Eric J. Lerner

THE BIG BANG NEVER HAPPENED

Eric J. Lerner received an undergraduate degree in physics from Columbia University and attended the University of Maryland for graduate studies. An independent researcher and writer, he has published numerous articles on science and technology in many publications, including *Discover*, *Science Digest*, *Aerospace America*, and *Spectrum*. He received the Award of Excellence from the Aviation/Space Writer's Association.

THE BIG BANG NEVER HAPPENED

Eric J. Lerner

Vintage Books
A Division of Random House, Inc.
New York

FIRST VINTAGE BOOKS EDITION, NOVEMBER 1992

 Published in the United States by Vintage Books, a division of Random House, Inc., New York, and simultaneously in Canada by Random House of Canada Limited, Toronto. Originally published in hardcover by Times Books, a division of Random House, Inc., New York, in slightly different form in 1991.

Permissions acknowledgments may be found on pages 439–440.

Library of Congress Cataloging-in-Publication Data
Lerner, Eric J.
The big bang never happened / Eric J. Lerner.—1st Vintage Books ed.
p. cm.
Originally published: New York: Times Books, 1991.
Includes bibliographical references and index.
ISBN 0-679-74049-X (pbk.)
1. Big bang theory. I. Title.
[QB991.B54L47 1992]
523.1′8—dc20 92-50100
CIP

The front cover depicts the action of plasmas, electrically conducting gases, in the cosmos: magnetic plasma filaments at the center of the galaxy (National Radio Observatory). The back cover illustrates the earth's plasma generator (*left*) (NASA) and gigantic plasma filaments traced by concentrations of galaxies in this map of the universe within one billion light-years of earth (*right*) (R. B. Tully and J. R. Fisher). Such huge conglomerations couldn't have been formed in the time since the Big Bang.

Manufactured in the United States of America
10 9 8 7 6

To Hannes Alfvén, Pioneer

ACKNOWLEDGMENTS

While I have personally contributed to the development of plasma cosmology, this book is overwhelmingly about the work and ideas of others. The main concepts of the plasma universe and the alternative to the Big Bang are primarily the work of Hannes Alfvén, as is the "cosmological pendulum," the idea that cosmology alternates between mythical and scientific approaches. The notion that the development of science and the political and economic development of society are linked has been developed and elaborated by many, especially the historian V. Gordon Childe. Some of this story may be familiar to the reader from Carl Sagan's excellent television series and book *Cosmos*. As I detail in Chapter Seven, the other key conception of the new view of the cosmos—time as an evolutionary process—is mainly the work of Nobel laureate Ilya Prigogine.

The one major concept that, as far as I know, *is* original is the idea that the changes in cosmology, thermodynamics, and particle physics are part of a single, interconnected scientific revolution. In addition, some of the specific ideas in this book, especially those in the second half of Chapter Six, dealing with recent developments in plasma cosmology, come from my own research. The social implications of this scientific revolution, as set out in Chapter Ten, also derive from my own opinions and analysis.

Throughout the book I have tried to indicate,

wherever necessary, the many cases where I integrate ideas from others' work. Since this book is predominantly an overview of a broad development in science, my debts are extensive.

One source, however, should be singled out. In my analysis of the political implications of the struggles of the early Christian Church, and their connection to the rise of medieval cosmology, I am particularly indebted to the work of Elaine Pagels of Princeton University and her first-rate book, *Adam, Eve and the Serpent*. In other cases, the bibliography should give the reader both a guide to my sources and a means of pursuing topics in greater depth.

Many people helped me immensely in the writing of this book. My gratitude to Hannes Alfvén, to whom this book is dedicated, is especially great. He not only developed the basic ideas that this book explains and elaborates, he has in many lengthy conversations clarified the ideas and the history of their development, and given me invaluable suggestions for the development of this work. He has, as well, unfailingly encouraged my own scientific research. My friend and colleague Tony Peratt has also been of great help in clarifying many scientific points, in his enthusiasm for this book, and in his fostering of the debate that this book describes.

Many people encouraged me to write this book, especially John Dinges and Andrea Eagan, who, over an excellent Chinese dinner, first convinced me that there might indeed be interest in such a book. The way was also prepared by Randy Rothenberg of the *New York Times* and Paul Hoffman and Robert Kunzig of *Discover* magazine, who first allowed me to bring this cosmological debate to the attention of a broad audience. (The latter two suggested the book's title.)

Equally, this book would never have been published without the willingness of my editor at Times Books, Hugh O'Neill, to take in hand a risky and controversial enterprise. Thanks go to Derek Johns, Hugh's successor at Times Books, and my copy-editor, Ted Byfield, both of whom greatly improved my manuscript with their careful editing and suggestions and to Derek's successor, Steve Wasserman, for guiding the book to publication. My successful navigation of the tricky waters of the publishing industry was helped by my very able agent, Rick Balkin.

In writing this book, I was aided by conversations with many colleagues and experts. In particular, I want to thank Timothy Eastman for his discussion on the development of plasma astrophysics, Dr. Ilya Prigogine for his extremely helpful conversations about his own work, Dr. Conrad Hyer and Dr. Nahum Sarna for their insightful remarks on the relation of cosmology and religion, and Dr. Alan Krisch for his update on his crucial experiments.

I want especially to thank James Skehan, director of Weston Geological Observatory, Boston College. I am grateful to him both for the stimulating discussions on theology and science and for his detailed and valuable critique of the first draft.

My thanks, too, to those who encouraged me throughout the writing of this book and discussed its concepts with me, especially, but not only, my good friends Robin and Greg Williams, Carolyn and Richard Pollak, Wendy Solof and Terry Brownschilde, and Mary and David Heller. It was David who pointed out to me that Edgar Allan Poe was actually the first to propose the Big Bang.

Authors customarily thank those who prepare their manuscripts, but Sue Wolfarth deserves more than customary thanks. My own typing produced a manuscript that had one or more mistakes per word, few of which were corrected by my proofreading. Sue transformed this manuscript into English typescript, and corrected my grammar as well.

Finally, I want to thank my wife, Carol, for bearing with me during my years-long pregnancy with this, my first book, for listening to the basic arguments innumerable times, and for carefully reading and criticizing my manuscript. And I thank my two daughters, Kristin and Erin, for patiently hearing a thousand times that "Daddy has to work on his book."

TECHNICAL NOTE

In this book very small and very large numbers are sometimes written in exponential notation. Here, 10^{15} means 1 followed by fifteen zeroes and 10^{-15} means $1/10^{15}$. Large distances are described in terms of light-years (ly)—the distance light travels in a year (about 10 trillion kilometers). Some papers referred to in this book commonly use another unit, the parsec (pc), which is about three light-years.

Metric units are used throughout. A centimeter is .4 inch, a meter is about a yard, and a kilometer is .6 mile. A gram is $1/30$ ounce and a kilogram is 2.2 pounds. One degree K is $9/5$ degree F. Zero K, or absolute zero, is -273°C.

CONTENTS

PREFACE TO THE VINTAGE EDITION

Four hundred years ago Galileo broke the bonds that had entangled science with religion. Defying his fellow scientists' near unanimous commitment to Ptolemy's finite, earth-centered universe, Galileo defended Corpernicus's unlimited, sun-centered cosmos. He argued that observation, not scientific or religious authority, must be the test of cosmological theory. Science and religion must be separate, he declared: "Religion teaches how to go to heaven, not how the heavens go."

But now, four centuries after the Scientific Revolution, we seem to have come full circle. "Historic Big Bang Discovery May Prove God's Existence" reads the headline of an Associated Press story dated April 25, 1992. Leading cosmologists are quoted as saying that recent astronomical discoveries "are like looking at God," that they prove the reality of the Big Bang—a scientific version of the Biblical story of Creation. Cosmology again seems to be entangled with religion, at least in the headlines and in the minds of some cosmologists.

To be sure, these newspaper headlines have told a confusing story. In January 1991 the headlines boldly stated that the idea of an explosive birth of the universe, the Big Bang, was dead: "Big Bang Theory Goes Bust" read one in the *Washing-*

ton Post. But in April 1992 another headline in the *New York Times* reported "Astronomers Detect Proof of Big Bang—profound insight on how time began." What accounts for this sudden turnaround in the heavens? According to the reports, this decisive proof of the Big Bang, this "scientific discovery of the century, of all time," this key evidence of the Creation and of the Deity, was the discovery of tiny ripples in the intensity of the microwave background, a sort of universal radio hiss. Thus, if we are to believe the reports, the finding of tiny fluctuations in the background radiation overshadows in importance the discovery of nuclear energy, DNA, antibiotics, the theory of relativity, and the quantum theory of matter, among other more minor scientific ideas.

But reality is different from headlines. In fact, the overwhelming mass of scientific evidence still contradicts the Big Bang, as this book endeavors to show. As of this writing—May 1992—the Big Bang remains in just as deep trouble as ever, with even wider divergence from observation than when the first edition of this book was completed in late 1990. The blizzard of press releases that accompanied the discovery of these fluctuations by the Cosmological Background Explorer (COBE) Satellite are not mere objective statements of fact but a salvo in the developing cosmological debate, a debate that is steadily growing and that has profound implications for science, and indeed for society.

In the year and a half since this book was written, the evidence against the Big Bang has grown stronger, and the COBE results, far from "proving" the theory, have not in any way resolved the problems raised by other discoveries. The key problem, as I describe in Chapter One, is that there are objects in the universe—huge conglomerations of galaxies—that are simply too big to have formed in the time since the Big Bang, objects whose age is greater than the age Big Bang cosmologists assign to the universe itself. These conglomerations stretch over a billion light-years of space and were first discovered in 1986. In January 1991, while the first edition of this book was at press, a team of astronomers led by Will Saunders of Oxford unveiled a survey of galaxies that confirmed beyond all doubt the existence of these conglomerations, termed supercluster complexes. The survey, based on data from the Infrared Astronomical Satellite (IRAS), showed how prevalent these large structures are. Since no version of the Big

Bang predicted the existence of such vast structures, cosmologists viewed the new finding with alarm. It was this discovery that led to the widespread headlines in early 1991 that the Big Bang theory was dead or at least in great doubt.

This alarm was with good reason. By measuring the speeds that galaxies travel today, and the distance that matter must have traveled to form such structures, astronomers can estimate how long it took to build these complexes, how old they are. The answer to the latter is: roughly 60 billion years. But the Big Bang theory says that the universe is between ten and twenty billion years old. The existence of objects "older than the Big Bang" is a direct contradiction to the very idea that the universe emerged suddenly in a great explosion.

This "age of the universe" crisis is rapidly worsening because the theoretical estimate of that age is shrinking by the month. Astronomers have known since the 1920s that the farther away a galaxy is from us, the faster it seems to be moving away. From this basic fact, astronomer George Lemaître first proposed that, at one time, all matter was squeezed together and exploded outward in a giant explosion—the Big Bang. (As we shall see in Chapter Six, this is by no means the only possible explanation.) Big Bang theorists therefore argue that by measuring the distance to galaxies, and their velocities today, we can determine the time since the Big Bang and the age of the Universe.

Now, measuring distances to galaxies is difficult. Some "standard candle" that is of a known brightness must be used so that, from its apparent brightness here on earth, the distance to the galaxy can be determined. In the past year, many different such estimates have seemed to converge on an answer—the time since the Big Bang, according to these observations, is at most thirteen to sixteen billion years. While this may seem like a long time, for astronomers it is uncomfortably short. Astronomers agree that they know enough about the stars to measure their ages when they are gathered together in globular clusters—spherical balls of hundreds of thousands of stars in our own and other galaxies. The oldest such clusters in our own galaxy are at least fifteen to eighteen billion years old—close to or beyond the maximum that Big Bang estimates of the age of the universe allow.

The matter is worse than that, however. As will be explained in Chapter One, cosmologists have predicted a density for the universe that is a hundred times greater than the density that astronomers observe from counting galaxies. This hypothetical "dark matter" is essential to the Big Bang. But so much matter would, in the Big Bang theory, slow down the expansion of the universe. In the past, the expansion would have been faster, and thus the age of the universe even shorter—some eight to eleven billion years. So not only are the great supercluster complexes some five times older than the "age of the universe"—even humble stars in our own galaxy are some four to seven billions years too old!

What has been the response of cosmologists to this age crisis? Characteristically, there has been *no* consideration of the idea that the Big Bang theory itself might be wrong. Instead, there have been two general approaches that maintain the faith. On the one hand, many Big Bang proponents simply say, "Yes, it's true that we can't explain the large-scale structures—but this is a mere detail that doesn't affect the validity of the Big Bang itself." This is much like a fundamentalist saying, "Yes, it appears that mountains are millions of years old, but this is a mere detail that doesn't affect the idea that the earth is six thousand years old." It is simply an abandonment of the idea that scientific hypotheses can be tested against observation.

The second, and increasingly popular approach, is to add new hypotheses—something Big Bang cosmologists are fond of doing (see Chapter Four). The latest idea is somehow to push the Big Bang farther back in time by maintaining that expansion was slower in the past. Cosmologists theorize that a cosmological expansion force of unknown origin is speeding up the expansion. But such an accelerating force, aside from being entirely plucked out of the air, created conflicts of its own with observation.

Not only has the age crisis worsened in the past year, but an entirely new problem has arisen for the Big Bang. The *only* quantitative predictions of the Big Bang are the abundance of certain light elements—helium, lithium, and deuterium (the heavy form of hydrogen). The theory predicts these abundances as a function of the density of matter in the universe. In the past, these predictions seemed to accord reasonably well with observation, and this was considered a key support for the theory (see page 153). But beginning in April 1991, a growing number of observations showed that these predictions too were wrong. There is less helium in the universe than the theory predicts, and far less deuterium and lithium (Fig. 1). One can fit the amount of helium observed with one assumed density, deuterium with another, and lithium with a third, but no single amount of matter comes out right for all three. In particular, if helium is right (no more than 23 percent of the universe), then deuterium is predicted to be eight times more abundant than is observed (sixteen rather than two parts in one hundred thousand).

This is another fundamental challenge to the Big Bang, for with these light elements out of agreement with the theory, there is no single piece of data that theorists can point to as confirming the theory. Of course, again there have been efforts to fix things up. Perhaps nearly all the deuterium was burned up in stars so only one-eighth is left, some cosmologists argue. Perhaps there were little lumps in the Big Bang, so that different amounts of elements were created. But none of these fixes can account for all the data.

The COBE observations, announced in April 1992, had absolutely no impact on any of these problems. COBE detected fluctuations of one part in one hundred thousand in the smooth cosmic background radiation. According to Big Bang theory, these fluctuations are relics of similarly subtle variations in the density of matter soon after the Big Bang. Such fluctuations, the theory states, gradually attract matter around them to become large structures in the universe today. But this in no way explains how the structures could have grown fast enough, nor how the universe could be younger than some of its own stars, nor why the light element abundances are all wrong.

Nor did Big Bang theorists even accurately predict the magnitude of the fluctuations. Original Big Bang predictions in the 1970s said that fluctuations of one part in a thousand would be needed for matter to condense into any structures at all, even relatively small ones like galaxies. (This is one hundred times larger than the fluctuation that COBE found twenty years later.) When these larger ripples they predicted were not found, theorists decided that matter must be one hundred times denser than observation indicated, so that a stronger gravitational force could speed the growth of structures (see page 33). This was the famous "dark matter." But with this dark matter, predictions became flexible enough to fit nearly any result. In the months

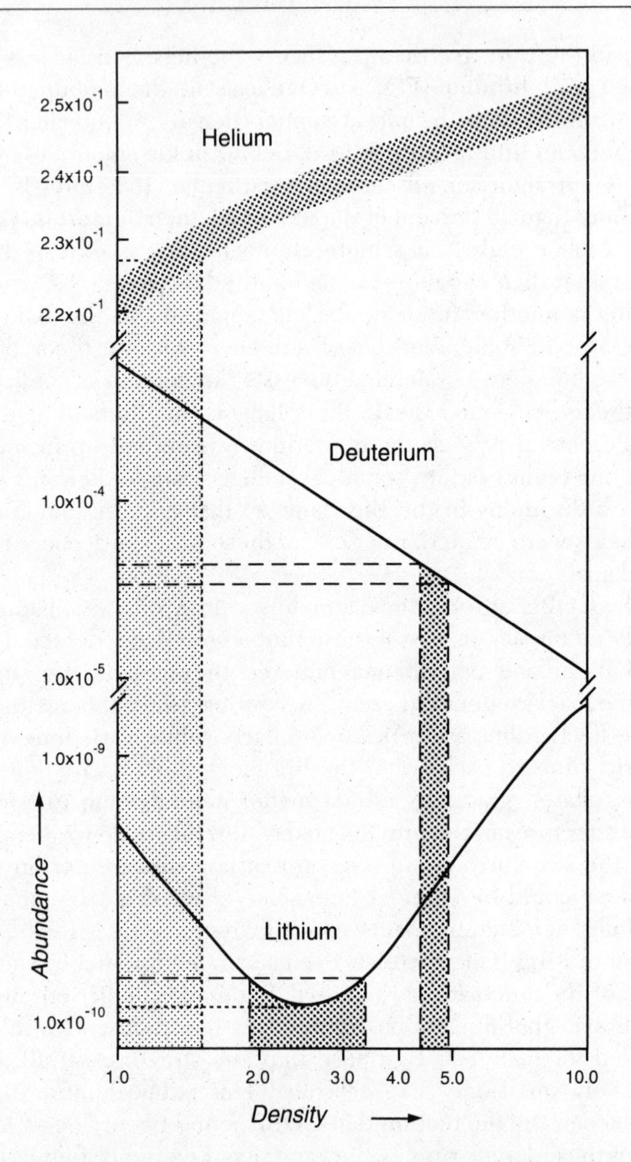

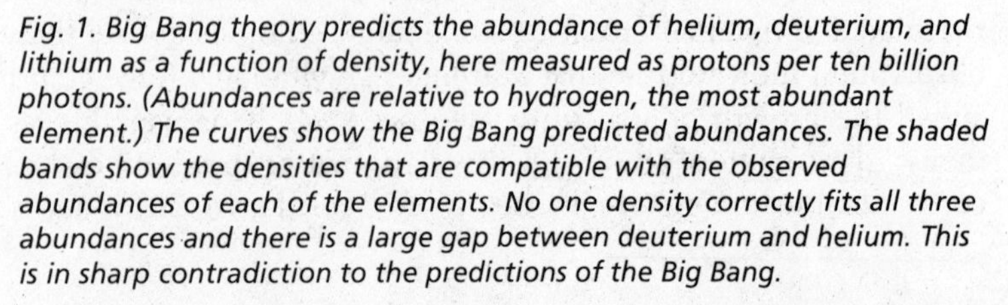
Fig. 1. Big Bang theory predicts the abundance of helium, deuterium, and lithium as a function of density, here measured as protons per ten billion photons. (Abundances are relative to hydrogen, the most abundant element.) The curves show the Big Bang predicted abundances. The shaded bands show the densities that are compatible with the observed abundances of each of the elements. No one density correctly fits all three abundances and there is a large gap between deuterium and helium. This is in sharp contradiction to the predictions of the Big Bang.

before the COBE results were announced, Big Bang predictions ranged from fluctuation of a few parts in a hundred thousand to a part in ten million—a hundred times *smaller* than the *COBE* results. Since no COBE result could contradict this shotgun pattern of predictions, none could confirm them either.

The results didn't even prove that the cosmic background is indeed an echo of the Big Bang. Other scientists, including myself and Dr. Anthony Peratt of Los Alamos National Laboratory, have hypothesized that the background is the glow from a radio fog produced in the present-day universe. Irregularities in this fog would produce fluctuations of just about the size observed, as we predicted prior to these results. And other observational evidence backs up the idea that such a fog exists between the galaxies (see page 276).

Then why was there such a celebration of the COBE findings? To most cosmologists, who have spent their lives elaborating the Big Bang theory, it has become an article of faith, not a hypothesis to be proved or disproved by the evidence. After two years in which every new observation produced a new contradiction, the COBE results, which did not contradict the theory (indeed *could not* have), were seized upon as a way to defend the faith. Cosmologists loudly proclaimed that none could now question their theory.

The press took the cosmologists, the existing authorities, at their word. None seem to have doubted the overblown claims, questioned exactly how these ripples dispelled all the theory's problems, or asked any of the dozens of critics of the theory to comment. In an uncertain time, journalists were all too willing to report that the authorities had the cosmos well in hand, that final truths were now known, that science and religion spoke with one voice.

This new entanglement of science, authority, and faith, this attempted Scientific Counterrevolution, is dangerous to the whole scientific enterprise. If the wildest theoretical claims are accepted on the word of scientific authority alone, the link with observation is broken. And if appeals to authority extend to Scripture, if one accepts that proof of the Big Bang is proof of one variety of Judeo-Christian doctrine, then attacks on this scientific theory become heresy, as Galileo's attacks on Ptolemy were deemed four hundred years ago. This is a return to a cosmology built on faith, not observation, a trend that is a major theme of this book.

Fortunately, this is not the only trend in cosmology. The publication of the first edition of this book in May 1991 has considerably sharpened the cosmological debate and brought this debate to the attention of a broad audience outside the narrow confines of cosmology itself. The idea that there is a scientific alternative to the Big Bang has now been discussed on the editorial page of the *New York Times*, in popular astronomy magazines like *Sky and Telescope*, on scores of radio stations, and on several TV news shows. In the past, Big Bang cosmologists have simply ignored the theory's critics. Now they are reluctantly beginning to debate with these critics. Perhaps most important, Big Bang supporters have had to take the challenge we pose seriously in their own scientific circles. At a recent seminar by a leading cosmologist at Los Alamos National Laboratory, the speaker began by raising this book and assuring his audience that the Big Bang was still valid. When I gave a seminar on the failure of Big Bang cosmology and the plasma alternative at Princeton University, several leading researchers and their flock of graduate students attended. Significantly, in the discussion that ensued, there were few defenses of the Big Bang, and the cosmologists' comments focused on their criticism of plasma cosmology. When I remarked on this, one Big Bang supporter shrugged and said, "We all know that the Big Bang has many problems. But if there is no alternative, we must stick with it."

Today, this debate is only beginning to be reported in the popular press and in the scientific journals. Yet it is nonetheless occurring and growing. This book is a report on that emerging debate, its roots, and its consequences. And since, as history abundantly shows, people's views of the universe are bound up with their views of themselves and of their society, this debate has implications far beyond the realm of science, for the core of the cosmological debate is a question of how truth is known. Must we rely on experts, whose pronouncements, no matter how seemingly absurd, are accepted on faith, or do we trust in the evidence of the senses, in our observation of the world? This question is also at the center of today's social events. As I write, there is not a government east or west that today enjoys the confidence of its people or that can credibly promise them any improvement in their future. The global decline of production and standards of living, begun twenty years ago, has accelerated. To extricate society from this whirlpool, must we rely on "the experts" who, east and west, call insistently for policies that benefit the few and sacrifice the many? Or can we rely on our own judgment to take into our own hands—the hands of those who work —the direction of society, and of the economy that supports that society? How these questions are answered will shape not only the history of science, but the history of humanity.

Eric J. Lerner
May 1992

THE BIG BANG NEVER HAPPENED

INTRODUCTION

When leading scientists publicly predict that science will soon reach its ultimate goal, that within a decade everything will be explained, you can be sure that they are wrong. A century ago, one of the leading scientists of the day, Lord Kelvin, stated that the future of physics lay "in the last decimal place." All the main problems, he declared, had been solved, only further accuracy was needed. Yet within two decades, the discovery of radioactivity, the theory of relativity, and the development of quantum mechanics had thoroughly transformed physics and profoundly changed humanity's view of the universe.

Today we again hear renowned scientists, such as Stephen Hawking, claiming that a "Theory of Everything" is within their grasp, that they have almost arrived at a single set of equations that will explain all the phenomena of nature—gravitation, electricity and magnetism, radioactivity, and nuclear energy—from the realm of the atoms to the realm of the galaxies and from the beginning of the universe to the end of time. And once again, they are wrong. For quietly, without much fanfare, a new revolution is beginning which is likely to overthrow many of the dominant ideas of today's science, while incorporating what is valid into a new and wider synthesis.

The Big Bang theory of cosmology—the idea that the universe originated in a single cataclysmic explosion some ten or twenty billion years ago—

was popularized in the fifties and sixties, and has become central not only to astronomy, but to all current theories of the basic structure of matter and energy as well. Yet in the past few years, observation after observation has contradicted the predictions of this theory. Rather, such observations are far more consistent with new theories based on the idea that the universe has existed for an infinite time—without beginning or end.

As yet, such alternative theories, known as "plasma cosmology," have been developed by only a relatively small group of physicists and astronomers, the most notable being Swedish Nobel laureate Hannes Alfvén. But as the evidence mounts, more and more scientists are questioning their basic, long-held assumptions.

The emerging revolution in science extends beyond cosmology. Today the study of the underlying structure of matter, particle physics, is intimately tied up with cosmology—the structure of the universe, theorists argue, is the result of events in the first instants of time. If the Big Bang hypothesis is wrong, then the foundation of modern particle physics collapses and entirely new approaches are required. Indeed, particle physics also suffers from an increasing contradiction between theory and experiment.

Equally important, if the Big Bang never occurred our concept of time must change as well. Instead of a universe finite in time, running down from a fiery start to a dusty, dark finish, the universe will be infinite in duration, continuously evolving. Just such a concept of time as evolution is now emerging from new studies in the field of thermodynamics.

The changes in these three fields—cosmology, particle physics, and thermodynamics—are merging into a single global transformation of how science views the universe, a transformation comparable to that which overthrew the Ptolemaic cosmos and initiated modern science.

This book is a first effort to describe that emerging revolution and its implications. Since it gives the view of what is at the moment a minority of the scientific community, the ideas presented here are far different from, and contradictory to, the most common beliefs about cosmology and fundamental physics. Yet what I describe here is not a fringe view, a Velikovskian fantasy. It is a summary of work presented in thousands of papers published by leading technical journals, work that, although not yet widely accepted, is beginning to be widely discussed. In the winter of 1988, for example, Alfvén was invited to present his views to the Texas Symposium on Relativistic Astrophysics, one of the most important conferences of cosmologists.

My aim is to explain these new ideas to the general reader, one who is interested in the crucial issues of science but who has no special training in the subject. I believe that if the issues are presented clearly, readers will be able to judge the validity of the arguments involved in this debate.

The ultimate test of scientific theories is observation, and I will emphasize how observations conflict with, or support, various cosmological ideas. But this debate involves more than just two views of the universe and its origins: it is a struggle between two different ways of learning about the universe. One, the method of learning from observation, is used by the vast majority of scientists today and by those who are proposing the new ideas in cosmology. The other method, advocated by mainstream cosmologists and particle theorists, is the deductive method, mathematically deducing how the universe *must* be.

Both methods date back millennia, and over time they have alternately dominated the study of the universe and its origins. To understand the present debate in cosmology, we must understand something of this long history, how the ideas themselves—a universe without a beginning, a universe created from nothing at a single moment—came into existence. For the only real way we have of judging these methods is by their results—the consequences they had for the development of science, and for the development of society.

This history, then, involves more than the history of cosmology, or even of science. One of the basic (although far from original) themes of this book is that science is intimately tied up with society, that ideas about society, about events here on earth, affect ideas about the universe—and vice versa. This interaction is not limited to the world of ideas. A society's social, political, and economic structures have a vast effect on how people think; and scientific thought, through its impact on technology, can greatly change the course of economic and social evolution.

So now, as in the past, the evolution of society and the evolution of cosmology are intertwined, one affecting the other. This interaction must be understood before one can comprehend what is happening in cosmology today. Otherwise it is a mystery how certain ideas develop, come to the fore, and are then abandoned, how the vast majority of cosmologists can arrive at conclusions so clearly contradictory to observation.

Today Big Bang theorists see a universe much like that envisioned by the medieval scholars—a finite cosmos created *ex nihilo,* from nothing, whose perfection is in *the past,* which is degenerating to a final end. The perfect principles used to form this universe can be known only by pure reason, guided by authority, independent of observation. Such a cosmic myth arises in periods of social crisis or retreat, and reinforces the separation of thought and action, ruler and ruled. It breeds a fatalistic pessimism that paralyzes society.

By contrast, the opposing view, plasma cosmology, is empirical, a product of the scientific method of Galileo and Kepler. Its proponents see an infinite universe evolving over infinite time. The universe can be studied only by observation—there is no final answer in science and no final authority. This approach, binding together thought and action, theory and observation, has proved, over the ages, to be a weapon of social change. The idea of progress in the universe has always been linked with the idea of social progress on earth.

THE STRUCTURE OF THE BOOK

The first part of this book explains the ongoing debate in cosmology. Chapter One begins with the evidence that the Big Bang theory is wrong, and that alternative theories, based on the study of electrically conducting gases, called plasmas, are probably right. I then take a long step back to trace the history of the cosmological debate. Chapter Two shows how the basic concepts of both the empirical and the deductive methods arose in ancient Greece and how they were tied up with the conflict between free and slave labor. The deductive method's disregard for observation and practical application of science originated with the slave master's disdain for manual work, while the empirical method's system is based on free craftsmen and traders combining theory and observation.

In the first swing of the cosmological pendulum, the deductive method became dominant, leading to the static and finite universe of Ptolemy. The central idea of modern cosmology, the origin of the universe from nothingness, then arose not from Genesis but from the ideological battles of the third and fourth centuries A.D., as Roman society disintegrated and the basis was laid for feudalism. The Church fathers Tertullian and St. Augustine introduced the doctrine of creation *ex nihilo* as the foundation of a profoundly pessimistic and authoritarian world view, a cosmology that denigrated all earthly endeavor and condemned material existence as "created from nothing, next to nothing," inevitably decaying from a perfect beginning to an ignominious end. This cosmology was to serve as the philosophical and religious justification for a rigid and enthralled society.

Chapter Three describes the next long swing of the pendulum—the centuries of struggle that led to the scientific revolution. The rise of a new and more profound empirical method went hand in hand with the rise of a new view of the universe—infinite in space and time, without origin or end—and with the rise of a new society, one based on free labor. By the middle of the nineteenth century, the scientific view of the universe was that of an unending process of evolution, as the revolutionaries of the eighteenth and nineteenth centuries saw an unending process of social evolution and progress.

The Big Bang and twentieth-century cosmology constitutes a startling return to the discredited medieval concepts, as Chapter Four details. The deep social crisis of the present century gave credence to the old philosophical view of a decaying universe, degenerating from its perfect origins, and to the deductive method. It is from these primarily philosophical premises, rather than from observation, that present-day cosmology developed. For this reason, as we will explore in Chapter Four, the repeated conflicts between theory and observation that have dogged the Big Bang never led to its abandonment.

However, the challenge to the Big Bang did arise from observation. Chapters Five and Six describe how plasma cosmology grew out of the laboratory study of conducting gases and had its roots in the advancing technologies of electromagnetism. As observations have extended outward from the earth and the solar system to the galaxies and the universe as a whole, the predictions of plasma cosmology have been increasingly confirmed.

The second part of the book deals with the implications of a universe that is infinite in space and time, continuously evolving. In Chapter Seven I examine how new discoveries in the nature of time show that such a cosmos can exist indefinitely without "running down." In fact, the universe is characterized neither by decay nor by a random, aimless meandering or by the automatic progress of late-nineteenth-century concepts. The cosmos, and indeed any complex system, progresses only through a series of crises whose outcomes are not predetermined and can lead, over the short run, either to new advances or to retrogression. Progress, the acceleration of evolution, is a long-term tendency of the universe, but it is far from a smooth and mechanical process.

Chapter Eight looks at the equally profound problems that arise with the conventional ideas of matter if the Big Bang is refuted. Not only the most recent theories but much of the underlying structure of physical theory suffers from crucial inconsistencies that remain to be resolved.

Finally, in Chapters Nine and Ten, we look at the impact an infinite cosmos has on religion and society. As in the sixteenth century, the two approaches to cosmology today imply profoundly opposing reactions to a deepening crisis.

PART ONE

THE COSMOLOGICAL DEBATE

1 THE BIG BANG NEVER HAPPENED

It's impossible that the Big Bang is wrong.

—Joseph Silk, 1988

Down with the Big Bang.

—Editorial title, *Nature*, 1989

Cosmologists nearly all agree that the cosmos came into being some ten or twenty billion years ago in an immense explosion, the Big Bang. Our mighty universe, they believe, began in a single instant as an infinitely dense and hot pointlike ball of light, smaller than the tiniest atom. In one trillion-trillionth of a second it expanded a trillion-trillionfold, creating all the space, matter, and energy that now make up the galaxies and stars.

The present universe, the ashes of that explosion, is a strange one, as cosmology describes it. Most of it is dark matter, exotic particles that can never be observed. It is dotted by black holes, which suck in streams of dying stars, and it is threaded by cosmic strings, tears in the fabric of space itself. Our universe's future, cosmologists tell us, is grim: it is doomed either to end in a spectacular Big Crunch, collapsing into a universal black hole, or to expand and decay into the nothingness of an eternal night.

This striking cosmic vision, built up over the past twenty-five years by hundreds of theoreticians and explained in dozens of books, has sunk deeply into popular consciousness. Many have pondered what meaning life can have in a universe doomed to decay, unspeakably hostile and alien to human purposes.

Without doubt, the current concept of the universe is fantastic and bizarre. Yet despite the efforts and firm beliefs of so many cosmologists, it is also almost certainly wrong.

The validity of a scientific concept is not determined by its popularity or by its support among the most prominent scientists of the day. Many a firmly held doctrine, from the geocentric cosmos of Ptolemy to the phlogistic theory of heat, has enjoyed the nearly unanimous support of the scientific community, only to be swept away later.

In 1889 Samuel Pierpont Langley, a famed astronomer, president of the American Association for the Advancement of Science, and soon to be one of the pioneers of aviation, described the scientific community as "a pack of hounds . . . where the louder-voiced bring many to follow them nearly as often in a wrong path as in a right one, where the entire pack even has been known to move off bodily on a false scent."[1]

The only test of scientific truth is how well a theory corresponds to the world we observe. Does it predict things that we can then see? Or do our observations of nature show things that a theory says are impossible? No matter how well liked a theory may be, if observation contradicts it, then it must be rejected. For science to be useful, it must provide an increasingly true and deep description of nature, not a prescription of what nature must be.

In the past four years crucial observations have flatly contradicted the assumptions and predictions of the Big Bang. Because the Big Bang supposedly occurred only about twenty billion years ago, nothing in the cosmos can be older than this. Yet in 1986 astronomers discovered that galaxies compose huge agglomerations a billion light-years across; such mammoth clusterings of matter must have taken a *hundred* billion years to form. Just as early geological theory, which sought to compress the earth's history into a biblical few thousand years crumbled when confronted with the aeons needed to build up a mountain range, so the concept of a Big Bang is undermined by the existence of these vast and ancient superclusters of galaxies.

These enormous ribbons of matter, whose reality was confirmed during 1990, also refute a basic premise of the Big Bang—that the universe was, at its origin, perfectly smooth and homogeneous. Theorists admit that they can see no way to get from the perfect universe of the Big Bang to the clumpy, imperfect universe of today. As one leading theorist, George Field of the Harvard-Smithsonian Center for Astrophysics, put it, "There is a real crisis."

Other conflicts with observation have emerged as well. Dark matter, a hypothetical and unobserved form of matter, is an essential component of current Big Bang theory—an invisible glue that holds it all together. Yet Finnish and American astronomers, analyzing recent observations, have shown that the mysterious dark matter isn't invisible—*it doesn't exist.* Using sensitive new instruments, other astronomers around the world have discovered extremely old galaxies that apparently formed long before the Big Bang universe could have cooled sufficiently. In fact, by the end of the eighties, new contradictions were popping up every few months.

In all this, cosmologists have remained entirely unshakable in their acceptance of the theory. Many of the new observations have been announced in the most prominent journals and discussed at the biggest astronomers' meetings. In some cases, the observers are among the most respected astronomers in the world. Nonetheless, cosmologists, with few exceptions, have either dismissed the observations as faulty, or have insisted that minor modifications of Big Bang theory will reconcile "apparent" contradictions. A few cosmic strings or dark particles are needed—nothing more.

This response is not surprising: most cosmologists have spent all of their careers, or at least the past twenty-five years, elaborating various aspects of the Big Bang. It would be very difficult for them, as for *any* scientist, to abandon their life's work. Yet the observers who bring forward these contradictions are also not at all ready to give up the Big Bang. Observing astronomers have generally left the interpretation of data to the far more numerous theoreticians. And until recently there seemed to be no viable alternative to the Big Bang—nowhere to go if you jumped ship.

But now an entirely different concept of the universe has developed, although it is not yet known to many astronomers. It begins from the known fact that over 99 percent of the matter in the universe is plasma—hot, electrically conducting gases. (In ordinary gases, electrons are bound to an atom and cannot move easily, but in a plasma the electrons are stripped off by intense heat, allowing them to flow freely.) Extrapolating from the behavior of such plasma in the laboratory, plasma cosmologists envision a universe crisscrossed by vast electrical currents and powerful magnetic fields, ordered by the cosmic counterpoint of electromagnetism and gravity.

The phenomena that the Big Bang seeks to explain with a mysterious ancient cataclysm, plasma theories attribute to electrical and magnetic processes occurring in the universe today. These are similar in kind, if not magnitude, to processes seen in the laboratory and used in such mundane technology as neon lights and microwave ovens. Instead of working forward from a theoretically conceived beginning of time, plasma cosmology works backward from the present universe, and outward from the earth. It arrives at a universe without a Big Bang, without any beginning at all, a universe that has always existed, is always evolving, and will always evolve, with no limits of any sort.

As yet, plasma cosmology has attracted only a little attention among astronomers, in part because it was formulated by plasma physicists, who attend different conferences and publish in different journals. This situation is rapidly changing. As more contradictions of the Big Bang emerge, some astronomers, in particular observers with little investment in a single theory, have begun to look with interest at the new ideas. They are starting to ask questions and tentatively to measure the old and new cosmologies against each other. No longer is the Big Bang unquestioningly accepted by leading journals outside of cosmology. The widely read British journal *Nature*, for example, in August of 1988 ran a lead editorial entitled "Down with the Big Bang," which described the theory as "unacceptable" and predicted that "it is unlikely to survive the decade ahead." A new cosmological debate has begun.

THE COSMIC TAPESTRY

The challenge to the Big Bang begins with new observations that undermine the basic assumptions of conventional cosmology. Perhaps the most important of these assumptions is the idea that the universe is, at the largest scales, smooth and homogeneous. If such a smooth universe is dominated by gravity alone—a second important assumption—then, according to Einstein's theory of gravitation (general relativity), the universe as a whole must either contract to, or expand from, a single point, a singularity.

But we seem to have a "clumpy" universe, which would not warp all of space or cause it to expand or contract. Each clump would just dimple the space around it. Galaxies are clumped into vast supercluster complexes, which stretch across a substantial part of the known universe.

These objects, by far the largest ever seen, were discovered in 1986 by Brent Tully, a University of Hawaii astronomer and one of today's leading optical astronomers. Tully found that almost all the galaxies within a distance of a billion light-years of earth are concentrated into huge ribbons of matter about a billion light-years long, three hundred million light-years wide, and one hundred million light-years thick.

His discovery, while stunning, was perhaps to have been expected. For centuries, astronomers have been discovering ever-larger clumps of matter in the universe, and ever-larger stretches of space between them (Fig. 1.1). Since the seventeenth century, astronomers have known that most of the universe's mass is concentrated in glowing stars like our sun, dense objects separated by light-years of nearly empty space. A hundred and twenty years ago, astronomers realized that groups of a hundred billion or more stars form the great pinwheels we see as galaxies, and that *these* are separated by *larger* empty expanses. In the thirties, as telescopes penetrated more deeply into space, observations showed that even galaxies are grouped together into clusters, some containing a thousand galaxies.

Then, in the early seventies, it became clear that these spherical clusters are strung together into larger filaments termed superclusters. While galaxies are a mere hundred thousand light-years across and clusters not more than ten million or so, a

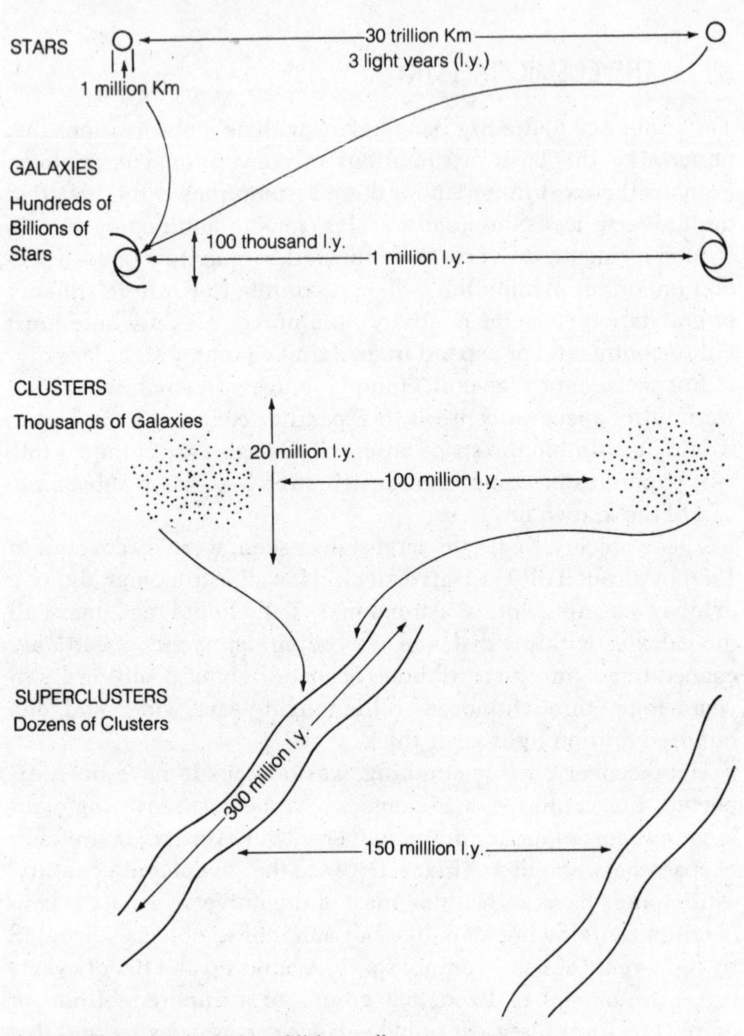

Fig. 1.1. The relative scales of "clumpy" space.

supercluster might snake through a few hundred million light-years of space.

Astronomers, excited by these latest observations, began to plot the locations of galaxies on the sky to see what patterns might appear. One group, led by Dr. P. J. E. Peebles of Princeton, used a supercomputer to plot nearly a million galaxies; the

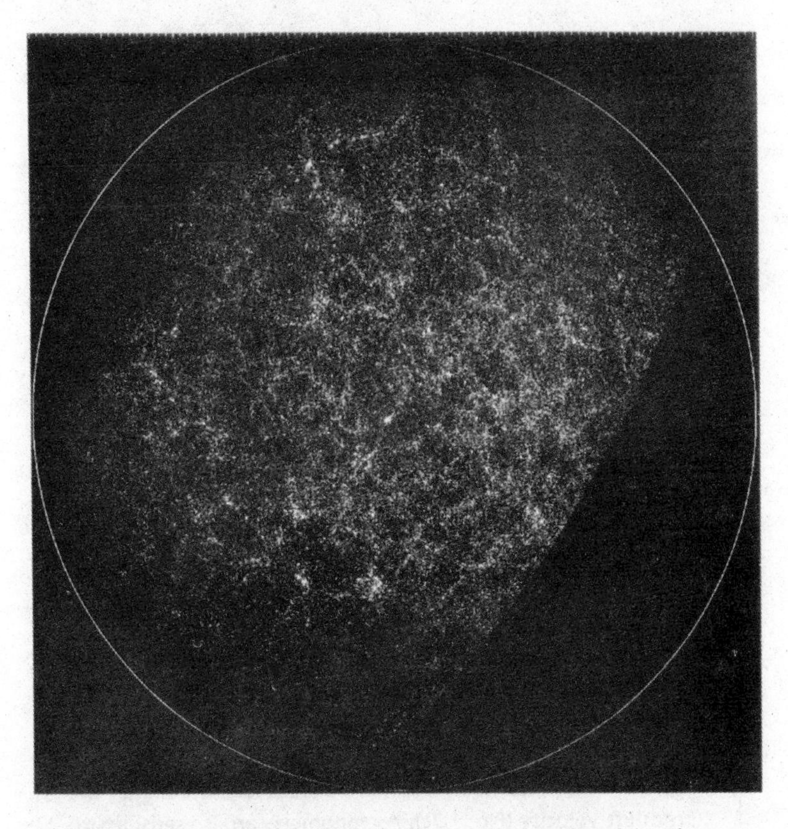

Fig. 1.2. The Cosmic Tapestry. Each dot represents a single galaxy. The million galaxies shown here (those visible from Lick Observatory) cluster into delicate filaments (P. J. E. Peebles).

result is a lacy filigree of interwoven threads, a pattern one astronomer dubbed "the Cosmic Tapestry" (Fig. 1.2).

But this was only a pattern in two dimensions, projected against the sky; to see where galaxies are really clustered in space, one needed to plot them in three dimensions. This was quite possible. Since the thirties, astronomers have known a way to measure the distance to galaxies—the Hubble redshift (see box). They had found that the farther away a galaxy is, the more its light shifts to the red end of the spectrum, just as if it were moving away from earth. On the one hand, this became the basis of the idea that the universe is expanding, an idea that led to the

Big Bang theory. On the other, it gave astronomers a powerful tool—by measuring the light from a galaxy one could calculate its distance from earth.

MEASURING THE DISTANCE TO A GALAXY

As an object travels farther away, its light shifts to the red end of the spectrum, just as a train whistle's pitch drops as it passes. Light waves (or sound waves) on the receding side of the object are more spread out than on the approaching side. A longer wavelength means a shift to the red (Fig. 1.3a). The redshift can be used to measure an object's velocity.

When light from a distant galaxy is put through a prism or grating, it produces a spectrum with characteristic dark lines. Comparing the frequency or color of the dark lines with those produced by heated gases on earth, astronomers in the twenties found that the galaxy lines shifted to the red, implying that the galaxies are receding at high velocity (Fig. 1.3b). Astronomer Edward Hubble found that the dimmer a galaxy is, and thus presumably the more distant it is, the higher the redshift velocity (Fig. 1.3c). Astronomers can use redshifts to measure distance far beyond the limits of other methods.

In the seventies, Brent Tully and J. R. Fischer developed another method of determining distance. They found that the intrinsic brightness of a galaxy was proportional to the fourth power of the rotational velocity (Fig. 1.3d). Because the rotational velocity could be measured from earth by comparing the redshifts on each side of a galaxy, the intrinsic brightness can be calculated. Knowing how bright the galaxy appeared in the sky would then give its distance.

a)

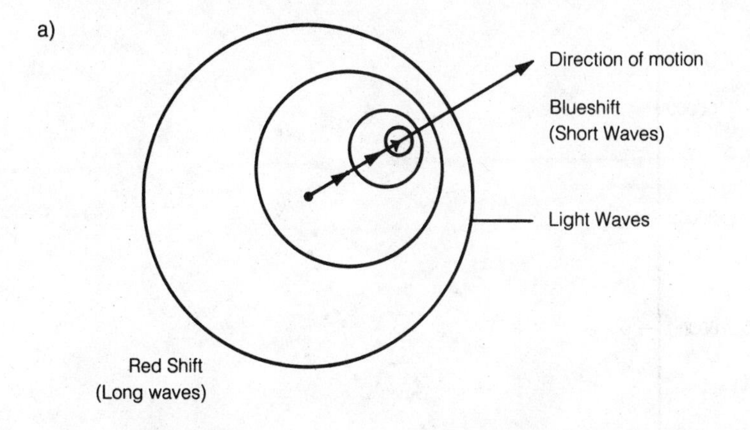
Direction of motion
Blueshift
(Short Waves)
Light Waves
Red Shift
(Long waves)

b)

GALAXY IN
RED SHIFTS

VIRGO

URSA MAJOR

CORONA BOREALIS

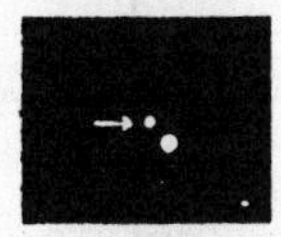
BOOTES

HYDRA

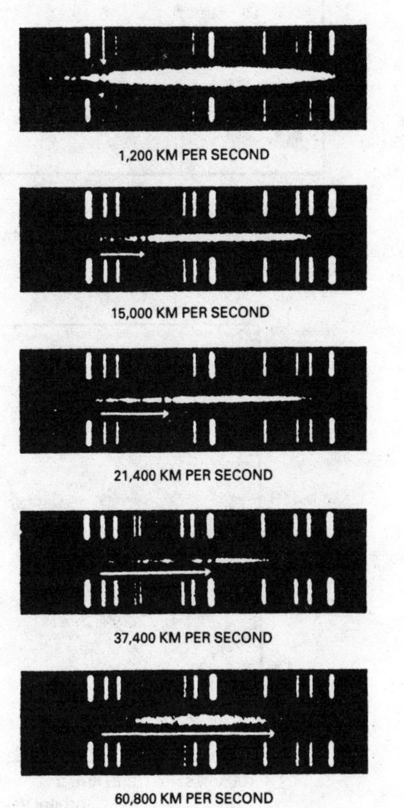
1,200 KM PER SECOND
15,000 KM PER SECOND
21,400 KM PER SECOND
37,400 KM PER SECOND
60,800 KM PER SECOND

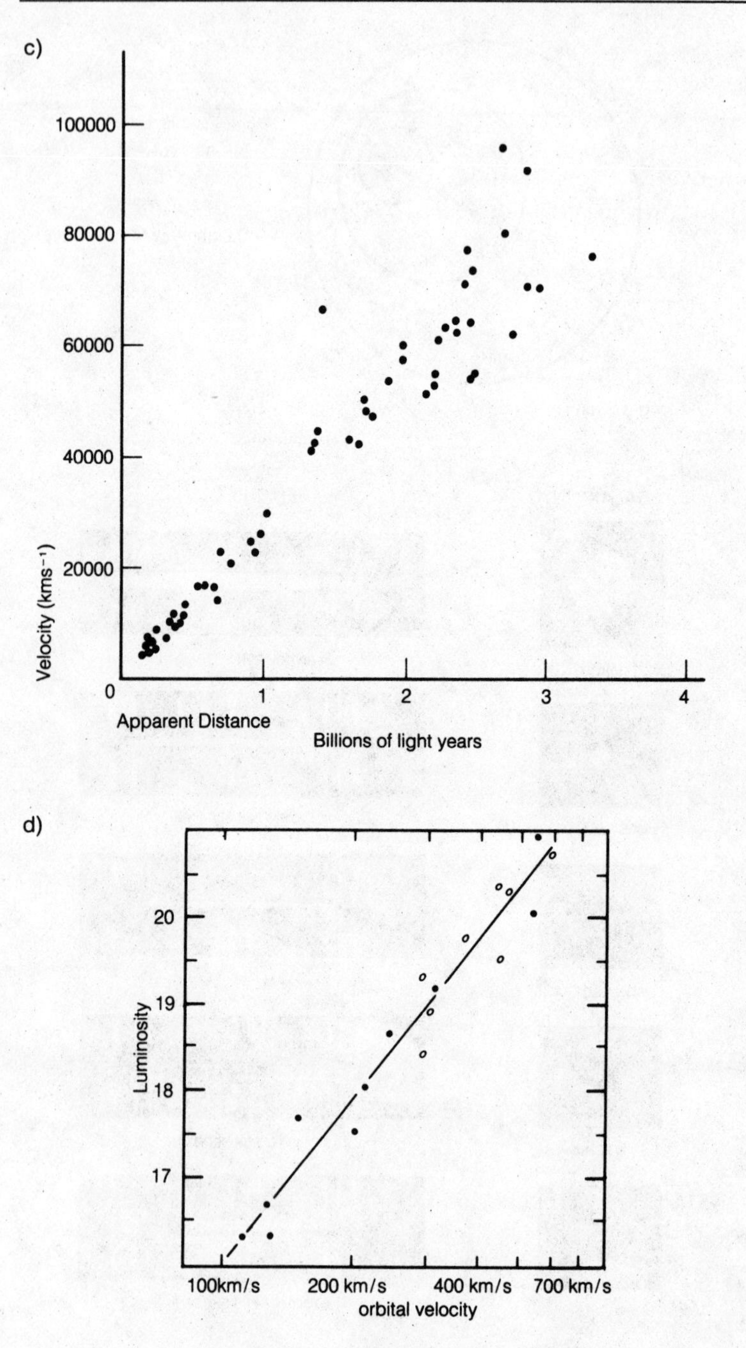
c)
100000
80000
60000
40000
20000
Velocity (kms⁻¹)
0
1
2
3
4
Apparent Distance
Billions of light years
d)
20
19
18
17
Luminosity
100km/s
200 km/s
400 km/s
700 km/s
orbital velocity

Dr. Tully and his colleague J. R. Fischer set out to use the distance measurements of two thousand nearby galaxies to create a three-dimensional atlas of our part of the universe. They were among the best qualified for the task, since they had themselves uncovered a complementary way of measuring distance to a galaxy, based on a link between how fast it spins and how bright it is.

After years of plotting and analyzing the data they had their map—the *Atlas of Nearby Galaxies*. Remarkably, they found the patterns in the sky were entirely real. With less than two dozen exceptions all of the thousands of galaxies are strung like Christmas lights along an interconnecting network of filaments—a glowing cat's cradle in the sky (Fig. 1.4). The filaments themselves, only a few million light-years across, extend across hundreds of millions of light-years, beyond the limits of Tully and Fischer's maps.

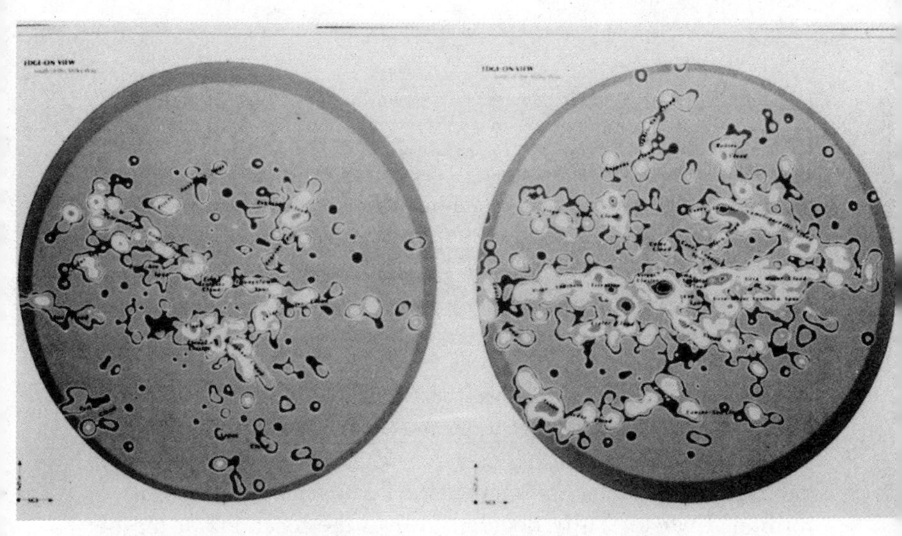

Fig. 1.4a. Tully and Fischer's maps show that galaxies within one hundred million light-years of earth are concentrated into filaments. The right-hand view is the view to the north and the left to the south (in both cases our galaxy is at the center of the map). The radius of the sphere mapped is 120 million light-years. Nearly all the galaxies lie along a few filaments, each less than seven million light-years across (R. B. Tully and J. R. Fischer).

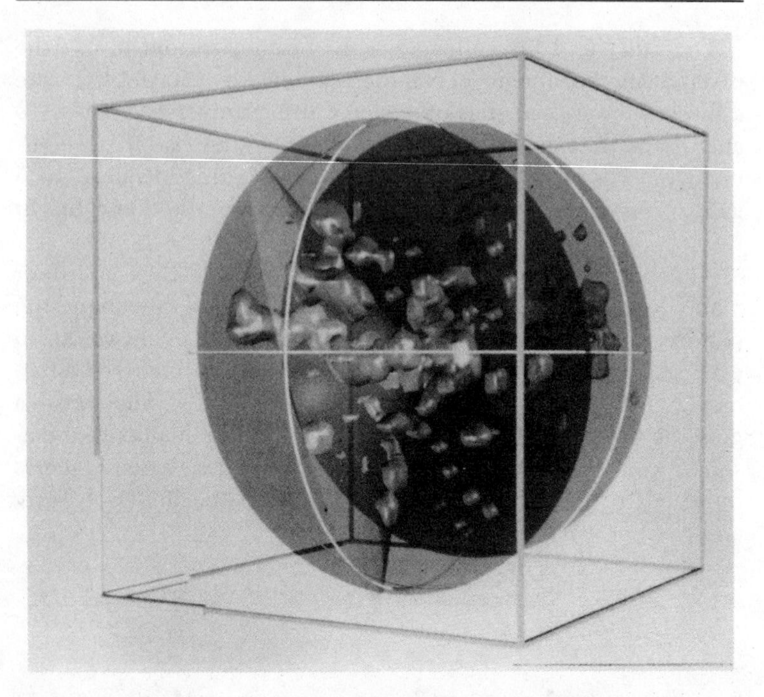

Fig. 1.4b. On a larger scale, clusters of galaxies are also concentrated into vast supercluster complexes. Here a sphere one billion light-years in radius is mapped, again with our galaxy at the center. Colors indicate the density, in this three-dimensional computer-generated map, with the densest regions being yellow and pink, slightly less dense regions being green (see back of book jacket). Nearly all the clusters are in the dense green and yellow columns, which take up only a fraction of the total volume mapped. Note the long filament, about one hundred million light-years across, and over a billion light-years long, snaking its way out to the left. The pink cone carves out a region of space that is not completely mapped.

How far beyond? Tully wanted to make a bigger map—out to a billion and a half light-years from earth. For that huge distance he couldn't use individual galaxies. Modern telescopes can see galaxies out that far, but there are far too many—a couple of million. Instead, Tully decided to map the locations of the big clusters of galaxies, clusters identified forty years earlier by astronomer George Abell.

The pattern of the clusters, to Tully's surprise, outlined the vast ribbons, each one made up of dozens of supercluster filaments. Tully identified about five "supercluster complexes," each containing millions of trillions of stars. The density of clusters within the ribbon was about twenty-five times that ouside of them. Moreover, several stretched to the boundaries of Tully's new map and beyond, and all of them seemed to lie in parallel planes—as if stacked in space as part of some still vaster structure.

TOO BIG FOR THE BIG BANG

The supercluster complexes directly contradict the homogeneity assumed by the Big Bang. This homogeneity has always been a problem, since it's clear that the universe is so clumpy: how did it get that way if it started out so smooth? The general answer has been that there were very tiny clumps in the early universe; through gravitational attraction these clumps gradually grew bigger and bigger, forming stars, galaxies, and clusters.

Of course, the bigger the clump, the longer the time to form. For stars, a few million years is enough, for galaxies one or two billion years are needed. Clusters take even longer. By the time superclusters were discovered, there was an obvious difficulty, and in the eighties cosmologists were hard at work trying to overcome them. Tully's objects made the situation impossible—they were just too big to have formed in the twenty billion years since the Big Bang.

It's not hard to see why. By observing the redshifts of galaxies, astronomers can see not only how far away they are, but roughly how fast they move relative to one another—their true speed, ignoring the Hubble velocities that increase with distance. Remember, redshifts indicate how fast an object is moving away from us. Redshifts increase with distance, but also with an object's own speed, relative to the objects around it. It's possible to sort these two velocities out, using other distance measurements, such as the one Tully and Fischer devised. It turns out that galaxies almost never move much faster than a thousand kilometers per second, about one-three-hundredth as fast as the speed of light.

Thus, in the (at most) twenty billion years since the Big Bang, a galaxy, or the matter that would make up a galaxy, could have moved only about sixty-five million light-years. But if you start out with matter spread smoothly through space, and if you can move it only sixty-five million light-years, you just can't build up objects as vast and dense as Tully's complexes. For these objects to form, matter must have moved at least 270 million light-years. This would have taken around eighty billion years at one thousand kilometers per second, *four times longer* than the time allowed by the Big Bang theorists.

The situation is really worse than this, because the matter would first have to accelerate to this speed. Even before this, a seed mass big enough to attract matter over such distances would have to form. So an age of one hundred billion years for such complexes is conservative. Simply put, if Tully's objects exist, the universe cannot have begun twenty billion years ago.

The initial reaction of most cosmologists to Tully's observations was to reject them altogether. "I think Tully is just connecting the dots in claiming to see these clusters of clusters," Marc Davis, a Berkeley cosmologist, commented dismissively. But that position has become increasingly untenable. During 1987 Tully carefully analyzed his data, proving that it is extremely unlikely that the clustering could have come about as a chance arrangement of random scattered clusters, or as a result of flaws in his calculations.

In 1990 the existence of these huge objects was confirmed by several teams of astronomers. The most dramatic work was that of Margaret J. Geller and John P. Huchra of the Harvard Smithsonian Center for Astrophysics, who are mapping galaxies within about six hundred million light-years of earth. In November of 1989 they announced their latest results, revealing what they called the "Great Wall," a huge sheet of galaxies stretching in every direction off the region mapped. The sheet, more than two hundred million light-years across and seven hundred million light-years long, but only about twenty million light-years thick, coincides with a part of one of the supercluster complexes mapped by Tully. The difference is that the new results involve over five thousand individual galaxies, and thus are almost impossible to question as statistical flukes.

Still larger structures were uncovered by an international team of American, British, and Hungarian observers, including David Koo of Lick Observatory and T. J. Broadhurst of the University of

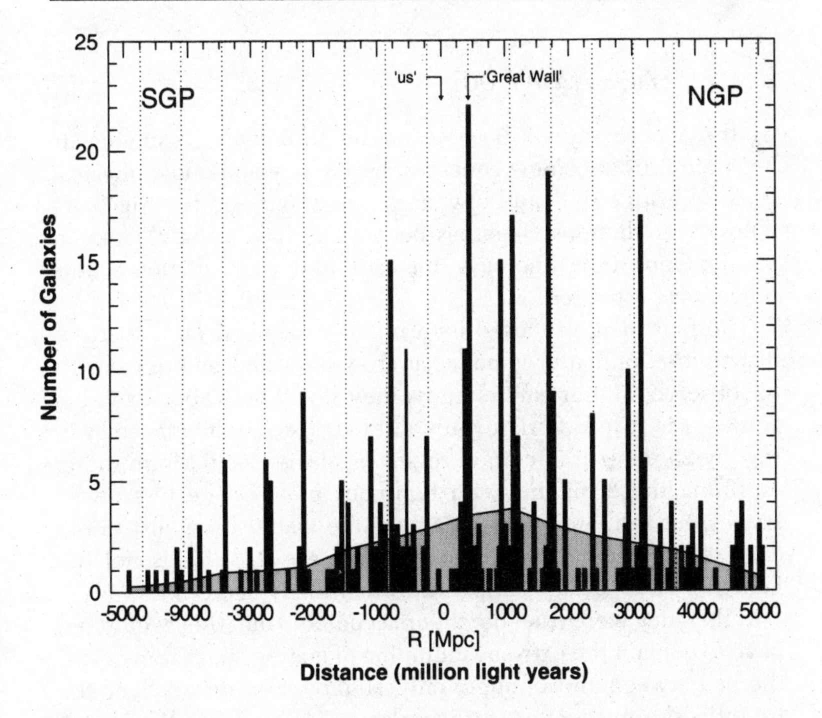

Fig. 1.5. A plot of the number of galaxies versus distance from earth in two small pieces of the sky. Distance increases with the increasing redshift of light from the galaxies. The galaxies are clumped in narrow peaks separated by voids about 700 million light-years across.

Durham, in England. The team looked very deeply into space in two opposing directions, scanning only narrow "wells" in space. To their surprise they found galaxies clustered in thin bands, evenly spaced some six hundred million light-years apart like the rungs of a titanic ladder (Fig. 1.5). The entire pattern stretched across a quarter of a diameter of the observable universe, a distance of over seven billion light-years. The galaxies seemed to be moving very slowly relative to one another—no more than five hundred kilometers per second. At that speed, the gigantic void-and-shell pattern appears to have taken at least 150 billion years to form—seven or eight times the number of years since the Big Bang allegedly took place.

SEEKING A WAY OUT

As these observations became harder to dispute, cosmologists began to introduce new concepts, based on wholly new physical laws, to bridge the gap between observations and the Big Bang theory's predictions. This has become an increasingly common phenomenon in cosmology—for each new contradiction a new process is postulated.

The first idea, proposed by a number of theorists, is that the distribution of matter is not accurately indicated by the galaxies we observe. Matter isn't clumpy, they say, it only appears to be. If matter is spread fairly evenly through space, but were denser, say, by 25 percent in certain regions, galaxies would form there, outlining these regions with luminous bodies. The less dense spaces, though, aren't *truly* empty—the matter there just didn't coalesce, for some reason, so we can't see it. (This is not the famous "dark matter," simply diffuse ordinary matter.)

If this idea were true, the theorists pointed out, they would not have to explain the extreme clumping of matter; the matter is still there, between the clumps, only slightly less dense than the brightly shining matter in the galaxies of the Great Wall or of Tully's complexes.

This theory is entirely ad hoc—that is, it was invented to bridge the gap between theory and observation. There is no reason to believe that there is a lot of gas in the voids, or that galaxies would not form in this gas. But more to the point, the "biased galaxy formation" theory is contradicted by observation.

Astronomers can deduce fairly accurately how much matter is actually concentrated into such objects as the Great Wall because such massive objects attract everything around them. By observing the velocities of galaxies around such objects, it is possible to "weigh" them. This is exactly what one astronomer, E. Shaya of Columbia University, did in 1989. Using Tully's maps of the region within 150 million light-years of earth, Shaya used the observed galactic velocities to measure matter density, assuming that all of it is concentrated in the regions traced by galaxies—that is, assuming *no* dim matter. He calculated that the average matter density is about one atom per ten cubic meters of space.

The question is, is this *all* the matter there is, or can there be additional, diffuse matter that isn't detectable by its gravitational attraction? It turns out that the Big Bang theory itself can predict the amount and density of ordinary matter. One of the two key predictions of the Big Bang is the abundance of helium and of two rare light isotopes—deuterium (heavy hydrogen) and lithium. These predictions depend on the density of the universe—the denser the nuclear soup, the more lithium and the less deuterium and helium would be produced.

Astronomers can measure the abundance of these elements quite accurately by observing the spectra of light from stars and other galaxies; from this they can calculate how much there really is—about 24 percent for helium, one part in one hundred thousand for deuterium, and one part in ten billion for lithium.

For theory to match observation, the overall matter density must be around one atom per ten cubic meters—just what Shaya obtained by "weighing" the matter concentrated in the clusters of galaxies. So if the Big Bang theory of element creation is right, there *can't* be any matter left over to fill up the voids, and the "biasing" idea is wrong. On the other hand, if we accept the idea that there is a great deal more ordinary matter than we see, the basic predictions of the Big Bang as to how much helium, lithium, and deuterium are produced are wrong. As a result of such contradictions, the popularity of this notion has drastically declined.

Other ideas have also fallen by the wayside. For example, Dr. Jeremiah Ostriker of Princeton University and others proposed the idea of the cosmic string—infinitely thin, infinitely dense objects, but stretching in length from one side of the observable universe to the other. While this remarkable string could thread the finest needle, it would be difficult to sew with, since it moves at nearly the speed of light, and a meter of the stuff weighs about as much as the moon.

A cosmic string, because of its immense mass, might pull matter from a huge distance, forming the long ribbons of the superclusters. Unfortunately, even cosmic strings could not help to overcome the main problem, the amount of time it takes to form supercluster complexes. They have another serious disadvantage —there is absolutely no evidence that they exist outside the blackboards and computers of cosmologists. They are hypothetical entities, predicted by theories that have no experimental verification.

And what about the problem of the apparent age of the supercluster complexes? "Perhaps matter moved faster in the past than it does now," speculate cosmologists, "so large objects could be built up quicker." So one unknown process accelerates matter to high speed, blowing it out of the voids, while another unknown process conveniently puts the brakes on, slowing the matter down to the observed sedate speeds before the galaxies form.

But enormous velocities would be needed to form the Great Wall and the supercluster complexes in the time since the Big Bang—about 2,000 km/sec for the Great Wall, 3,000 km/sec for Tully's complexes, and a speedy 5,000 km/sec to hollow out the voids observed by the American-British-Hungarian team. If this matter is now moving at only 500 km/sec the energy tied up in its motion had to be dissipated. Just as a car's brakes convert energy of motion into heat, which is radiated into the air, so the vast energy of the primordial matter would have to be radiated away. Matter colliding at several thousand kilometers per second would radiate very intense X-rays. And there is indeed a universal X-ray background, but the amount of energy in it is *one hundred times* less than what would be released by braking the speeding matter. So, where is this energy?

Theorists speculate that a *third* unknown process might convert this high-energy X-ray radiation to some other sort of radiation. Astronomers have observed only one type of radiation intense enough to contain the enormous amount of energy which would result from the hypothetical "braking" of matter—the cosmic microwave background. This even bath of microwaves, radio waves each measuring about a millimeter long, comes from every part of the sky and is considered the key piece of evidence that there was a Big Bang. According to conventional cosmology, the background is the dilute afterglow of the titanic explosion that created the universe. It reflects the state of the universe only a few hundred thousand years after the Big Bang. If the large-scale structures were created after this time, the energy released in slowing the speeding matter would show up in the background radiation.

Radiation can be described by its spectrum, a curve that shows how much power the radiation has at various frequencies. The Big Bang theory predicts that the cosmic background radiation must have a black-body spectrum—that is, the spectrum of an object in thermal equilibrium, neither absorbing nor giving up heat to its surroundings. Obviously, if the origin of the background radiation is an explosion involving the entire universe, it *must* be in equilibrium—there are no surroundings to get energy from or give it to.

The black-body spectrum is described by a simple mathematical formula that was worked out by Max Planck at the beginning of the century. Plotted on a graph, it rises slowly to a peak as frequency increases, and then falls off rapidly. This shape is the same no matter what the temperature of the object emitting the radiation is; only the frequency of the peak and its power change as the temperature changes.

After the discovery of the background radiation, astronomers used radio telescopes to measure its spectrum at shorter and shorter wavelengths. In every case the measurements fit the black-body curve predicted by the theory. This was considered a great confirmation of the Big Bang.

But, as the problem of large-scale structure became evident, cosmologists hoped that at short wavelengths the observed spectrum would differ slightly from a black-body. They predicted that it would have a little bump indicating the release of energy after the Big Bang—the energy needed to both start and stop large-scale motions. Since the earth's atmosphere absorbs the shorter-wavelength microwaves, radio telescopes would have to be lifted above the atmosphere in balloons, rockets, or satellites. In 1987 a Japanese rocket bearing an American instrument designed by Paul Richards and his colleagues at Berkeley finally succeeded in measuring the short-wavelength spectrum at three frequencies, and indeed they detected an excess of radiation over the predicted black-body. The catch was that the excess was too much of a good thing. It was *so* big, one-tenth of the total energy of the background, that it could not be accounted for by the slowing down of matter or by anything else. Instead of helping Big Bang theory, the new data just brought another headache to the theoreticians.

As a result, cosmologists eagerly awaited the first results from the Cosmic Background Explorer (COBE) Satellite. COBE, launched by a NASA Delta rocket in November of 1989, carried three extremely sensitive instruments. An infrared spectrometer was expected to produce definitive results on the spectrum of the background, since it would measure it at over one hundred wavelengths between one hundred microns and ten millimeters, with .1 percent accuracy. Theorists hoped that COBE would find a smaller excess radiation, perhaps one-third of what Richards had found.

But again they were disappointed. Preliminary results from COBE were announced in January of 1990 at the American Astronomical Society meeting: to everyone's surprise, the instrument detected *no* variation from a black-body spectrum (Fig. 1.6). There was *no* release of energy in excess of about 1 percent of the energy in the background itself, no more than one-tenth of that measured by Richards. Since the COBE instruments are highly sensitive and carry their own calibrations with them, it seemed clear that Richards's results were simply wrong.

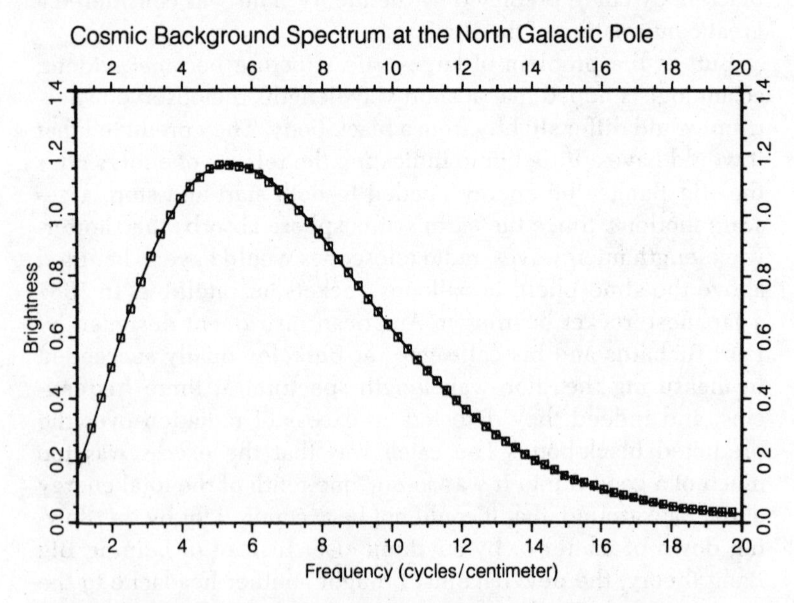

Fig. 1.6. COBE's measurements of the Cosmic Background spectrum (squares) showed no variation from the black-body spectrum (curve).

Now initially the cosmologists thought that this was just great—the black-body curve predicted by the Big Bang was exactly right. When the results were announced at an Astronomical Society meeting, there was actual cheering (not a common event at scientific conferences!). But after a few hours, theorists realized that this was actually bad news: if the excess radiation observed by Richards was too hot for the Big Bang, the lack of *any* excess observed by COBE is too cold. Since there is no variation from a black-body spectrum, there is no energetic process vigorous enough either to create, in twenty billion years, the large-scale structures astronomers have observed or to stop their headlong motion once they were created.

Dissipating the energy from the Great Wall's formation in twenty billion years would create a 1 percent distortion in the background spectrum. For Tully's structures 2 percent would be needed, and for the structure discovered by Koo and colleagues, 5 percent of the energy in the background would be needed. The COBE results ruled out such large energy releases. Thus the microwave spectrum is "too perfect." The close correspondence to the black-body curve, seen as confirmation of the Big Bang theory, at the same time rules out any way of forming the large-scale structure of the universe from the Big Bang.

The structures could not have formed *before* the epoch of the microwave background either. According to Big Bang theory, any concentration of matter present at that time would show up as hotter and brighter spots in the intensity of the background radiation. But even prior to COBE, ground-based observation had ruled out fluctuations from point to point of more than one part in thirty thousand. COBE confirmed these results. If the large-scale structures existed before the background formed, major fluctuations at least a thousand times larger should have been observed.

Again, this smooth perfection of the background, the same in all directions, has been cited as key evidence of the Big Bang and of the homogeneity of the early universe. Yet this very perfection makes it impossible for the theory to explain how today's clumpy universe could have come to be. So there is simply no way to form these objects in twenty billion years.

Nor can the Big Bang be moved back in time. The estimate that the Big Bang occurred ten or twenty billion years ago is based on measuring galaxies' distance from us, and the speed at which galaxies appear to be receding from one another. If galaxies receding at half the speed of light appear to be about five or ten billion light-years away now, cosmologists reason, they were all much closer ten or twenty billion years ago. So to move the Big Bang back hundreds of billions of years, cosmologists must hypothesize a bizarre two-step expansion: an initial explosion to get things going, a pause of a few hundred billion years to allow time for large objects to form, and a resumed explosion to get things going again, so that they only *appear* to have started twenty billion years ago.

Here the questions multiply like rabbits. But the underlying problem is basic to science. A theory is tested by comparing predictions derived from it with observations. If a theorist merely introduces some new and arbitrary modification in his theory to fit the new observations, like the epicycles of Ptolemy's cosmos, scientific method is abandoned.

Yet Big Bang theory is supported in great part by arbitrary, hypothetical entities, such as cosmic strings. As Tully puts it, "It's disturbing to see that there is a new theory every time there's a new observation."

Despite the many new hypotheses, there remains no way to begin with the perfect universe of the Big Bang and arrive at the complex, structured universe of today in twenty billion years. As one COBE scientist, George Smoot of the University of California at Berkeley, put it, "Using the forces we now know, you can't make the universe we know now."

THE DARK MATTER THAT WASN'T THERE

The problem of large-scale structure is itself a serious challenge to the Big Bang, but it is not the only one: a closely related problem is the evidence that dark matter does not exist.

Dark matter is perhaps the strangest feature of conventional cosmology. According to most cosmologists nearly 99 percent of the universe is unobservable—dark, emitting no radiation at all. The universe we do see—stars, galaxies, and all—is only 1 or 2 percent of the total. The rest is some strange and unknown form of matter, particles necessitated by theory but never observed.

This curious concept was introduced a decade ago and has since become a fundamental part of the modern Big Bang cosmology.

Long before the question of supercluster formation emerged, cosmologists realized that there is a difficulty with forming even objects such as galaxies. As we've seen, Big Bang theory assumes that these objects grew by gravitational attraction from tiny clumps, called fluctuations, in the early universe.

As early as 1967 Peebles and Joseph Silk had concluded that such primordial fluctuations should show up as fluctuations in the brightness or temperature of the microwave background. If matter was unevenly distributed at the time the microwave background originated, around a million years or so after the Big Bang, then the background produced by that hot matter would not be isotropic (uniform), but would have irregular hot spots, or "anisotropies." By 1970 they had calculated that this variation in temperature should amount to five or six parts per thousand.

At the time, measurements were not sufficiently accurate to test this prediction. But in 1973 observers showed that the anisotropy must be no more than about one part in a thousand. Throughout the seventies, observers continually lowered the limits of the anisotropy, and theorists modified their theories to make new predictions below these limits. Unfortunately, by 1979 it had become clear that this game could not continue, since there was no anisotropy at even one part in ten thousand—and *every* theory required at least a few times that amount.

The theorists realized that there was just too little matter in the universe. The less matter, the less gravity, and hence the more slowly little fluctuations would grow into large galaxies. Thus if the fluctuations were very small to start with, more matter was needed to make them grow faster.

Astronomers had a pretty good idea of how much matter we can see. They simply counted the galaxies. Knowing how bright stars of a given mass are, they could calculate roughly how much mass there is in a given volume of space, hence the density of the universe—something like one atom for every ten cubic meters of space.

Cosmologists found that this was not enough. They needed a *hundred times* more. They calculated that for galaxies to have formed as a result of the Big Bang, there must have been so much matter in the universe that its gravitation would eventually halt its expansion. But that required a density of about ten atoms per cubic meter. Cosmologists decided to represent the density of the universe as a ratio to the density needed to stop the expansion, a ratio they termed "omega." If there were just enough matter to stop the expansion, omega would equal 1. It appeared, however, that omega was really about .01 or .02—only a few one-hundredths of the matter needed to stop the expansion of the universe, and far too little to magnify the fluctuations fast enough to form galaxies.

This is where the dark matter came in. *If* omega is really 1, or close to it, then gravity would act so swiftly that even a tiny fluctuation could have grown to galaxy size in the time since the Big Bang. So theorists simply *assumed* that this was true (if it wasn't, the whole theory would collapse). But the observers could not see nearly this much matter, with either optical or radio telescopes. Since it *had* to exist but *couldn't* be seen, it could only be one thing—unobservable, "dark." Dark matter was "the little man who wasn't there."

But that's not all: dark matter had to be quite different from ordinary matter. As mentioned earlier, one of the two key predictions of the Big Bang was the abundance of helium and certain rare light isotopes—deuterium (heavy hydrogen) and lithium. These predictions also depend on the density of the universe. If the dark matter was ordinary matter, the nuclear soup of the Big Bang would have been overcooked—too much helium and lithium, not enough deuterium. For theory to match observation, omega for ordinary matter, whether dark or bright, had to be around .02 or .03, hardly more than could be seen.

If it wasn't ordinary matter, what could the dark matter be? Around 1980 worried cosmologists turned to the high-energy particle physicists. Were there any particles that might provide the dark matter but wouldn't mess up the nuclear cooking? Indeed, there just might be. Particle physicists provided a few possibilities: heavy neutrinos, axions, and WIMPs (Weakly Interacting Massive Particle—a catchall term). All these particles could provide the mass needed for an omega of 1, and they were almost impossible to observe. Their only drawback was that, as in the case of cosmic strings, there was no evidence that they exist. But unless omega equaled 1 (thus lots of dark matter), the Big Bang theory wasn't even self-consistent. For the Big Bang to work, omega *had* to be 1, and dark matter *had* to exist.

So, like the White Queen in *Through the Looking Glass* who convinced herself of several impossible things before breakfast, cosmologists decided that 99 percent of the universe was hypothetical, unobservable particles. But cosmologists were comforted that there was some evidence that some dark matter could exist. And if some, why not more?

SEARCHING FOR DARK MATTER

The evidence was in studies of the rotation of galaxies, and of the motions of galaxies in groups and clusters. Galaxies rotate like pinwheels and move through galactic clusters in looping orbits. By measuring the redshifts of stars or gas clouds in galaxies, or of galaxies in clusters of galaxies, astronomers could deduce the speed of rotation of the galaxies *and* the speeds of the galaxies themselves. Now if the galaxies and galactic clusters were held together by gravity, as astronomers assumed must be the case, the mass of a galaxy or a cluster could be found from Newton's law of gravity. The greater the velocities of the stars in a galaxy, or of galaxies in a cluster, the stronger the force needed to hold them in orbit; the stronger the gravity, the more mass there must be producing the gravitational attraction. This is like measuring the strength of an Olympic hammer-thrower by measuring how fast he can whirl the hammer around without letting it go. The faster the hammer whirls around, the stronger the hammer-thrower (Fig. 1.7).

Astronomers found that there seemed to be more mass in galaxies, measured in this way, than could be accounted for by the stars. There also seemed to be more mass in clusters than in the galaxies that made them up—five to ten times more. Perhaps, astronomers thought, this extra mass is the dark matter.

Unfortunately, there was only enough to bring omega up to .1, far too little to "close the universe" and solve the various problems confronting the Big Bang theory. But, cosmologists reasoned, at least there is some dark matter, so perhaps there is more

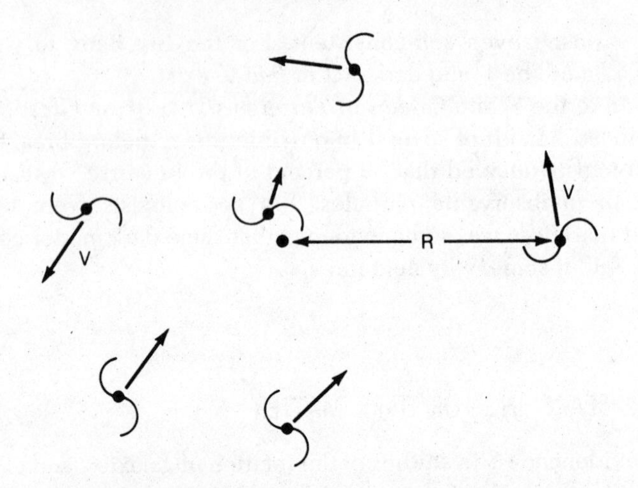

Fig. 1.7. Measuring a cluster's mass. Using redshifts, astronomers can measure the velocities of galaxies moving in the gravitational field of a cluster. They can also measure the distance (R) of each galaxy from the cluster's center. Knowing both numbers, they can estimate the mass of the galaxy—the bigger the cluster and the faster its galaxies, the higher its mass. To be exact, the mass is just the product of the square of the velocity times the radius of the objects, divided by G, the universal gravitational constant.

dark matter, more exotic, evenly spaced throughout the universe, not even revealing itself by its gravity.

This was, to be sure, a very slender thread to hang a theory of the entire universe on—and in 1984, that thread was cut.

Mauri Valtonen of the University of Turku, Finland, and Gene Byrd of the University of Alabama teamed up to take a critical look at this evidence for dark matter. They started with galactic clusters, where they knew there was a potential complication. The redshift of the galaxies was being used for two purposes: first, to measure the distance to the galaxies and thus to see *if* they were even part of the cluster; and second, to measure their velocities *within* the cluster. There was a potential for error: a galaxy nearer to us than the cluster to which it appeared to belong could be mistaken for one in the cluster that is moving toward us, while one farther away could be misidentified as a cluster galaxy moving away (Fig. 1.8). It would then be an "interloper"—appearing to be part of the cluster, but actually being far behind it. If these interlopers (which are not in fact part of the cluster) are included in calculations, *their velocities would drive up the apparent mass of the cluster, creating apparent mass where there is none*—"missing" mass. To go back to the hammer-thrower, the error would be the same as watching a film of the athlete and accidentally measuring the speed of a flying hammer in the background, rather than the speed of the hammer he is actually holding. If the background hammer was far faster, the strength of the athlete would be overestimated, just like the mass of the cluster. Valtonen and Byrd found a telltale sign that this was happening.

Astronomers had observed the curious fact that in virtually every cluster of galaxies the brightest galaxy seemed to be mov-

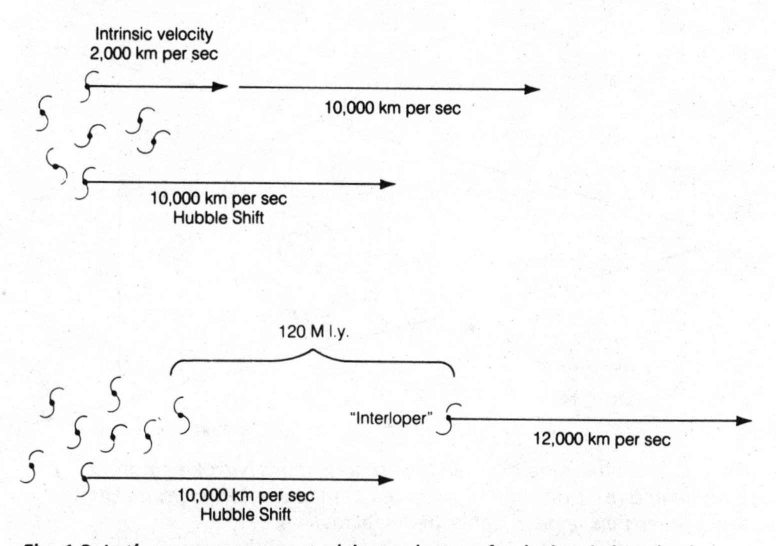

Fig. 1.8. Let's assume we are studying a cluster of galaxies six hundred million light-years (Mly) away. Its average redshift indicates a Hubble expansion velocity of 10,000 km/sec—that is, it's receding from us at that rate. Now we see in the same line of sight a galaxy with a redshift of 12,000 km/sec. We can assume it is part of the cluster and shares its 10,000 km/sec Hubble velocity—and thus that the 2,000 km/sec difference is its orbital velocity relative to the center of the cluster. Or it might have very little velocity and be located 720 Mly from us—so that the whole 12,000 km/sec is the galaxy's Hubble expansion velocity. In that case it is 120 Mly behind the cluster and not part of it at all (a typical cluster is only 10 or 12 Mly across).

ing away more slowly than the cluster it belonged to—that is, the brightest galaxy's redshift was always less than the average redshift of the cluster as a whole.

Valtonen and Byrd showed that this should be expected if some of the galaxies apparently in the cluster are really interlopers, not actual cluster members. Since the "cone" of our vision widens with distance, there will be more interlopers *behind* the group than in front of it (Fig. 1.9)—and they'll be redshifted relative to the true center of the cluster. If, as seems reasonable, the brightest galaxy (because it's largest) is generally near the center, its redshift will be *less* than the average of all the galaxies thought to be in the group, including the predominantly background interlopers.

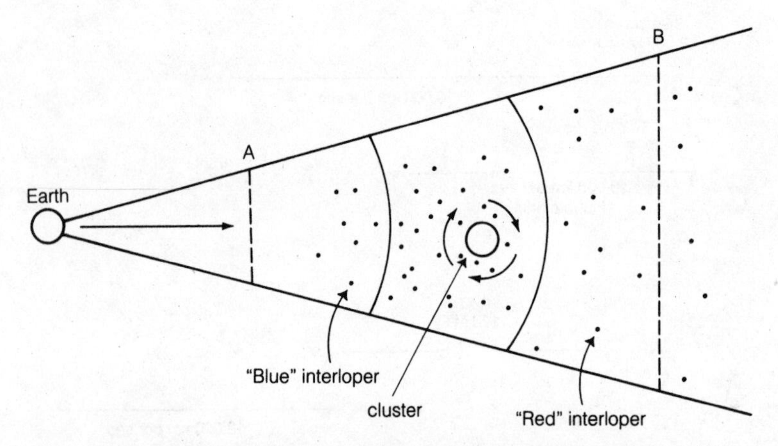

Fig. 1.9. Since the cone of vision toward a cluster is narrower in front (A) than behind (B), there will be more red than blue interlopers, making the average redshift appear higher than it actually is.

There was another reason, the two astronomers found, that the cluster mass might be overestimated. Clusters tend to be dominated by a pair of extremely heavy elliptical galaxies. Astronomers believe these galaxies grew to be as much as a thousand times more massive than our own galaxy by gravitationally swallowing smaller neighbors. But Byrd and Valtonen, using computer simulations, discovered that small galaxies might suffer a different fate: they might be caught in the pair's gravitational field and be thrown away from the cluster at high speed.

Here was another source of error. If astronomers included escaping galaxies as members of the cluster, thinking them still bound to it by gravity, again they would overestimate the gravity of the cluster and therefore its mass, just as the hammer-thrower's strength would be overestimated if the speed of the hammer was measured *after* he had let go of it. If astronomers included *both* the galaxies that had been flung away from the cluster *and* the interlopers in their calculations, the cluster's mass would be greatly exaggerated. In fact, Valtonen and Byrd found that these two errors would account for *all* of the "missing mass": in pairs of galaxies, groups of galaxies, and clusters there is *no* dark matter. And when they examined the motions of small nearby companions, they found the galaxies themselves weighed just as much as the visible matter composing them.

Valtonen and Byrd's results have now received important confirmation from Columbia's Shaya. Shaya measured the velocities and positions of hundreds of galaxies in a broad region, in effect weighing all the matter in the clusters at once. He found a value of omega, .03, very close to the value of .02 found by Byrd and Valtonen. Again, there is just no room for dark matter—about half the matter is in galaxies and their bright stars, another half in glowing gases tightly bound into the clusters and superclusters, gas that can be observed by radio telescopes.

These results have been published in leading journals, yet have stirred little discussion and no attempts at refutation. They completely eliminate any evidence for dark matter—what you see in the universe is what there is. The implication is that the many papers written about axions, heavy neutrinos, cold dark matter, and hot dark matter are entirely without any real foundation. But without dark matter, the Big Bang theorists say, no galaxies, stars, or planets can form. As a scientist on the COBE team, John Mather, quipped, "If these theories are right, we shouldn't be here."

THE PLASMA ALTERNATIVE

The test of scientific theory is the correspondence of predictions and observation, and the Big Bang has flunked. It predicts that there should be no objects in the universe older than twenty billion years and larger than 150 million light-years across. There are. It predicts that the universe, on such a large scale, should be smooth and homogeneous. The universe isn't. The theory predicts that, to produce the galaxies we see around us from the tiny fluctuations evident in the microwave background, there must be a hundred times as much dark matter as visible matter. There's no evidence that there's *any* dark matter at all. And if there is no dark matter, the theory predicts, no galaxies will form. Yet there they are, scattered across the sky. We live in one.

Dozens of new papers on the Big Bang are published every month, but less than a handful question its basic validity. With so many scientists assuming that it's right, abandoning it is not that easy. "The Big Bang could fail altogether," conceded Harvard's George Field. "It's a question of taste as to when you jump ship and go off into the unknown. I myself am conservative and I'll stay with it for now."

Historically, few theories in science have been abandoned without a clear alternative in sight. For decades, there has been no evident alternative cosmology. Now there is one: plasma cosmology. Its pioneer is Hannes Alfvén, a Swedish Nobel laureate and the virtual founder of modern plasma physics.

To Alfvén, the most critical difference between his approach and that of the Big Bang cosmologists is one of method. "When men think about the universe, there is always a conflict between the mythical and the empirical scientific approach," he explains. "In myth, one tries to deduce how the gods must have created the world, what perfect principle must have been used." This, he says, is the method of conventional cosmology today: to begin from a mathematical theory, to deduce from that theory how the universe *must* have begun, and to work forward from that beginning to the present-day cosmos. The Big Bang fails scientifically because it seeks to derive the present, historically formed universe from a hypothetical perfection in the past. All the contradictions with observation stem from this fundamental flaw (as we shall see in greater detail in Chapter Four).

The other method is the one Alfvén himself employs. "I have always believed that astrophysics should be the extrapolation of laboratory physics, that we must begin from the present universe and work our way backward to progressively more remote and uncertain epochs." This method begins with *observation*—observation in the laboratory, from space probes, observation of the universe at large, and derives theories from that observation rather than beginning from theory and pure mathematics.

According to Alfvén, the evolution of the universe in the past must be explicable in terms of the processes occurring in the universe today: events occurring in the depths of space *can* be explained in terms of phenomena we study in the laboratory on earth. Such an approach rules out such concepts as an origin of the universe out of nothingness, a beginning to time, or a Big Bang. Since nowhere do we see something emerge from nothing, we have no reason to think this occurred in the distant past. Instead, plasma cosmology assumes that, because we now see an evolving, changing universe, the universe has always existed and always evolved, and will exist and evolve for an infinite time to come.

There is a second critical difference in the two approaches to cosmology. In contrast to the Big Bang universe, the plasma universe, as Alfvén calls his conception, is formed and controlled by electricity and magnetism, not just gravitation—it is, in fact, incomprehensible without electrical currents and magnetic fields.

The two differences are related. The Big Bang sees the universe in terms of gravity alone—in particular, Einstein's theory of general relativity. Gravity is such a weak force that its effects are evident only when one is dealing with enormous masses—such as the earth we live on. Only very powerful gravitational fields, far more powerful than earth's, show the principal consequence of general relativity—the curvature of space by gravitating bodies—as anything other than a tiny correction to Newton's laws. The exotic effects of such powerful fields, central to conventional cosmology, cannot be either studied or applied on earth. Moreover, the exotic particles created in the Big Bang are impossible to generate on earth even in the most powerful of particle accelerators. Thus for the Big Bang there is a complete separation between the celestial and the mundane, between what is important here on earth, in technology, and what is important in the cosmos. Cosmology has become the purest of pure science, devoid of connection or application to the humble day-to-day world.

But the electromagnetism that is the basis of plasma cosmology is also the basis of our thoroughly technological society: electricity and magnetism are applied every instant to run our factories, televisions, cars, and computers. Plasmas are studied not only to learn about the universe but to study how radio and radar waves are propagated, how computer screens can be more brightly lit, how cheaper power can be generated. Plasma cosmology derives, of necessity, from the interplay between the problems of astrophysics and those of technology, between the celestial and the mundane.

The plasma universe is not only studied differently from the universe of the Big Bang, it also *behaves* differently. "I have never thought that you can get the extremely clumpy, heterogeneous universe we have today from a smooth and homogeneous one dominated by gravitation," Alfvén says. But plasma becomes inhomogeneous naturally. From the thirties Alfvén's scientific career has been devoted to studying and explaining the manifold ways in which plasma, electrical currents, and magnetic fields work to concentrate matter and energy, to make the universe the complex, dynamic, and uneven place that it is.

▪ PLASMA WHIRLWINDS

As a boy in Sweden, Alfvén was fascinated by the spectacular displays of the northern lights, the moving curtains of filaments and spikes. "Our ancestors called them 'the Spears of Odin' and they look so close that they might fall on your head," he jokes. As a young scientist he learned that the Norwegian physicist Kristian Birkeland had explained the aurora as the effect of electrical currents streaming through plasma above the earth. In his own experiments in nuclear physics labs, Alfvén saw the same lacy filaments: "Whenever a piece of vacuum equipment started to misbehave, there they were," he recalls. They were there, too, in photographs of solar prominences and of the distant Veil and Orion nebulas (Fig. 1.10).

Many investigators had analyzed the laboratory filaments before, so Alfvén knew what they were: tiny electromagnetic vortices that snake through a plasma, carrying electrical currents. The vortices are produced by a phenomenon known as the "pinch effect." A straight thread of electrical current flowing through a plasma produces a cylindrical magnetic field, which

Fig. 1.10. Filamentary structure is evident in the Orion nebula. The nebula is a mass of heated plasma surrounding stars.

attracts other currents flowing in the same direction. Thus the tiny current threads tend to "pinch" together, drawing the plasma with them (Fig. 1.11). The converging threads twine into a plasma rope, much as water converging toward a drain generates a swirling vortex, or air rushes together in a tornado. The filaments are plasma whirlwinds.

Almost any plasma generates inhomogeneity, pinching itself together into dense, swirling filaments, separated by diffuse

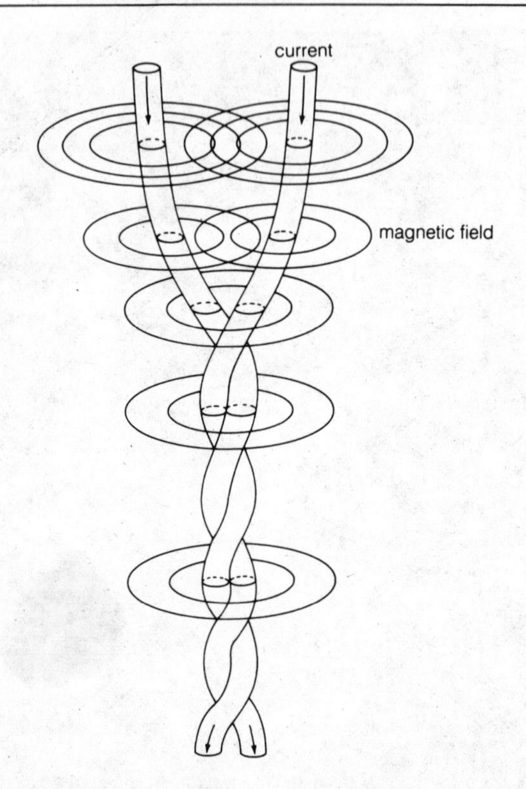

Fig. 1.11. An electrical current creates a magnetic field around it, while a magnetic field bends an electrical current. These effects permit parallel currents in a plasma to attract each other and twist into a plasma vortex filament with magnetic fields and electrical currents in the same helical pattern.

voids. Alfvén believes that the filaments seen in the laboratory, in the sun, in nebulas, are all one phenomenon.

Magnetic fields and currents can concentrate matter and energy far faster and more effectively than can gravity. The magnetic force of a plasma thread increases with the velocity of the plasma. This leads to a feedback effect: as threads are pulled into the vortex, they move faster, which increases the force on the threads of current and pulls them still faster into the filament. In addition, a contracting mass tends to spin faster and faster, like an ice skater who pulls in his or her arms. This generates a centrifugal force which fights the contraction. Magnetic filaments can carry away this excess spin, or angular momentum, allowing further contraction, while gravity cannot.

Over a period of decades, Alfvén and a small group of colleagues applied concepts learned from the laboratory study of plasma to the mysteries of the heavens. He proposed new theories to explain cosmic rays, solar flares and prominences, and the origin of the solar system—and met initially with fierce opposition or indifference. Yet as the years passed, the idea that space is alive with networks of electrical currents and magnetic fields filled with plasma filaments was confirmed by observation and gradually accepted—often after most scientists had forgotten who first proposed the theories, and after Alfvén himself had long since turned to other problems.

The turning point came in the late sixties, when space probes explored the solar system. "Having probes in space was like having a cataract removed," says Alfvén. "We could see things never seen before, just as Galileo could with his telescope." The early probes showed that filaments do exist near earth, where currents flow along the lines of the geomagnetic field and create the aurora as they strike the atmosphere. Later, in the seventies, the Pioneer and Voyager spacecraft detected similar currents and filaments around Jupiter, Saturn, and Uranus.

Currents and filaments are now known to exist throughout the solar system, and astronomers have come to accept Alfvén's theories about the origin of the solar system and the electromagnetic origin of cosmic rays.

A FILAMENTARY UNIVERSE

By the late seventies many scientists studying the solar system were convinced that electrical currents and magnetic fields do indeed produce a complex, highly inhomogeneous filamentary structure in space, just as Alfvén had theorized. For Alfvén, however, a description of the solar system was only a first step. Plasmas should look similar no matter how big or small they are. "If we can extrapolate from the laboratory to the solar system, which is a hundred trillion times larger," he asks, "then why shouldn't plasma behave the same way for the entire observable universe, another hundred trillion times larger?"

In 1977 he applied his concepts to the next order, the galaxies, proposing a new way to explain the violent outbursts of energy that occur in their cores. Conventional wisdom ascribes their highly concentrated outbursts to black holes, bizarre objects with a gravitational field so intense that light itself cannot escape it. Alfvén had a less exotic concept based on laboratory experience with electrical systems.

In his theory, a galaxy, spinning in the magnetic fields of intergalactic space, generates electricity, as any conductor does when it moves through a magnetic field (the same phenomenon is at work in any electrical generator). The huge electrical current produced by the galaxy flows in great filamentary spirals toward the center of the galaxy, where it turns and flows out along the spin axis. This galactic current then short-circuits, driving a vast amount of energy into the galactic core. The galaxy "blows a fuse": powerful electrical fields are created in the nucleus which accelerate intense jets of electrons and ions out along the axis.

Again, few astrophysicists took Alfvén's description of electrical currents and magnetic fields of galactic strength seriously. But the new theory soon received support. In 1979 Tony Peratt, a plasma physicist and former student of Alfvén's, began to see things in the lab that seemed to confirm Alfvén's theory. Working at San Diego's Maxwell Laboratory with machines that produced powerful electrical currents in plasma, he saw the current develop vortex filaments, which twisted up into what looked like

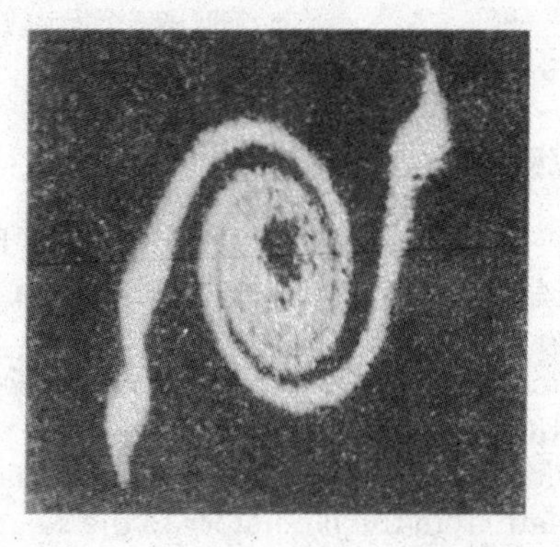

Fig. 1.12. Spiral filaments of current and glowing plasma, a few millimeters across, are formed in the lab, resembling the mighty spiral galaxies of space (A. Peratt).

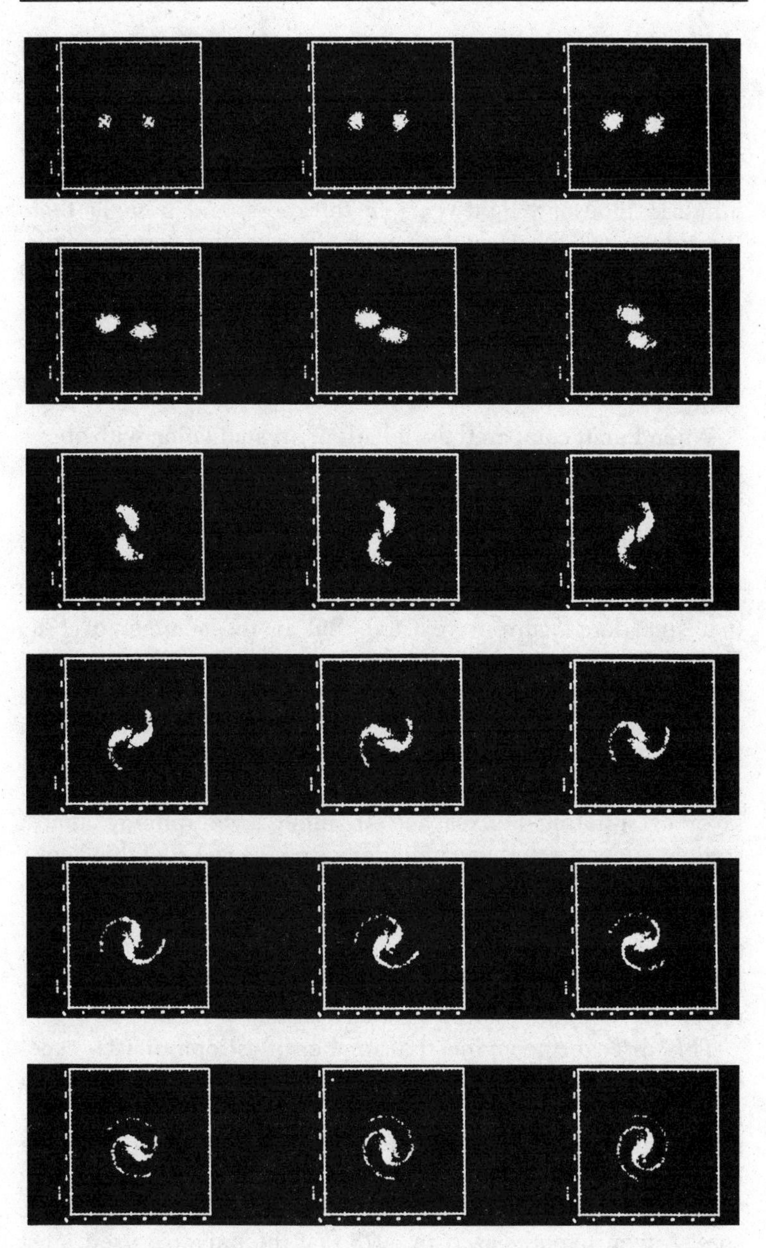

Fig. 1.13. A computer simulation shows how two currents in space (seen here in cross section) interact through their magnetic fields to produce a spiral galaxy (A. Peratt).

tiny spiral galaxies (Fig. 1.12)—a phenomenon that, Peratt later learned, had first been observed in the fifties. Curious about these tiny plasma "galaxies," he used a recently developed computer program to simulate the action of plasma on a galactic scale.

In his model he created two filaments of current, each a hundred thousand light-years in thickness, and brought them together to see what would happen. The results were dramatic: the two filaments merged, generating the graceful forms of spiral galaxies (Fig. 1.13). As Alfvén had predicted, the simulation showed currents streaming along slender filaments toward the galactic core, from which intense bursts of radiation were emitted.

When Peratt compared the details of his simulation with observations of real galaxies, there was excellent agreement: "I found in photographic atlases of galaxies examples of just about everything I saw in simulations—the shapes, the radio emission, all were the same as in the computer."

Astrophysicists either ignored the work or remained skeptical that such large currents existed. But in the summer of 1984 Farhad Yusef-Zadeh of Columbia University, and colleagues at the Very Large Array radio telescope in Zoccoro, New Mexico, discovered large-scale magnetic vortex filaments at the heart of our own Milky Way galaxy. Hundreds of light-years long, they were a textbook example of Alfvén and Peratt's vortices: an outer layer of spiraling helixes and an inner layer running almost straight along the axis of a cylinder (as on the jacket of this book), the whole pattern arcing out of the plane of the galaxy straight up into its axis of rotation. Their magnetic field strength, at least a few ten-thousandths that of the earth's surface, was also just what Peratt's simulations predicted—and far above what most astrophysicists thought possible on such a scale.

This discovery convinced a number of astrophysicists, especially those already familiar with the work on solar system plasma, of the reality of current filaments in space. The alignments and shape of the galactic filaments simply could not have been created by gravity.

Following up his 1977 work on magnetic storms at the galactic core, Alfvén hypothesized in 1978 that the universe itself must have an inhomogeneous, cellular structure. In any plasma, from laboratory to intergalactic scale, filaments form naturally. Currents moving in the same direction attract each other, and small currents formed by the random motion of the plasma merge and grow into bigger currents. Given enough time, currents and filaments of any magnitude, up to and including supercluster complexes, could form—in fact, *must* form.

Peratt, in creating his computer models, had also hypothesized that galaxies themselves are created by still vaster filaments, which then provide the magnetic fields that drive galaxies to generate currents. Peratt knew from experiments that such filaments were typically ten thousand times longer than they are wide; thus the galactic filaments, one hundred thousand light-years across, should be about a billion light-years long. From the standpoint of plasma physics, galaxies should be strung along such filaments, groups of which would, in turn, organize into still larger ropes. This is, of course, exactly what Tully, Fischer, and others later observed while compiling their maps. As one astronomer, Margaret Haynes, commented on the twisting filaments of galaxies she and her colleagues had discovered, "The universe is just a bowl of spaghetti." Moreover, in 1989 a team of Italian and Canadian radio astronomers detected a filament of radio emissions stretched along a supercluster, coming from the region between two clusters of galaxies. Electrons trapped in a magnetic field emit radio radiation, so their finding provided indirect evidence of a river of electricity flowing through the empty space. The estimated size of the current, some five or ten million trillion amperes, was exactly that predicted by Peratt's model. The existence of filaments at the supergalactic scale—explicitly predicted by a small group of plasma theorists—was confirmed by observation.

▪ WITHOUT A BEGINNING

Plasma interactions can, given a few hundred billion years, form the supercluster complexes. For Alfvén and the slowly growing band of plasma theorists like Peratt, time is no problem. If one starts from the present and attempts to go backward in time, there is no reason to assume that there ever was a Big Bang or that the universe had any beginning.

To challenge the Big Bang, however, plasma cosmology must

account for the three observed phenomena that the Big Bang claims as evidence: the helium abundance, the microwave background, and the Hubble expansion. The first two phenomena can be explained by the same cause—massive stars generated in the formation of galaxies. In 1978 Cambridge astrophysicist Martin J. Rees had proposed that such stars would, in a few hundred million years, produce the 24 percent helium now observed: having transformed part of their hydrogen fuel into helium, they would explode into supernova, distributing the helium through space. Later, smaller stars would then form out of this helium-enriched gas. The energy the massive stars produced would be absorbed by interstellar dust, which would then emit the microwave background.

Conventional cosmologists raised a number of objections to this proposal, the most serious of which concerns the isotropy of the microwave background. If developing galaxies had generated the background radiation, we should be able to see warm spots in the background where the newly formed galaxies were. Looking out in space means looking back in time, so the early galaxies would still be observable as clumps in the microwave background—yet the microwave background is completely smooth. What process could smooth out the microwave background?

Radio astronomers had long known that electrons, trapped in magnetic fields, emit radio waves and microwaves—the process by which, as Alfvén first pointed out in 1950, celestial objects produce radiation. (This is, in fact, the principle behind a microwave oven—electrons are forced to move in a circle by magnetic fields.)

Any object that emits radiation can also absorb it, however—this is the key to the microwave background. My own work is relevant here. In the early eighties I had been working along parallel lines to Peratt, detailing plasma theories of galactic nuclei. I asked myself: Why can't electrons in intergalactic magnetic fields absorb microwave radiation and then *reemit* it? (The idea occurred almost simultaneously to Peratt and his colleague Bill Peter.) Since there would be no relation between the direction the radiation was traveling when it was absorbed and its direction as a reemission, the microwaves would be scattered. After a few scatterings, the radiation would be "smoothed out"—much as the water droplets of a fog scatter light into a near-uniform glow.

I realized, however, that a magnetic field much stronger than the average field between the galaxies would be needed. What could provide such a strong field? The obvious candidate was the jets emitted from galactic nuclei. These had powerful magnetic fields and energetic electrons, and had been observed in detail. My calculations showed that a thicket of millions of such filaments would act like a radio fog, scattering the hot spots into a smooth microwave background. Moreover, they would be nearly invisible themselves, since their radiation would be the same everywhere, just as fog droplets are hard to see in a fog. But they would have easily observable effects: they would absorb radio waves from more distant objects, which could be observed only by peeking through the random holes in this intergalactic thicket of filaments. Indeed, distant objects should appear to have less radio radiation than nearer objects.

In the fall of 1990, I published an article showing that this is exactly what happens. Galaxies that are equally bright infrared emitters, equally "hot," are fainter and fainter radio sources the farther they are from earth. This is clear observational evidence that *something* is absorbing radio waves, including microwaves, as they travel between the galaxies. But even more significant, this shows that the conventional explanation of the microwave background must be wrong. Such absorption would distort the black-body spectrum of the background if it really was the faint echo of the Big Bang. Since the spectrum, as COBE showed, is not distorted, the radiation must instead come from nearby, from the intergalactic medium itself. In equilibrium, such an absorbing medium would produce just the black-body spectrum observed.

Thus not only is there an alternative explanation for the microwave background, which naturally explains its energy, smoothness, and spectrum without the Big Bang, but the observational evidence is incompatible with a Big Bang origin for the cosmic background radiation. There is no contradiction between the smoothness of the background and the lumpiness of the universe. The background is smooth for the same reason a dense fog looks smooth in all directions, not because the universe itself was ever that smooth.

And finally, what about the Hubble expansion? As Alfvén has commented, a Big Bang will certainly produce an expansion, but an expansion does not require a Big Bang. "This is like saying that because all dogs are animals, all animals are dogs," he quips. There are, in fact, a number of possible explanations of the Hubble relation other than the Big Bang. None as yet is confirmed, or even fully worked out, but it is clear that there are alternatives.

One of the simplest suggestions, that of Alfvén and his former teacher the late Oskar Klein, begins from the known fact that when matter is produced, antimatter is also produced. Antimatter and matter have opposite electrical charges—for example, antiprotons are negatively charged, while protons are positively charged. When they combine, they annihilate one another with an enormous release of energy. Since, in the laboratory, matter and antimatter are always created in equal amounts, Alfvén and Klein reasoned that this must be true for the universe as a whole.

In general, plasma processes separate the large regions of matter and antimatter so that they don't mix. However, Alfvén and Klein hypothesize that many billions of years ago the small corner of the infinite universe that we can observe started to contract, under the influence of its own gravity. When it was about a tenth its present size, matter and antimatter started to mix, annihilating each other and generating huge quantities of energetic electrons and positrons. Trapped in magnetic fields, these particles drove the plasma apart over hundreds of millions of years. The explosions were gentle enough not to disrupt previously formed filaments of plasma, so these far more ancient objects still exist today, in expanded form—just as designs printed on a balloon persist while it is inflated.

The explosion of this epoch, some ten or twenty billion years ago, sent the plasma from which the galaxies then condensed flying outward—in the Hubble expansion. But this was in no way a Big Bang that created matter, space, and time. It was just a big bang, an explosion in one part of the universe. Alfvén is the first to admit that this explanation is not the only possible one. "The significant point," he stresses, "is that there are alternatives to the Big Bang."

SCIENCE AND IDEOLOGY

In later chapters I will discuss the scientific issues raised so far in greater depth. But from this brief survey a few conclusions are clear. The Big Bang arose initially as an explanation for the Hubble expansion—the relation of the redshifts and distances of the galaxies. The observations of the past several years have put that theory into grave doubt, contradicting all its predictions as well as its basic assumptions. A plausible alternative, plasma cosmology, has arisen, and its predictions have been systematically confirmed by observation. Moreover, it provides simple explanations for phenomena that the Big Bang cannot consistently explain—the inhomogeneous and filamentary structure of the universe, the abundance of helium, and the microwave background.

One would think that these developments would reopen a debate over the correct explanation of the Hubble expansion, and redirect theoretical work from the nuances of hypothetical creatures such as dark matter and cosmic strings to an examination of the validity of cosmology's basic assumptions. However, cosmologists have either ignored or dismissed plasma theory—few have even bothered to read about it. To P. J. E. Peebles, a pioneer of the Big Bang, Alfvén's ideas are "just silly." His colleague at Princeton, Jeremiah Ostriker, comments, "There's no observational evidence that I know of that indicates electric and magnetic forces are important on cosmological scales."

In part the problem is the increasing specialization of science. The average scientist, not only the cosmologist, reads certain journals, attends certain conferences, meets with basically the same groups of specialists year in and year out. Plasma cosmology was developed not by astronomers or theoretical cosmologists but by plasma physicists, who publish in electrical engineering and related journals, not in the magazines that most astronomers read.

To be sure, this is not entirely by choice. Alfvén, as well as far lesser known plasma physicists, have repeatedly had their papers rejected by the astrophysical journals because they contradict conventional wisdom. Again, this is not a problem unique to cosmology. "When scientists are specialized," Alfvén comments, "it's easy for orthodoxy to develop. The same individuals who formulate orthodox theory enforce it by reviewing papers submitted to journals, and grant proposals as well. From this standpoint, I think the Catholic Church was too much blamed in the case of Galileo—he was just a victim of peer review."

The system of peer review—having all papers and grant proposals controlled by a small group of "leading specialists"—has had a profoundly conservative effect on all branches of science, since theorists in particular are reluctant to admit the truth of papers that contradict their decades of work. However, while peer review would explain the dismissal of the plasma alternative, it cannot explain the reactions to the new observations which have been written by leading astronomers and published in the leading cosmological journals. For a decade now the accumulating contradictions have met not with a reexamination of basic assumptions, but with boilerplate hypotheses. Just as the medieval astronomers added epicycle after epicycle to Ptolemy's spheres in order to match his geocentric theories with observed planetary movement, so today cosmologists add dark matter to cosmic strings to inflation, papering over the yawning crevices in their theory. "It's impossible that the Big Bang is wrong," Joseph Silk, a leading astrophysicist at Berkeley, states flatly. "Perhaps we'll have to make it more complicated to cover the observations, but it's hard to think of what observations could ever refute the theory itself."

This attitude is not at all typical of the rest of science or even the rest of physics. In other branches of physics the multiplication of unsupported entities to cover up a theory's failure would not be tolerated. The ability of a scientific theory to be refuted is the key criterion that distinguishes science from metaphysics. If a theory cannot be refuted, if there is *no* observation that will disprove it, then nothing can prove it—it cannot predict anything, it is a worthless myth.

There is more than science involved here. While the Big Bang as a scientific theory is less and less supported by data, its prominence in our culture has increased. The scientific press has taken it as unquestionable truth, a touchstone of the scientific outlook. In a recent test of the scientific literacy of the American and British public, two questions were used to "test acceptance of the scientific world-picture." People were asked to agree or disagree with two propositions: "The universe began with a huge explosion" and "Human beings developed from earlier species of animal." Disagreement with Big Bang theory was equated with rejection of evolution, and scientific illiteracy. (Evidently Dr. Alfvén would have flunked this particular test!)

The ideas of modern cosmology have, as well, become increasingly tied to theology. In books like Paul Davies's *God and the New Physics*, which now fill the science shelves at bookstores, scientists and popularizers argue that the theories of the Big Bang lead to a proof of God's existence or at least to knowledge of why the universe came into existence. From these bases we can hope, in Stephen Hawking's words, "to know the mind of God."

To the interested layman, much of this seems strange indeed. The Big Bang theory starts with some peculiar premises—the universe was once smaller than the head of a pin, and there was a beginning to time. The natural response is: What came before that? Some cosmologists, such as Hawking, answer with even weirder ideas: perhaps, they speculate, tiny pulsations in the space around us, even within us, are at every instant giving birth to submicroscopic universes, tiny bubbles of space-time, that then pinch off from our universe to form another universe. From every point, even the tip of one's nose, quadrillions of universes are forming every second. Ours is only one among them, formed presumably from the tip of someone's nose in another, more ancient universe.

Stranger still, many scientists proclaim that, through these mathematical calculations and theorizings, they are approaching *the* solution for all the fundamental mysteries of the universe. In his recent book, *A Brief History of Time*, Hawking expressed the belief, shared by many of his colleagues, that perhaps within a decade cosmology will discover a Theory of Everything, a small set of equations that describe all of physics—gravitation, electromagnetism, and nuclear forces—equations so simple and elegant that, as one joked, "they can fit on a T-shirt." From these simple equations, the true reality of the universe will flow by logical deduction, not only the queer zoo of cosmic strings and multiple universe, but also galaxies, stars, planets—everything.

Whatever one's religious views, such speculations seem an odd way of arriving at a sure knowledge of the existence or non-existence of God. And it seems odder that they prevail while cosmologists ignore the observations that seem to cut at the base of this tower of theoretical fancy.

How did such a state of affairs develop in the first place? For an answer, we must take a long step back. To describe the cosmological debate in the eighties and nineties is like coming into a room in the middle of a heated argument. One needs to know how the argument developed to understand and judge it today: opposition to the Big Bang did not start in the seventies, and the Big Bang itself did not spring into being when it was popularized in the fifties and sixties.

This is more than a scientific discussion about observations and theories. The Big Bang rests on a pair of assumptions that form the core of conventional cosmology's method: the universe came into existence at a specific moment, created from nothing, and we can learn about creation and the universe as a whole by developing exact mathematical theories—that is, by our own reason, by logical deduction. We can, as Hawking and others argue, determine how the universe *must* have been formed, by sheer logical necessity, by what laws it *must* be governed, and we can then divine its "true" properties from that necessary beginning. The mathematical laws we develop are the *essence* of the universe, the reality behind all the phenomena of the visible cosmos —"the mind of God."

Plasma cosmology, however, assumes that we learn about the universe by observing processes that act in nature today. From the patterns we discern, we can derive generalizations that allow us to guess how these same processes led to the present configuration of the universe. Because today nothing comes from nothing, the reasonable hypothesis is that this has always been true— the universe, in some form, has always existed.

These two approaches to knowing the universe are not new. The ideas underlying the Big Bang and plasma cosmology have their roots not in the present century or even the present millennium, but in Greece of the fourth, fifth, and sixth centuries B.C., and in Rome of the fourth century A.D. Over the centuries these two concepts have battled, each, in turn, dominating for a while.

Alfvén has called the alternation of these two broad ideas the "cosmological pendulum," an oscillation between a mythological and a scientific approach. For Alfvén, the cosmology of today is based on the same mythological views as that of the medieval astronomers, not on the scientific tradition of Kepler and Galileo. But this pendulum does not swing in the ethereal reaches of pure ideas. Since antiquity, how people have looked at the universe has been intertwined with how they viewed their society and the needs of that society. People have projected their social ideas onto the universe and have used their cosmology, their ideas about the heavens, to justify their practices on earth. The battles between these two views of the cosmos have been linked to the most crucial questions of society and history: Is progress, the continual betterment of human life, possible? Must there always be rulers and ruled, or should those who work decide what work is to be done? In the Middle Ages, for example, the hierarchy of heavenly spheres of the Ptolemaic world was used to justify the hierarchy of king, nobles, priests, and commoners. And in the 1600s those battling for democracy used the Copernican system as a model of their ideals of equality under universal law.

To understand the cosmological debate today, we must therefore trace the origins of each side's assumptions and methods, and the historical contexts from which these ideas arose.

NOTES

1. Samuel P. Langley, "Address as retiring President of the A.A.S.," *American Journal of Science*, Series 3, Volume 37, pp. 1–23, 1889.

2 A HISTORY OF CREATION

In the beginning, only the ocean existed, upon which there appeared an egg. Out of the egg came the sun-god and from himself he begat four children: Shu and Tefnut, Keb and Nut. All these, with their father, lay upon the ocean of chaos. Then Shu and Tefnut thrust themselves between Keb and Nut. They planted their feet upon Keb and raised Nut on high so that Keb became the earth and Nut the heavens.

—EGYPTIAN MYTH, C. 2500 B.C.

As long as humans have walked the earth, they have wondered how things came to be as they are. How the world began has always been central to knowing what it is and what we are, just as one needs to know a person's history to know him.

From the start people have tied their ideas of the origin of the world to their way of learning about the world. The first way men had of learning about the universe was from authority—the stories of priests, from myth. Doubtless our cave-dweller ancestors had their own ideas and myths, but the earliest creation myths that we know of were written down in Mesopotamia and Egypt about 2500 B.C. In these stories creation is a magical-biological reproduction: gods emerge from a primeval ocean and mate with one another to produce additional deities—the earth, the sky, the heavens, and the oceans. Egyptian paintings show the sky god Shu standing on the earth god Keb and holding up the heaven goddess Nut (Fig. 2.1).

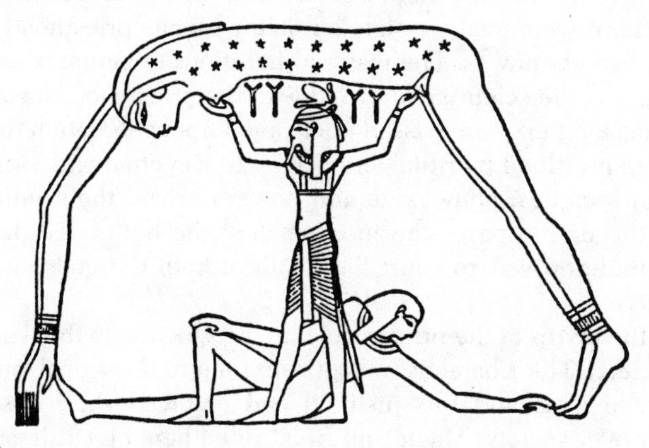

Fig. 2.1.

These priestly myths reflect how these societies were run. The first civilizations were based on large-scale irrigation agriculture, organized by a centralized priesthood headed by an all-powerful and divine king or pharaoh. The creation stories tell of these societies' origins, how their people had organized the lands between the Tigris and Euphrates rivers and in the Nile valley, had literally separated the earth from the waters by channeling swamps into canals, which superseded the chaos of agriculture dependent on fickle and sparse rains. The priests gave the credit for this vast social enterprise to the gods, the "ancestors" of the pharaohs.

According to these myths, this task was accomplished not with reason, planning, and hard work (as actually occurred), but by fertility-based magic. This too reflects how these societies functioned, for they had little use for reason. This may seem strange, since the earliest civilizations developed some of the most essential inventions of the human mind—metallurgy, writing, arithmetic, geometry, and astronomy.

Yet, once these agricultural improvements were instituted by neolithic farmers or by the first priesthoods that organized the irrigation works, these societies persisted without further technical advance for over fifteen hundred years. The social organization set up to create the irrigation works—a king and priesthood directing the work of millions of peasants—itself prevented further progress. As historian V. Gordon Childe points out, the peasants who grew food and the artisans who worked in royal workshops were totally isolated from the literate priesthood, who held absolute power. The material traditions of peasants and artisans, and the scientific knowledge of the priesthood, separated from each other, were passed unchanged from generation to generation, mystified by ritual and magic. In Egyptian and Mesopotamian societies, knowledge and power derived their authority from the divine past—the more ancient the better. To deviate from tradition was to court the wrath of both the gods and the pharaoh.

So the myths of the priests gave divine sanction to the working of society. The pharaohs and priests inherited magical powers from the gods and this justified and enforced their absolute power over society. Magic and ritual ruled here on earth, and so it must have been in the heavens, in the beginning.

▪ THE BEGINNINGS OF SCIENCE

> The formation of the world began with a vortex, formed out of chaos by Energy. This vortex started at the center and gradually spread. It separated matter into two regions, the rare, hot, dry and light material, the aether, in the outer regions, and the heavier, cooler, moist material, the air, in the inner regions. The air condensed in the center of the vortex, and out of the air, the clouds, water and earth separated. But after the formation of earth, because of the growing violence of the rotary motion, the surrounding fiery aether tore stones away from the earth and kindled them to stars, just as stones in a whirlpool rush outward more than water. The sun, moon and all the stars are stones on fire, which are moved round by the revolution of the aether.
>
> —ANAXAGORAS, c. 430 B.C.

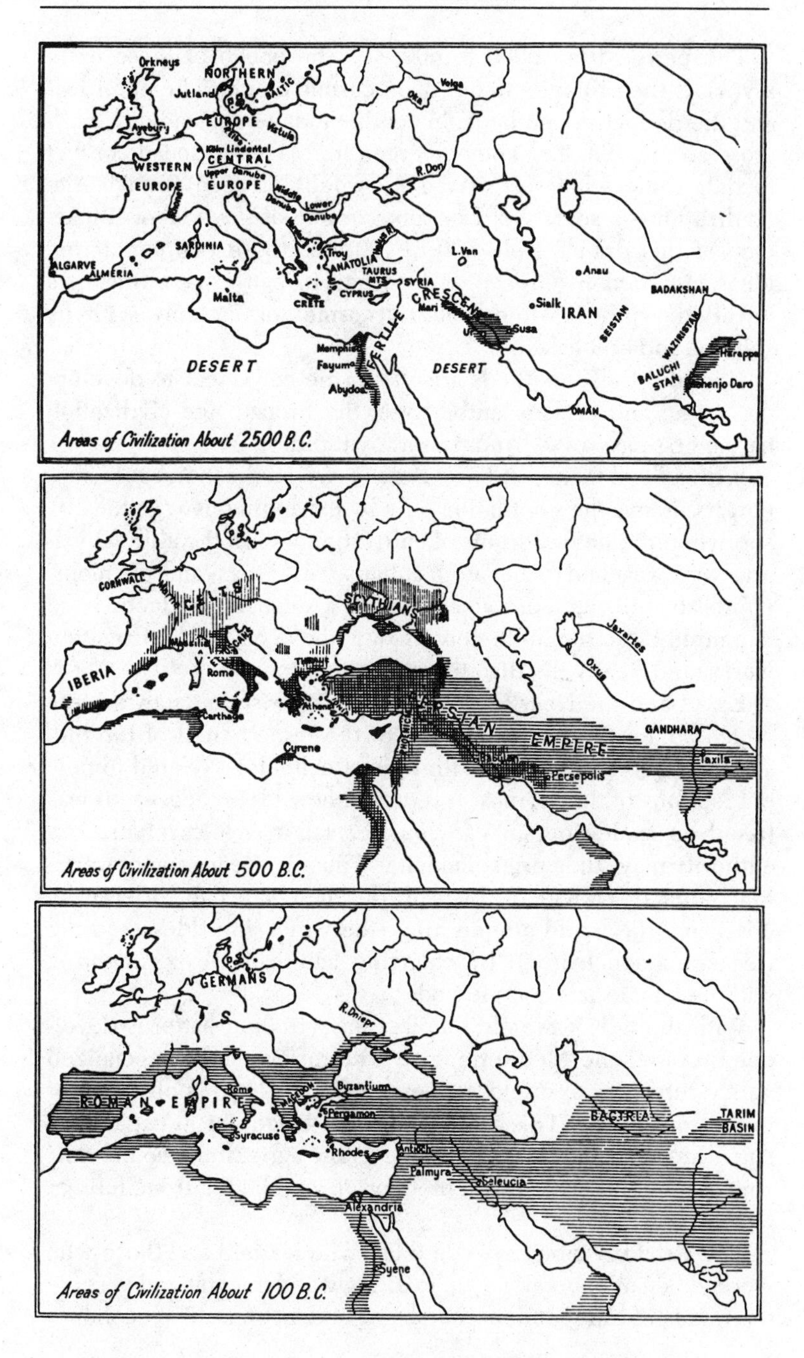

Areas of Civilization About 2500 B.C.

Areas of Civilization About 500 B.C.

Areas of Civilization About 100 B.C.

The period from 2500 to 600 B.C., the epoch of these early myths, is the prologue to our main cosmic drama. The empirical and the deductive methods, in conflict today in cosmology, both arose around 500 B.C. They emerged from a fierce social conflict to determine what sort of society would succeed Bronze Age civilization—a society of free labor or one of slave labor. In Act One of this drama, from 600 to 100 B.C., these two conflicting ideas are born in Greece and engage in battle, resulting in a lopsided synthesis—the fantastic cosmos of Ptolemy with its spheres and epicycles.

How did these methods arise? For the new ideas to develop, the magic and priestly authority of the Bronze Age civilization had to first pass away. And pass away it did.

With a fixed technical base, Bronze Age civilization could support its increasing population only by expanding geographically, and when the natural limits of cultivation within the alluvial valleys were reached, it began to collapse. All kings' and pharaohs' efforts to squeeze more wealth out of a stagnant production led to a rapid depletion of the population, the decay of the irrigation works, and finally the disintegration of society.

Egypt and the Near East, however, gave rise to a new society which sprang into existence out of the ruined shell of the old. The new society brought with it new technology related to new perceptions of the cosmos. It required new ideas, because it was based on trade and, in part, on free labor. While reliance on authority may suit a priesthood, it is a poor guide for an enterprising trader or craftsman. Instead, the merchant had to learn by observing the world around him—the winds and tides. And the free craftsmen learned by changing nature, by experimenting with new materials and methods.

With the fall of the old empires, new trading patterns developed around the Mediterranean, patterns based on specialized agriculture—for example, farmers in Greece trading olive oil for grain from Egypt. This allowed further geographical expansion, since areas unsuited to intensive grain agriculture could now contribute their specialized products to, and be fed from, foreign lands.

The strict division between those who learned and those who worked began to break down: learning was democratized to serve the needs of independent merchants and artisans. The economi-

621–543 B.C.	Thales	
570–510 B.C.	Pythagoras	
428–349 B.C.		**PELOPONNESIAN WAR**
427–347 B.C.	Plato	
330–320 B.C.		**ALEXANDER'S CONQUEST**
310–230 B.C.	Aristarchus	
160–100 B.C.	Hipparchus	
120–180 A.D.	Ptolemy	
150–220 A.D.	Clement	
155–222 A.D.	Tertullian	
410 A.D.		**FALL OF ROME**
354–430 A.D.	Augustine	
350–420 A.D.	Pelagius	

cal Phoenician alphabet superseded the elaborate hieroglyphics of the ancient priests.

Nowhere were the changes so thorough as in the trading colonies established by Greeks in Ionia, on the eastern shore of the Aegean (now a part of Turkey). As is generally the case with colonies, the inherited social patterns were left behind when the more adventurous set up shop in new, previously unpopulated lands.

By 700 B.C. the Ionian trading cities, increasingly dependent on trade in specialized agriculture and craft products like textiles, had thrown off the earlier subordination to the great landowners of mainland Greece. They established new societies of traders, craftsmen, and freeholding peasants—the first limited attempts at democracies and republics. They needed new ideas to run such new societies—the old gods were outmoded.

This change is evident in the Ionian conception of the universe and its origin. Around 580 B.C. Thales, a native of the trading and textile center Miletus, first asserted that the world was formed by natural processes which could be *observed* in the world. He secularized the old creation myths. The world did begin with water, he taught, but it evolved out of water by natural means, as marshland and dry earth can be reclaimed from the sea. Man's alteration of nature, observable day by day, provided fuel for a new philosophy that supplanted the mythic past.

While Bronze Age priests had seen an unchanging society ruled by the unchanging cycles of the seasons, the Ionians saw a society in the midst of convulsive changes as aristocratic landholders, merchants, artisans, and peasants battled for power. Having experienced tumultuous overthrows of government Heraclitus concluded that the universe was in constant flux, like a fire, ever changing.

After Ionia was conquered by the Persians, the new ideas spread to Athens in mainland Greece. Here some of the most striking theories of early cosmology were born. Anaxagoras, a native of Ionia and later a friend of the Athenian leader Pericles, derived his theory of origins from close observation of nature. Seeing how whirlpools in nature order the chaotic flow of water and separate materials of different densities—mud and wood are drawn to the center, while stones and pebbles are flung outward —he reasoned that such vortices, driven by a primeval power *(nous)* could separate the air from the earth. The sun and stars are rocks larger than all of Greece, torn loose from earth, flung outward, and heated by friction to their present fiery state. Stars, he correctly guessed, are suns too far away for us to feel their heat.

From simple observation of nature—whirlpools, the glowing hot metal of the blacksmith's forge, the distant light of merchants' signal fires—Anaxagoras hypothesized a naturalistic theory of cosmic origins which was essentially correct.

In Anaxagoras' view, the universe is infinite, populated by a host of different worlds—many of them inhabited. There is no difference between the heavens and the earth, no finite earth surrounded by an unknowable heaven. Instead, all operate by the same principles that can be seen in everyday life, in the workings of nature and technology.

Thus by the fifth century B.C. the germ of the empirical scientific method had developed. The method was based on observations of nature in the here and now and led to extrapolations of those observations to parts of the universe distant in space or time. Necessarily such a method led to a theory of the world's origin through the same processes observable today—natural processes without divine intervention. Because such a cosmos evolves and changes, it can never have a start in time, a creation from nothing—since such events are never seen to occur. Instead, it is unlimited in space and time, for there are no limits to what can be observed and learned.

As the early magical view of the universe was tied to a stagnant society, where work was subordinate to unchanging authority, so the new empirical view was tied to a progressive society, where freer labor could and did continuously improve its technique. The basic idea of science, a changing cosmos governed by natural laws, was inspired by a society molded by technical and social progress. Without that progress, science lost its necessity and its inspiration, while without scientific observation, the basis for technical progress withered.

GOD AS MATHEMATICIAN

> Was the world always in existence and without beginning or created, and had it a beginning? Created, I reply, being visible and tangible and therefore sensible; and all sensible things are created. . . . Which of the patterns had the Artificer in view when He made the world—the pattern of the unchangeable or of that which is created? If the world is indeed fair and the Artificer good, it is manifest that He must have looked to that which is eternal—Thus, when all things were in disorder, God created in each thing all the measures and harmonies which they could possibly receive. For in those days nothing had any proportion except by accident nor did any of the things which now have names deserve to be named at all, neither fire nor water nor the other elements. All these the Creator first set in order and out of them he constructed the universe.
>
> —PLATO, *TIMAEUS*, c. 370 B.C.

Ancient Greece originated not one but two new ways of understanding the cosmos and its origins. The other method envisioned a world ruled neither by magic nor by natural process, but by number—knowable not by authority or by observation, but by pure reason. Like other methods, this reflects the development of society. For Greece was transformed not only by new technology, but also by money. It contained not only free craftsmen and peasants, but also slaves and slaveholders.

Around 600 B.C. an economic revolution began with the development of coinage. Before this, the only money was bullion weighed out at the time of a transaction. Small farmers and artisans were forced to barter. The introduction of coins that petty producers could use brought the mass of the population the benefits of money. As V. Gordon Childe puts it, the craftsman was no longer condemned to eat his wages. He could buy products with his earnings, opening up new markets for fellow artisans. The small farmer was free to follow the aristocrats into specialized farming for an export market.

But money was a mixed blessing, and in its wake followed usury, mortgages, and debt slavery. Money was power. Anything could be reduced to abstract numbers: the value of a pot, a jar of oil, a plot of land, a slave, could all be expressed by exact numbers of coins, as could the wealth and worth of any citizen. Numbers seemed to have magical powers. Money invested at interest could even multiply itself without any effort on the part of the lender.

The Ionians had generalized the social experience of free craftsmen and merchants into a view of the cosmos (the word "cosmos" itself is based on an earlier word meaning the coming together of all the Greek clans). Other Greek thinkers extrapolated from the power of numbers in society to the idea that numbers rule the universe as well. Pythagoras, living in the sixth century B.C., between Thales and Anaxagoras, witnessed the effects of money on the Greek states. To Pythagoras the pure relations of numbers in arithmetic and geometry are the changeless reality behind the shifting appearances of the sensible world. In contrast to the Ionians, Pythagoras taught that reality can be known not through sensory observation, but only through pure reason, which can investigate the abstract mathematical forms that rule the world.

Anaxagoras was to emphasize the similarity of the heavens and the earth, but Pythagoras contrasted them. The heavens, he taught, are the realm of pure number, where objects move in perfect, unchanging circles, the realm that can best be perceived through pure reason. The earth, realm of sense and appearances, is where human souls are condemned. Our only release from our earthly body, "the tomb of the soul," is withdrawal from the world to dispassionate contemplation of reason and mathematics.

The split or dualism in Pythagoras' thinking, between thought and action, reason and the senses, so foreign to that of the Ionians and their successors—as Childe and others have pointed out—was closely linked to the rise of slavery. As the money economy developed, so did chattel slavery. Such an institution, based on the sale of slaves, is impossible without the free exchange of money. Slavery threatened either to enchain the small producers themselves or to undercut their livelihood, and so devalued productive activity. To Greek slaveholders, work was something done by slaves, thus in itself degrading—*banausic.* Only detached thought is worthy. As slavery separated thought from action, so did the Pythagorean trend in Greek philosophy, glorifying abstract reason while denigrating physical observation. Slavery also undercut the development of the technology that required and fed observation: slaveowners don't need labor-saving devices.

Beyond its justification of social practice, Pythagorean philosophy substantiated itself with indisputable and vital discoveries in science. Despite their theoretical disdain for the senses, the Pythagoreans did in practice make accurate observations of nature. Pythagoras' fantastic theories of the heavens were based on his very real discoveries of the laws of musical harmonies and of the regular polyhedra (solid shapes whose sides are identical polygons).

Aspects of Pythagorean dualism were elaborated into a powerful and immensely important system of philosophy in the work of Plato, who was born half a century after Pythagoras' death. In the nearly twenty-five hundred years since, Plato's ideas have become so pervasive that they still influence science today.

Much of Plato's immense appeal through the ages lies in his championing of human reason. Rationality, not ancient authority, is Plato's guide to knowledge and morality, to the Highest Good. But at the same time, Plato emphasizes that knowledge comes through reason alone, diminishing the role of the senses and the earthly realm as a whole. According to Plato, the universe we see is based on ideal forms, which are imperfectly embodied in various objects—individual horses, for example, are embodiments of the ideal form of a horse from which their existence derives. Since these ideal forms are ideas, they cannot be perceived by the senses, but only uncovered by the use of reason, guided by a critical use of logic.

In the *Timaeus* Plato formulates a cosmology and creation story consistent with this concept of human knowledge. At the beginning of time, a beneficent creator used the eternal ideas or forms to mold preexisting, chaotic matter. (Like Pythagoras, Plato believed that the ultimate basis of these forms was mathematical and geometrical.) The creator molded matter into approximations of these ideal shapes, creating a universe ruled by eternal mathematical laws, laws which humans can deduce through reason. These eternal mathematical laws are the true reality while the changeable universe we see is mere appearance—the observation of nature is thus unreliable.

Plato emphasizes the *ethical* implication of this distinction: the ideal forms are the source of all good, while base, earthly matter is the source of the world's evils. The mundane, changeable world of everyday life cannot be used to understand the eternal, perfect, and unchangeable heavens. The most perfect motion, circular motion, occurs only in heaven, not on earth.

Plato thus developed another mode of thinking about the universe and creation. Against the traditional appeal to authority, Plato counterposes the power of human reason. But Plato attacks observation as a route to knowledge and strictly separates the worlds of thinking and doing, the spirit and the flesh, the heavens and earth. He thereby created a mathematical myth, a formidable barrier to the development of science.

Plato's belief in the supremacy of pure reason *necessarily* led him to formulate a myth to account for the rational origins of the universe, and thus led to the reestablishment of authority as the source of all knowledge. The observations that Plato so firmly rejected have the great advantage of objectivity. But a theory of the universe based on ideal mathematical forms relies on the *authority* of a priesthood of reason that can dictate *which* mathematical forms are the most ideal, most beautiful, most perfect: the ones which the creator chose at the beginning. The story of creation that is based on such priestly authority is just as much a myth as the Egyptian tales of Keb and Nut. Mathematics can make the myth more impressive, but no more objective.

It was thus not accidental that Plato's mathematical myth integrated the traditional Greek gods, who were the creator's helpers, and who guided the planets and stars, divine beings themselves, in their perfect paths. Plato's rational myth makes the earlier irrational mythology respectable again, and he produces a cosmology that again separates thought and action, the heavens and earth. In the process, he erects more obstacles for science: a world inhabited by a multitude of divine beings who control the stars and the winds is hardly a suitable subject for scientific inquiry.

Plato's concept of eternal mathematical laws is two-sided. The belief in such laws and the search for them has been immensely important to science. But scientists have had two contrasting views about what these laws are, views that have colored their investigations. One view follows Plato and believes that the laws truly rule the universe, that the universe is the embodiment of abstract mathematics, "the mind of God," knowable by reason alone. The other, quite different view, is that mathematical laws are *descriptions* of physical processes and patterns in nature—the reality is the process *described* by mathematics, the language of exact science. These different interpretations of mathematical laws have affected debates in cosmology to this day.

HEAVEN, EARTH, AND SLAVERY

Plato conceived his view of the universe as consistent with, and reinforcement for, his concept of the ideal society. In that society, outlined in his dialogue *The Republic,* all thought is to be done by philosopher-kings, aided by a small elite of guardians. No one else has political or social rights.

As ideas and matter, heaven and earth, are separated at creation, so guardians and philosopher-kings must be separated from those who work: slaves are to work without thinking, and philosopher-kings are to think without working. As the creator gave eternal mathematical laws to the universe, so the philosopher-kings give laws to society.

Plato's *Republic,* a rejection of Athenian democracy, was modeled on Sparta, where a small body of landholders ruled over a mass of rightless serfs, or helots. Sparta had defeated Athens in the thirty-year-long Peloponnesian War, begun in the year of Plato's birth, 428 B.C. Deprived of its colonies in the wake of defeat, Athens erupted in social conflict as rich landholders battled freeholders and artisans. To protect themselves from the growing demands for abolition of debts and land distribution, the landholders sought to combat political democracy and to erect a hierarchical society. Plato became the theoretician of this new society, rationalized in *The Republic* and justified by the cosmology of *Timaeus.*

The two ways of looking at the universe that arose in ancient Greece and which still battle today were from the start entangled with, and justified, two forms of society. Platonic dualism describes a cosmos knowable only to the pure reason of the few who then had the right to rule over the many, as the heavens rule earth, as the soul rules the body, as the master rules the slave. It was the worldview of the slaveholder. The alternative Ionian science assumed a world knowable by observation, where thought and work joined together. It was the worldview of the free craftsman and peasant. Knowledge was available to all and therefore power could not be the monopoly of the few. This science was the child of democracy and free labor and would, for the next two thousand years and more, be the constant enemy of authority and slavery.

▪ IDEAL AND OBSERVATION

While Greece generated two methods of learning about the cosmos, two cosmologies, it was a sort of mongrel synthesis of mathematical myth with observational method that triumphed. The reason again lay in social development. Neither authoritarian Sparta nor free Ionia became the model for the social evolution of the Mediterranean. Instead, slavery, free labor, and expanded trade all coexisted in the centuries that followed the fall of Athens.

Where Athenian imperialism had failed, Macedonian imperialism succeeded spectacularly. Beginning in 330 B.C., Alexander the Great conquered the area now occupied by Turkey, Syria, Jordan, Iraq, Iran, and Egypt, and he established colonies of Greek freeholders, artisans, and merchants. The free population increased, Mediterranean-wide trade flourished, and living standards rose.

Unlike the nobles of Sparta, the merchants of the Hellenistic world needed observations of nature to speed their ships across the Mediterranean and to ports in India. Even before Alexander's conquest, Plato's students had begun systematic astronomical observations in order to convert his ideas about perfect circular motions into an explanation of the observed motions of the planets. One such disciple, Eudoxus, created a system of moving spheres, with the earth at their center, which carried the planets, sun, and moon on their complex travels. This was the cosmological system of perfect motion that Aristotle then popularized (Fig. 2.2).

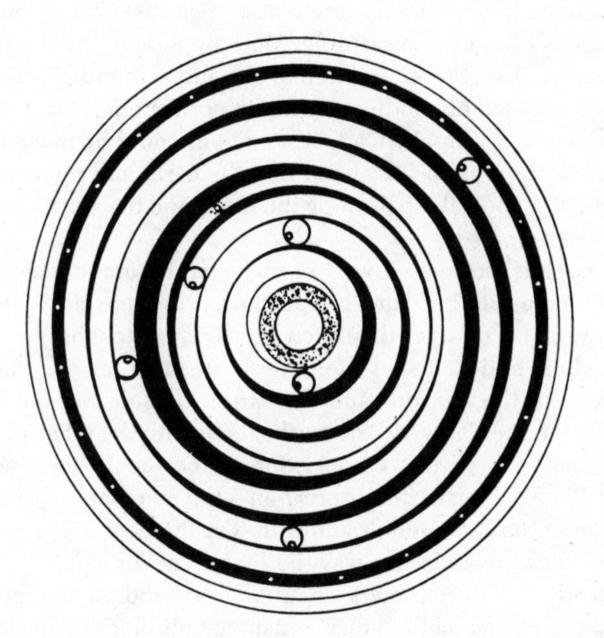

Fig. 2.2. The heavens, according to Eudoxus, were centered on the earth, with each planet carried on its own sphere. Later, Ptolemy added epicycles (small circles) and offset the spheres to better match observations.

Following Alexander's death, the various heirs to his empire and its provinces began a long scramble to assert their power—the impetus for systematic observation increased. The ruling Ptolemies (no relation to the astronomer) in Egypt established the Museum at Alexandria as a liberally endowed research library to generate and centralize such observations. Alexandrian astronomy, for example, used observation to solve practical problems of navigation.

From these observations startling theoretical results followed. Using Euclid's discoveries in geometry, Aristarchus of Samos estimated from observations that the sun was five million miles away and six times as large as the earth. Hipparchus, in the late second century B.C., refined these measurements and accurately obtained the true distances and sizes of the sun and moon. The sun, he realized, was a full hundred times larger than the earth in diameter, perhaps a million times as massive.

To Aristarchus, the idea that a much larger sun, even six or seven times larger, should circle at a great distance around the small earth didn't seem sensible. More important, the increasing accuracy of observations led him to conclude around 250 B.C. that the idea of heavenly bodies moving in perfect circles around the earth must be wrong. Instead, observations could be much better accounted for if it was assumed that the earth and planets orbit the sun, the moon alone orbiting the earth, and the earth spinning on its axis.

Aristarchus' correct views were rejected by other ancient scientists, including the later Hipparchus, who himself calculated that the sun is far larger than even Aristarchus thought. To accept Aristarchus' heliocentric system meant abandoning the Platonic hierarchy of the heavens and the earth: the sun, not the earth, would be motionless, and the earth, along with the other planets, would move in perfect circles. But neither could they wholly accept Plato's disdain for observation—the merchants that relied on them certainly couldn't afford to. On the contrary, they systematized observation in a way the Ionians never had.

Instead, the astronomers compromised, and in the process forged a scientific method that contained within it the tensions of ancient society; and which would, for better or worse, endure for millennia. On the one hand, the basic assumptions about the universe must come from pure reason, which can fathom the perfect mathematical laws of the heavens. On the other, observation serves to correct these basic mathematical laws in practice, modifying them as needed to "save the phenomenon" or to fit observations. So, rather than junking geocentricism and the Platonic philosophy that went with it, Hipparchus and his successors—notably Ptolemy—added new assumptions consistent with their mathematical ideas to close the gap between theory and observation. To the simple circular motion of heavenly objects they added lesser circular motions (epicycles), placed an orbit's center itself on an orbit (deferents), or added additional tilted spheres. The result was a fantastic clockwork, a neoplatonic Rube Goldberg invention.

Of course, such a scientific method can produce theories that fit observation. But a science that scrambles for a new, arbitrary entity that will jury-rig the agreement of theory and each new observation is no science at all. The ability to predict what has not yet been observed is the main use of science to mankind. The best science seeks simple patterns in nature, not because nature is in some mystical way "simple," but because only simple patterns are of much use in making predictions.

THE PROBLEM WITH EPICYCLES

If a scientist today plots a series of measurements on a piece of graph paper, he or she wants to know a mathematical formula that will predict new measurements, new points on the curve, which haven't yet been made. The formula thus obtained will, if confirmed by observation, save the work of making the new observations for each new situation—and may lead to new concepts about nature.

Now there are mathematical ways of simply taking any set of, say, twelve points and plotting them on a curve, while knowing nothing of the physical relationship that underlies the measurement. But such a curve has twelve independent factors, each one derived from a single measurement. The probability that such a curve will accurately predict even the thirteenth measurement, let alone the hundredth, is vanishingly small. However, if the scientist, by trial and error, or by observing the underlying physical relationship, hits on a formula with only one factor that needs to be fitted to, say, one point, and which then fits the other eleven, he or she can be confident that, by induction, the hundredth point is likely to be fit as well.

But if the scientist clings to a preconceived notion of what (by reason) the mathematical relationship *must* be, and introduces additional factors to save both theory and observations, no useful, predictive theory will result. It is in this way that the Platonic appeal to pure reason has again and again stood in the way of scientific and technical advance.

The synthesis achieved by 100 B.C. extolled the rationality needed for technical progress, progress essential to the survival of small freeholders, craftsmen, and traders. But it was dominated by a dualism that justified the powerful slaveholders who ruled ancient society. The separation of matter and spirit, action and thought, the glorification of pure reason, and of the authority of myth and tradition, became the ideological underpinnings of slave society, a society hostile to technical progress and scientific inquiry.

THE GENESIS OF GENESIS

In the beginning God created the heaven and the earth. And the earth was without form and void; and darkness was upon the face of the deep. And the Spirit of God moved upon the face of the waters. And God said, 'Let there be light'; and there was light. And God called the light Day and the darkness he called Night. And the evening and the morning were the first day. And God said, 'Let there be a firmament in the midst of the water and let it divide the water from the water.' And God made the firmament and divided the waters which were under the firmament from the waters which were above the firmament; and it was so. And God called the firmament Heaven and the evening and the morning were the second day. And God said, 'Let the water

> under the heaven be gathered together unto one place, and let the dry land appear'; and it was so. And God called the dry land earth and the gathering together of the waters he called Seas; and God saw that it was good.
>
> —GENESIS 1:1–10, c. 430 B.C.

By the end of Act One, around 100 B.C., one of the two central ideas of conventional cosmology—a universe knowable primarily by pure reason—had arisen in the struggles to create ancient society. In Act Two of the drama, from 100 B.C. to 400 A.D., the second great idea of modern cosmology—a universe created from nothing and decaying from its perfect origin—emerges with the fall of ancient society. The Ptolemaic synthesis splits apart into two contending visions of heaven and earth; again the underlying conflict is between free and forced labor.

One cosmology elaborates the dualism that separated heaven and earth, and abandons Platonic rationality in a renewal of magic and the occult. An earth "created from nothing and next to nothing" becomes the cosmological justification for a humanity rightly subjugated to absolute authority. The other, antidualist trend takes from Platonism only its appeal to reason, affirming the goodness of the material world and moving back toward the Ionian ideas of the unity of heaven and earth. In opposition to slave society this worldview champions the radical doctrines of human equality and freedom.

The antidualist ideas arose from the religious tradition of Judaism and early Christianity. Contrary to what is commonly supposed, the doctrine of creation *ex nihilo*—creation from nothing—does not come from the first chapter of Genesis. Genesis tells how God created the heavens and the earth by organizing a preexistent chaos, "the waters." "The earth was without form and void; and darkness was on the face of the deep. And the Spirit of God moved upon the face of the waters." As biblical scholar Nahun Sarna says, "Genesis is silent about where the initial chaos càme from mainly because the priests who wrote the account had no interest in that question. Their concern was how God created order in the universe."

While creation *ex nihilo* emerged from a dualistic tradition, the first chapter of Genesis is, to a large extent, a polemic against pagan dualism and its denigration of the material world. There are, to be sure, important similarities between the Genesis and Platonic creation stories (which were both written down around 400 B.C.). Both tell of a single creator who makes the cosmos according to a reasoned plan. But Plato's creator populates the heavens with divine beings—the sun, moon, and stars, the divinities of traditional Greek religion. By contrast, in Genesis the heavens are created in exactly the same manner as the earth and have no more perfection. Instead of sun and moon being divinities, Genesis demotes them to mere functional objects—a greater light and a lesser light. Whether creating the sun and the moon or the birds and the beasts, God pronounces each day's work equally "good."

Just as the heavens are not filled with divine perfection, neither is earth, matter, or mankind subordinated, as they are in Plato. Man is not created out of a separate and superior soul with an inferior body. God simply creates man and woman "in his own image." So humans, both male and female, partake in some measure in God's perfection, not just in their souls but in the whole unified being.

This crucial difference is not surprising given the history of the Israelites: with brief exceptions, their society of peasant farmers and shepherds was continually oppressed by one conquering group or another. In this soil the Platonic disdain for this world, the separation of thought and action, master and slave, did not take root.

THE EARLY CHRISTIAN COSMOS

The idea of an origin from nothing does not arise in early Christianity either. Christianity, originally a Jewish sect, emerged from the egalitarian traditions in Judaism. Jesus spoke to the humble of the world and attacked the dualism and inequality of pagan society. While dualism put an unbridgeable gap between heavenly perfection and earthly corruption, Jesus taught the attainability of perfection on earth: "Be ye therefore perfect, as is your Father in heaven." Indeed such perfection would arrive shortly in the Kingdom of God on earth.

Against the authority of either the Jewish priests or the pagan philosophers Jesus' parables appeal to the common sense of ordinary people. His analogies, based on common experience of weeds, trees, and houses, are antithetical to such abstruse concepts as creation *ex nihilo*. As the early Christians began actively proselytizing amid the Greco-Roman culture, steeped in Platonic ideas, Christianity remained opposed to dualism and its cosmology. Clement of Alexandria, a leading Christian theorist living at the end of the second century A.D., attacked the Platonic division between heaven and earth, freeman and slave, denouncing ancient society and the ideology that justified it. While Clement admires Plato's glorification of reason, he denies the Platonic view of matter as the origin of evil. Clement reaffirms the goodness of both man and matter, and locates the source of evil in ignorance and custom, which enslaves men with irrational ideas.

Neither Clement nor any other early Christian was in a modern sense abolitionist, but for Clement slavery was an unmitigated evil. It is "monstrous that human beings who are God's handwork should be subjected to another master." As Elaine Pagels shows in *Adam, Eve and the Serpent*, Clement and other Christians believed that "all people are begotten alike, with a capacity and ability for reasoning and emotion, without preference to age, sex or social status."

The early Christians developed no new cosmologies of their own, yet by attacking slavery and dualism they were eroding the main ideological, social, and economic obstacles that had impeded scientific advance in the preceding centuries. (The trade expansion that had emerged from Alexander's conquests had ended three centuries earlier, by 100 B.C., and Hellenistic society's dependence on slavery prevented any further advance in the technology of production.) Rather than furthering Platonic cosmology, the early Christians were casting it off. However, after Clement's time, around 200 A.D., a sudden shift in Christian thought led to the idea of a world created from nothing.

> And aught else besides Thee was there not, whereof Thou mightest create them, O God, One Trinity, and Trine Unity; and therefore out of nothing didst Thou create heaven and earth; a great thing, and a small thing; for Thou art Almighty and Good, to make all things good,

> even the great heaven and the petty earth. Thou wert, and nothing was there besides, out of which Thou createdst heaven and earth; things of two sorts: one near Thee, the other near to nothing; one to which Thou alone shouldest be superior, the other to which nothing should be inferior.
>
> —AUGUSTINE, *CONFESSIONS,* XII. 7, C. 400 A.D.

THE FALL OF ROME AND THE ORIGIN OF THE UNIVERSE

As in ancient Greece, the immense accumulations of wealth in the Mediterranean after 100 B.C. were based on conquest and imperialism. In the battle over Alexander's empire Rome emerged to swallow up the whole of the Mediterranean. The Roman legions enforced ruinous taxation, looting existing wealth but creating none. Slavery was massively extended and living standards dropped precipitously throughout the empire.

Although there was a period of recovery in the early first century A.D., by the end of the century Roman defeat at the hands of the Germanic tribes terminated the empire's expansion. With the supply of slaves cut off, Roman internal depredations increased: taxes soared, and the population began to decline. In Clement's youth a plague decimated the weakened population, and in his maturity the empire entered a period of crisis as revolt and persecution spread everywhere.

In this long epoch of increasing misery and oppression, Christianity became the only empire-wide opposition to Rome's slave system. The early Christian message of the universal brotherhood of all humanity, the antithesis of the legion's robber-rule, appealed to the enslaved and the poor. As wider sections of the population defected from allegiance to Rome, educated Christians formulated a potent antidualistic rationalist argument against the ideology that justified the empire.

But Christian opposition did *not* lead a holy war against pagan Rome. On the contrary, the chaos and savage persecutions of the third century created a sharp break with early Christianity. It bred among many Christian theologians an increasing pessimism, a withdrawal from the travails of this life, a retreat from rationality, and a resigned acceptance of existing society and ideas. In the third and fourth centuries, Christian thought split into two trends: one transformed Christianity from the chief enemy of existing society to its chief bulwark. The earlier oppositional stance became the faith of a heretical minority.

It was in this social context that the Christian contribution to cosmology, creation from nothing, comes into existence as a way of reconciling Christianity with existing society and its dualistic social structure. The doctrine was first formulated by the theologian-lawyer Tertullian, a contemporary of Clement, who converted to Christianity at the beginning of the third century.

Tertullian embraced Platonic dualism and rejected its rationality. Faith alone matters, not reason: "That the son of God died is to be believed because it is absurd," he wrote, "and the fact that he rose again is certain because it is impossible."

To Tertullian, as to the pagan neoplatonists, the material world is evil. But how could an omnipotent and good God have created an evil world? Tertullian's solution was the doctrine of creation from nothing. The material world is evil, Tertullian argues, because it had a beginning in time—the moment of creation. Things that have beginnings necessarily have ends, they are finite and subject to decay, therefore they are imperfect, hence the source of evil. By contrast, only God, who is eternal and infinite, can be wholly good and divine. His infinitude makes Him divine and separates Him from the finite material world.

Creation *ex nihilo* was for Tertullian what separated the finite and decaying earth from the infinite and divine heaven. We will see this powerful idea of the finiteness of the world arise repeatedly as a basic axiom of cosmology.

THE AUGUSTINIAN COSMOS

The idea of a universe created from nothing therefore arose as a way of reconciling Christianity's increasingly abstract God with a debased earth and society. In the tortured decades of the third century, Tertullian's denial of the world and rationality became the dominant trend in the Christian movement. It would be elaborated into a new cosmology as Church and empire merged in the following century.

Out of the chaos of the third century the emperors Diocletian and Constantine completely reorganized the Roman Empire. By this time large-scale trade had collapsed, and slavery had disintegrated too, as supplies dried up and the population fell. The reorganized empire was a society based on impoverished serfs, bound to the land and raising subsistence foodstuffs for powerful landlords: virtually the entire population was reduced to a level not much above slaves. It was a prefiguring of medieval society, but far more oppressive since the voracious demands of the imperial state were superimposed on the greed of the local landholders.

In this empire of universal compulsion there could be no competing loyalties. Christianity remained the only possible challenge—it had to be either extirpated or embraced by the imperial state. Diocletian took the road of repression with savage and widespread persecutions. When this failed, Constantine took the other road of merger, converting to Christianity and almost immediately subordinating the Church to imperial rule.

Many Christians revolted against the idea of an alliance with the empire they had fought so long. But many more saw the advantages to the Church of a new and powerful friend. The most prominent of these was Augustine, Bishop of Hippo, a city in North Africa. Augustine formulated the ideology of the new alliance of Church and State, an alliance that would shape the next thousand years of western history. He sought to reconcile Christians with imperial rule—converting the Church into a powerful buttress of secular authority—as well as to reconcile with Christianity the pagans who ran the empire.

The foundation of this doctrine was a new cosmological myth, and creation *ex nihilo* was central to that myth. To Augustine, as to Tertullian, creation *ex nihilo* necessitated the unbridgeable gap between heaven and earth, the extreme denigration of the material world "created out of nothing and next to nothing."

The infinite gap between the eternal, limitless perfect God and the transient, finite earth was a reflection of the empire that Augustine accepted as a needed prison for man's unruly will: an emperor with godlike powers, and subjects without the least hint of freedom.

But Augustine went far beyond Tertullian, by creating a cosmological justification for imperial rule. The suffering and oppression derive not only from the abstract finitude of creation, but also from a continuous process of decay from the perfection of the beginning. God has subjected humanity to an ever-mounting burden of evil as just punishment for Adam's sin. Augustine developed a new cosmology and political philosophy from the story of Eden.

The Adam and Eve story in the Bible, dated from four or five centuries before the first-chapter Genesis story, implies a more ambiguous attitude toward the goodness of the existing world.* Augustine transformed that ambiguity into an outright condemnation of the material world. To Augustine, nature as a whole, including humanity, was corrupted irredeemably by Adam's sin. Not only the pain of childbirth but all the suffering, disease, starvation, and misery of this world, and death itself, are God's just punishment for that sin which was conveyed by the sexual act to all subsequent generations. To Augustine newborns cursed with blindness or deformity are not suffering innocents but are being justly punished for Adam's sin; if they die unbaptized, they will burn eternally in hell.

From this initial sin came not only natural evils but political ones as well. Augustine says that after Adam's fall, man lacks the free will to avoid sin. Not only slavery, but all forms of rule of man by man, including the empire itself, are necessary institutions imposed on all humans, Christian and unbeliever alike, because all lack the will to govern themselves.

For Augustine, it is inevitable that these evils must not only exist but grow, as the world degenerates from the lost epoch of Eden. Earlier Christians had viewed human progress as held back only by the weight of pagan ignorance and custom. But Augustine viewed existing society as necessary, not a barrier to progress. To him, there was no such thing as progress. In his great work *The City of God,* Augustine interprets the fall of Rome to the Visigoths and other earthly catastrophes as a consequence of cosmic decline. The universe had begun at a moment in time, out of nothing, and it would end at a certain moment, returning to nothing. The evils of this transient world must be endured. Only by fixing one's eyes steadily on the next world could one hope for salvation.

* *Genesis* tells the creation story twice—once in Chapter 1, the seven days, and again in Chapter 2, the story of Eden. Scholars agree that the two were composed separately.

Like Tertullian, Augustine absorbed late Platonism's dualism and pessimism, its denigration of the senses. In a depraved world, Augustine argued, the senses are not to be trusted. Knowledge comes from the intellect alone, from the authority of the Church.

One consequence of Augustine's radical devaluation of the material world (and of his desire to reconcile pagans and Christians) was his easy acceptance of the pagan gods. The many gods worshiped by the pagans can exist as creatures of the one God who created all from nothing: Christians call them angels, Augustine wrote, "but if they [the Platonists] see fit to call such blessed and immortal creatures gods, this need not give rise to any serious discussion between us."

Thus by 400 A.D. Augustine had elaborated a cosmology strangely similar to the Big Bang: a universe created in an instant out of nothing, decaying from a perfect origin toward an ignominious end, populated by strange and miraculous creatures, and knowable only by the mind, not the senses. These fundamental conceptions arose as religious and philosophical justifications for a decaying and oppressive society.

To be sure, the origins of these assumptions do not necessarily invalidate any modern cosmological theory. But it is impossible to understand why these assumptions have again become so entrenched in cosmology without knowing their history. For as we will see in Chapter Four, the revival of these axioms in the twentieth century is again entangled with the development of society as a whole.

THE VICTORY OF AUGUSTINIAN COSMOLOGY

The integral connection between Augustine's cosmology, theology, morality, and politics was recognized by his opponents at the time. The most important among these was a British monk, Pelagius, who countered with an equally comprehensive worldview. The disease and pains of this world, death itself, Pelagius and his disciples taught, are not punishment for sins, but the result of nature's laws. Since nature exists independently of man, it is impossible that death, the pain of childbirth, deformity, and illness derive from Adam's fall: "The merit of one single person is not such that it could change the structure of the universe itself," exclaims Pelagius's disciple Julian. For the first time in a millennium, Pelagius and Julian revived the ancient Ionians' idea of a nature distinct from the human will, a nature whose working and processes, whose births and deaths, can be learned by observation.

Such a view of nature was linked still more directly than in Greek times with social issues. For if nature is ruled by processes that all can see, not by punishments devised by a capricious God, then the same must be true of human society. Human will is free and remains free for the Christian, who can alter his own impulses and cravings and subject them to his reason and morality. To Pelagius, the evils of society have causes and cures in this world, not the next. Pelagius extols human freedom and denounces the rich who impoverish the poor. The injunction to give up all riches is to be taken literally. Julian adds, "if there are no rich, then there will be no poor."

These differing views of the cosmos and society were not mere scholarly arguments among priests. In Egypt and all of North Africa, opposition to the empire, its social system, and the alliance of Church and State had fueled the Donatist movement in Christianity. While the northern barbarians invaded Italy, the Donatists led an open revolt against Rome. Donatist peasants and agricultural workers terrorized landlords, tax collectors, and creditors, liberating slaves and destroying rent rolls and land titles, unraveling the fabric of Roman rule. The Donatists, controlling the national churches of North Africa, organized a rebellion the imperial legions could not defeat.

It was at this point that Augustine, "the hammer of the Donatists," promulgated a doctrine to justify the persecution of heretics by Church *and* State, working together—the first inquisition. He mobilized the Church's own vast resources to hound the leaders of the Donatist heresy far more effectively than the decaying empire could. Equally important in this open civil war, he gave broad ideological justification to the fight against Donatism, winning over waverers and steeling the orthodox. His cosmology of a universe depraved, created from nothing and next to nothing, a humanity justly punished with slavery, gave moral sanction to the cruel and bloody work of the early inquisitors.

The Donatists were crushed by the combination of Catholic

inquisition and imperial force. When the invading Vandals, after sacking Rome, conquered North Africa in 430, the year of Augustine's death, they were able to take over as a going concern the vast landed estates and enserfed population of the empire.

With the collapse of the empire in the west, Augustine's cosmology was adopted by Christians in the following millennium. This view of a world created out of nothing, steeped in sin and misery, and rightly ruled by the harsh authority of Church and State, was perfectly fitted to the petrified society of the self-sufficient landholders, who needed neither merchants nor philosophers nor scientists. They required only a religion that would encourage serfs to accept their lot. Augustine's world, like that of the paganism the peasant previously knew, was a world with a yawning gap between heaven and earth, an earth peopled by demons and spirits, witches and devils. As Roman society retreated toward the level of primitive and impoverished agrarianism, so Augustine's cosmology retreated toward the magical, irrational world of myth.

Augustine's doctrine of creation *ex nihilo* became the orthodox wisdom in the west. By 400 A.D. the ancient cosmologies had evolved, at least in Europe, into a single doctrine. Dualism was triumphant, the role of the senses and the value of this world rejected. Knowledge came through the intellect alone, sternly guided by the authority of the church hierarchy. The finite universe, graded into celestial spheres, echoed the hierarchy of power on earth. It had come into being out of nothing and depended for its very existence on an inscrutable deity whose justice was, in Augustine's words, shown "in the agonies of tiny babies." This grim vision was, of course, hostile to the least vestige of science.

■

3 THE RISE OF SCIENCE

The eye can never have too much seeing, so the mind is never satisfied with sufficient truth.

—NICHOLAS OF CUSA, *On Learned Ignorance*, 1440

Religion teaches men how to go to heaven, not how the heavens go.

—GALILEO GALILEI, 1616

The result, therefore, of this physical enquiry is that we find no vestige of a beginning, no prospect of an end.

—JAMES HUTTON, *Theory of the Earth*, 1785

The second swing of the cosmological pendulum, Act Three of the cosmic drama from 400 to 1900 A.D., completely reversed this situation. Again, the finite and infinite cosmologies, the deductive and empirical methods battled, but this time the infinite universe prevailed. By the late nineteenth century the medieval world of lord and serf had disappeared and with it the medieval cosmology. The Ptolemaic spheres that carried the planets in their perfect circles, that were created in an instant, were replaced by an infinite, eternal universe, evolving by natural processes, a universe knowable by observation and experiment. The Ionian methods were again accepted wisdom. Scientists had gone far beyond them, creating a detailed history of the natural world, and a detailed description of its workings, which enabled society to generate a mighty technology. But before the rise of the scientific worldview could occur, the two central concepts of medieval cosmology had to be overthrown—the idea of a decaying universe, finite in time and space, and the belief that the world could be known through reason and authority.

This overturn would have been impossible within the old society. It could only happen as the new society of merchants, craftsmen, manufacturers, and free peasants, a society based on free labor, came to be. In turn, the new ideas became potent political weapons in the efforts to overthrow authoritarian power. The triumph of science was linked to the triumph of the system of free labor. Because of this link, cosmology became again, in the Renaissance and Reformation, something for which people killed and died.

THE PROGRESS OF HERESY

In the first millennium of this period, from 400 to 1400 A.D., there were three abortive attempts to develop a new scientific view of the universe. Each time, the resistance of the surrounding society defeated the efforts, but each time new concepts developed that paved the way for further advance.

The first step toward a new cosmology was not in the realm of science but in that of politics and theology. As long as the dualism that held sway went unchallenged, the very act of investigating nature was held valueless. And attacking the dualism of heaven and earth in the ancient world meant attacking the dualism of master and slave, ruler and ruled.

This challenge began during Augustine's own lifetime, when factions of the eastern churches took up the political and moral arguments of Pelagius or independently developed similar ones. In the east, both resistance to Roman rule and the traditions of free inquiry had deeper roots than in the western empire. The Church increasingly split along pro- and anti-imperial lines, a split that became final under the rule of Justinian a hundred years after Augustine.

465–542	Severus
480–560	Philloponus
630–690	**ISLAMIC CONQUEST**
965–1038	Ibn al-Haythan
980–1037	Ibn-Sina
1170–1253	Grosseteste
1214–1294	Roger Bacon
1348	**BLACK DEATH**
1401–1464	Nicholas of Cusa
1410–1495	Toscanelli
1452–1519	Leonardo
1473–1543	Copernicus
1588	**SPANISH ARMADA**
1525–1595	Digges
1550–1600	Bruno
1564–1642	Galileo
1571–1630	Kepler
1642	**ENGLISH REVOLUTION**
1642–1727	Newton
1775	**AMERICAN REVOLUTION**
1789	**FRENCH REVOLUTION**
1724–1804	Kant
1726–1797	Hutton
1733–1804	Priestley
1809–1882	Darwin

Justinian's rule, beginning in 527 A.D., was one of the harshest of any in imperial history. To finance his campaigns to reconquer Italy, Justinian piled tax on tax, provoking violent revolts both in Constantinople, the capital, and the provinces. The devastation wreaked by his depredations at home and his savage conquest of Italy and northern Africa paved the way for the great plague of Justinian in 542.

Opposition to the empire in the east crystallized around a faction of the Church called the Monophysites, led by Severus of Antioch. Since Church and State, theology and politics were one, the split first centered on a seemingly arcane theological question: What is the nature of Jesus? Are his divine and human natures separate, as the orthodox proimperial Diphysites contended, or is there only one nature, combining both divinity and humanity, as the Monophysites argued? Behind this seemingly abstruse debate lay the life-and-death politics of empire. Just as the human part of Jesus was subject to this divine will, argued the Diphysites, so the empire's merely human masses must be subject to the emperor, God's image on earth. For the Monophysites the unity of divine and human in Jesus symbolized the idea that *all* men share in the divine, and thus in the right of self-governance. Thus, the empire was without justification.

Most important for the future rise of science, Severus attacked for the first time the notion that soul and body are separate. Just as Christ's divine and human sides are combined into a single nature, Severus writes, a man's soul and body are a single entity: "Let us take the example of a man who builds a house: he draws up a plan, he decides on a location and what order to do things in, then he starts digging and sealing, hammering and sawing, roofing and painting. He has performed mental and physical work, but all the work sprang from the same source within the man, his will."

Severus's simple example, seemingly trivial to a modern mind, struck at the base of the dominant ideology. A man planning and building his own house assumes free labor, not slavery or serfdom. In the empire a master or lord would decide where the house was to go and his slaves, serfs, or tenants would be obliged to build it. Action and thought would be divided between ruler and ruled. But if a man is a unified whole, then he is capable of planning his own life and ruling himself. At the same time, there is no justification for the glorification of theory and the denigration of observation that aborted science. Knowing and doing are a single action.

Severus prepared the way for the rebirth of science in a second way by reviving causality—the idea that one event leads to other events, that man and nature can be understood as historical phenomena, autonomous of divine intervention. Science is impossible without this basic concept.

The Platonic-Augustinian worldview was anti-historical. God created the universe once and for all, creating all individuals for all times as good or bad, rulers or ruled. Cause and effect was excluded—things happened because God willed it, and thus science could not begin to take root.

In contrast, Severus sees human beings as *processes* whose individuality is based on their *history*—their parentage, their education, their actions and moral decisions—which shape them to be what they are. Evil arises historically, from people's relations with one another in society—rather than from the inherent sinfulness of man or matter. Therefore, to combat evil, society must be changed. This justified revolution. Severus's theology became a rallying cry for all those who wished for the destruction of Justinian's empire.

The idea that the roots of evil lie in the historical development of a society, that the laws of cause and effect can be applied to the affairs of men, has remained for fifteen hundred years one of the most subversive ideas in human history. But if the affairs of humanity can be understood in terms of historical causes and effects, then the *world as a whole* can be understood in the same way. The medieval cosmos of devils and angels, of a capricious divine will, is swept away and the ground cleared for the study of causes and effects—that is, for the development of science.

Severus thus revived the battle between an evolving world and one created once and for all. As we shall see, the ideas of an evolutionary universe and an evolutionary society have remained intertwined for the succeeding centuries down to the present.

THE REVIVAL OF OBSERVATION

Severus's philosophical notions implied a cosmos that unified heaven and earth, a universe not created by fiat, but developing by a historical process. However, with Severus these remained political and philosophical notions. John Philloponus, a Monophysite philosopher and a younger contemporary of Severus, applied the unity of soul and matter to cosmology. Reviving the Ionians' ideas, he argued that these same ideas apply to the heavens and the earth. Stars are neither divine nor perfect beings but material bodies on fire. The heavens are not unchanging but governed by the same changes as are earthly objects.

To counter the overwhelming authority of the Church, Aristotle, and Plato, John returned to the Ionian emphasis on observation and experiments. In supporting his assertion that stars are lighted by fire, John pointed to their obvious differences in color: this shows, he said, they cannot be simple bodies, made of pure ether (the rarefied medium filling the heavens), as dualism claimed, since we know that on earth different materials produce different colorations in fire—thus the stars must be composed of different materials, like those on earth. (John hit on the basis of spectrography, which centuries later allowed scientists to figure out what the stars are made of.)

John's approach, however, was limited: he refers to experiments and observation frequently, but he still subordinates them to philosophical inquiry. Equally significant, John did *not* question the finite, geocentric universe of late antiquity.

Even without these serious limitations, John's work could not have led to an early revival of science, for he lived during the collapse of Mediterranean civilization. The trade that demanded and supported scientific research was disrupted by Byzantine attempts at reconquest and its futile battles with Persia. John Philloponus represented the last flicker of ancient science. The next step was to be taken by a different society. That step was the development of a systematic experimental method.

THE ISLAMIC RENAISSANCE

The way for a revival of trade, and subsequently of science, was cleared by the Islamic conquest. By 613 A.D., when Muhammad first began to preach his new revelations, the once mighty Byzantine empire was an empty husk, holding no sway over the outlying territory of Arabia. Unlike Jesus, Muhammad stepped into a virtual political vacuum in which he was able to put his ideas of social justice (probably influenced by Monophysite doctrine) into immediate practice.

Those ideas of justice, based on the grievances of the merchants and tradesmen of the empire, restricted the depredations of tax gatherers and usurers, glorified fair dealing and trade, and created a sacred social obligation to devote a part of all wealth to social welfare—to help the impoverished believer. The morality of Islam seemed so obviously superior to that of the still ravenous and decayed empire that it won adherents with brushfire speed. By the time of his death in 632, Muhammad had unified the squabbling Arab tribes, and within ten years, Muslim armies fanned out to crush imperial troops, seizing Syria in 636, Iraq the following year, and then Mesopotamia and Egypt. The Muslims were greeted everywhere as liberators by the empire's alienated population, who had been rebelling, led by Monophysite and other anti-imperial groups. The new rulers slashed taxes by one-third or one-half and a slow recovery of trade and prosperity began throughout the Mediterranean.

With this came a gradual revival of support for science. By around 800, when the center of Muslim rule shifted to Iran, Muslim scholars, often working together with Monophysite and other Christian colleagues, were busy absorbing what remained of the ancient learning of the Greeks, as well as borrowing from India. But the Arabs did not merely pass on ancient knowledge. During the height of Arabic civilization around the year 1000, while Western Europe was still crawling out of the Dark Ages, they formulated for the first time the modern scientific method.

The most important person in this breakthrough was Ibn al-Haytham, known in the west as Al-Hazen. Primarily in the field of optics, he went beyond John Philloponus and all reliance on the speculative method of ancient natural philosophy. He started from systematic, repeated experiments, which were arranged to yield quantitative measurements, and from these he developed hypotheses expressed in mathematical form. These were inspired guesses as to the physical relationships underlying various sets of measurements. If a hypothesis was seen to fit the measurements, further experiments were devised to see if the proposed relationship could accurately predict new measurements.

Here are the basic ideas of the scientific method. Science begins from systematic observation and measurement, but it does not stop there, like a mere collector of information about nature. The creative act is to generalize from the data, to hypothesize a possible physical process and to describe the process in mathematical terms. Mathematics describes a relationship observed in nature, rather than claiming to *be* the underlying reality (as in Platonism or in conventional cosmology today). Finally, the hypothesis is judged not on its intrinsic logic or by debate, but solely by its ability to accurately predict further measurements.

Using this method Ibn al-Haytham demolished the old optics of Ptolemy and established the framework for a science of light. He refuted by quantitative experiments Ptolemy's logically derived laws of reflection and refraction and his ideas of "vision rays" from the eye.

While Ibn al-Haytham was perfecting the scientific method, other Islamic thinkers were beginning to jettison the ideas of the ancient world. The philosopher, physician, and scientist Ibn Sina (known as Avicenna) abandoned the various creation myths and studied geological formations to learn the origin of the present-day earth. He correctly concluded that nearly all land today was once under water, that sedimentary rocks were formed under water, and that the land was subsequently lifted by earthquakes.

Despite the great strides made by Islamic science, the Muslim renaissance faded by around 1100 A.D. The tenth-century thinkers had attacked an important part of the existing worldview, but had not formulated a comprehensive alternative. Like John Philloponus, with whom they were familiar, they developed observational method; but unlike him, none rejected the contrast of heaven and earth. The scientific method did not probe too deeply into matters that, in the Islamic east as well as in the Christian west, were so closely tied to religious orthodoxy. For the Muslim empires were just as closely linked to religious hierarchies as were the European feudal states. And while the Muslims encouraged trade, and, to a limited extent, manufacturing, political power rested with a landholding class whose power was centralized in the powerful caliphs.

The conflicts between the wealthy landholders, who exploited enserfed peasants and slaves, and tradesmen and manufacturers, who relied on free labor, broke out again and again in violent struggles. In the end, the power of the caliphs was gathered into the hands of the invading Turks, who crushed the budding merchant economies and dispersed the scientific institutions they had supported. Fundamentalists attacked philosophers like Ibn Sina as impious and heretical. The first serious effort to establish self-sustaining scientific enterprises had failed. The crucial breakthrough—a new scientific cosmology—would be achieved in the west.

CHURCH AND SCIENCE IN EUROPE

While the Islamic world was declining, Europe was recovering, and it was here that the further evolution of science occurred, because it was in Europe that free labor again started to develop, and with it the need for labor-saving inventions, thus for science. By 800 A.D. Arabic trade was already starting to stimulate revival in the west. By 900 industry on a scale not seen in the ancient period began to emerge in Flanders, where cloth manufacturing led to the formation of small industrial towns. As the population grew, the landless younger sons of serfs went to the towns to earn livings as peddlers, merchants, or artisans—free labor.

In the long recovery of the Middle Ages, European technology advanced far beyond that of the ancient world. Water power and animal power replaced the lavish use of human labor. This and other innovations permitted peasants to devote their energy to more sophisticated activities, so the European economy could support the bulk of its population, although enserfed, with incomparably better food, clothing, and housing than was commonly available in the ancient world.

Free labor was relatively scarce and costly for the new manufacturer, since most of the population was still tied to the land. Slave labor was even less available. Only in the east was the cultural level still low enough for the Europeans to capture slaves—Slavs. But with the rise of more developed technology, free labor, which was better skilled, became more productive and thus more economical.

As free labor reintroduced the incentive for rapid innovation and economic growth encouraged the expansion of trade, once again an interest in technology was kindled. And not by accident new learning became available from the east. Returning crusaders brought back to Europe the works of the Arabic scientists and philosophers, and translations, lost in the west, of the ancient Greeks.

Around 1200, Robert Grosseteste, a British monk, began assimilating the scientific method of Islam. In the following generation, Roger Bacon, a Franciscan friar, carried on the same work. For the first time in the west, Bacon asserted that the highest purpose of scientific work is its eventual practical application. In famous visionary passages, Bacon outlines a future filled with technological marvels—submarines, flying machines, and self-propelled vehicles. But his scientific theory was more utopian than practical.

Grosseteste's and Bacon's notions of science and its methods were more limited than those of the Muslims. Neither man performed significant experiments himself; instead, both wrote commentaries on questions of method. For them, experiments, though important, played a subordinate role to revealed truth. Bacon argues that, while revelation is the highest authority, experiment can be used to sort out which authority is valid, and which the result of misinterpretation or mistranslation. In no case can experiment contradict the core of authoritative truth. Yet Bacon's still timid contentions were chastised by Church authorities and he was placed under the surveillance of his Franciscan order.

While Bacon was attempting to absorb the new learning of the Arabs, Thomas Aquinas was integrating the translations of Greeks, especially Aristotle, into the still dominant Augustinian worldview. For Aquinas, faith and revelation remain the only source of knowledge, with reason as its handmaid. Where reason cannot determine the truth, as in, for example, the question of whether the world had a beginning or not, revelation—in this case, the Bible—provides the true answer.

The limited role allowed to medieval science reflects the limited role of medieval industry and commerce. The serfs paid their surplus produce to the lords, so the burghers and manufacturers depended on the nobility for their market. As a result, the scope of technological and economic expansion was limited by the old order.

But feudalism needed new lands to grow. As arable land became scarce around 1300, the nobility borrowed on a grand scale to finance their luxuries and wars, taxing their subjects to pay the debts. Peasant grain reserves were squandered, famine repeatedly swept over Europe, hunger gripped the filthy towns, and in 1348 feudal society collapsed in the grip of the Black Death, which carried off a third of Europe's people.

▪ THE FALL OF THE HIERARCHICAL COSMOS

The disintegration of feudal society in the mid-fourteenth century was the great event that cleared the way for the development of science and for modern society. In the next 250 years, the old cosmology crumbled and a new worldview triumphed.

Science could begin to develop only when the old society was weakened and its ideas discredited. Lacking the centralized states that enabled ancient society and the Muslim world to survive similar crises and to crush social opposition, the authority of the feudal lords fell apart in the wake of the Black Death. In France the Hundred Years' War, from 1337 to 1453, spread anarchy. In England Henry IV's usurpation of rule in 1399 led to a century-long series of dynastic wars in which the feudal nobility accomplished the singular service of wiping itself out.

The catastrophe of the Black Death undermined the ideological authority of both Church and State. The generations born after the plague saw it as God's judgment on the existent society. At the same time, the tremendous shortage of labor created by the plague (which repeatedly returned with decreasing ferocity) made serfdom unworkable in much of Europe. Runaway serfs could easily find untenanted lands, already cleared to farmland, or could find well-paid work in the towns.

The doctrines of Augustine and Aquinas, in which the people owed obedience to secular and ecclesiastical authority, no longer held sway. In England, the Oxford preacher John Wycliffe denounced the worldly Church as the robber of the people, and denied that its officers were necessary to mediate between man and God. Other movements throughout the European continent expressed similar beliefs.

Another preacher, John Ball, denounced all sources of authority—priests, nobles, kings, lords. "Matters cannot go well in England until all things are held in common, when there shall be neither vassals nor lords, when the lords shall be no more masters than ourselves." In 1381 the English peasants, enraged by the efforts of the nobles to reimpose the duties of serfdom, rose in revolt, led by John Ball and the veteran soldier Wat Tyler. Although it eventually was crushed, the Peasant Rebellion was the death knell of serfdom in England.

Other revolts shook the trading towns of Europe—the artisans seized Florence in 1379 and Liège in 1384. The successor to feudal Europe was not one society but two, in mortal conflict with each other. The free towns of artisans, merchants, and manufacturers, allied with the free peasants, warred with the great lords and bishops, kings, and popes.

As in early Greece the new trading cities around 1400 gave rise to revolutionary new ideas. For the first time in a thousand years the Augustinian finite universe was challenged—by a German-born bishop, Nicholas of Cusa. Born in 1401 and educated in Italy at the University of Padua, Nicholas became the great transitional figure between the worldview of the Middle Ages and that of the Renaissance. Like everyone in his time, he cast his thought as a continuation of tradition, yet his ideas initiated the fall of the entire cosmology and social outlook that had held sway since Augustine. Nicholas's ideas were in truth a rebirth of ancient Greek learning, but not that of Plato and Aristotle, which had indeed never been rejected by the medieval thinkers. It was instead a revival of the Ionian methods of exactly two thousand years earlier. Like Thales, Nicholas formulated a new way of viewing the universe.

In his major work, paradoxically entitled *On Learned Ignorance*, Nicholas returned to the central idea of Anaxagoras—an infinite, unlimited universe. In contrast to Ptolemy's finite cosmos circumscribed by concentric spheres with earth at their center, Nicholas argued that the universe has no limits in space, no beginning or ending in time. God is not located outside the finite universe, he is everywhere and nowhere, transcending space and time.

Nicholas's infinite universe is populated by an unlimited number of stars and planets, and, of course, has no center, no single immobile place of rest. The earth, he reasoned, must therefore move, like everything else in the universe. It appears at rest only because we're on it, moving with it. He cast aside the geocentric cosmos entirely.

Again, like Anaxagoras and John Philloponus, Nicholas demolished the great distinction of celestial and sublunar realms. The same material make up the earth and the stars: "To people elsewhere, the earth would appear to them as a noble star." While the earth is imperfect, there is no more perfection in the heavens: Plato's perfect circles are only approximations of real movement.

Nicholas linked this renewed cosmology to the continuing tradition of learning from experience and observation. He criticized Plato's theory of ideas as constructions of pure reason. Reason, he says, arrives at truth only by abstracting from, and organizing, the impressions of the senses.

Since reality is infinite in its complexity, knowledge can only be a series of better approximations, unifying larger realms of experiences. According to Nicholas, the human mind, though finite in its understanding, is infinite in its capacity for understanding and in its desire for truth. "The eye can never have too much seeing, so the mind is never satisfied with sufficient truth." In this sense, all learning is still ignorance—not because it is false, but because it will never arrive at the *final* truth. There can be *no* Theory of Everything.

This open-ended theory of knowledge is the greatest possible challenge to the old concepts of absolute truth, founded on pure reason and authority: if there is no final truth, there is no final authority. Nicholas extends this radical theory of knowledge even to the revealed truths of religion. In an age when the extermination of heretics was pursued with enthusiasm, and Muslim and Christian armies clashed, he extolled a vision of religious toleration. Christianity, Judaism, and Islam, he writes, are only partial, human perceptions of infinite religious truths. They can be unified on the basis of their common belief in God and a common view of what is moral and right on earth.

Nicholas's political thought embodied the new spirit emerging from the free cities. He rejected Augustine's and Aquinas's doctrines of human rule as punishment for original sin, instead echoing the Pelagian and early Christian doctrines of innate human freedom. Since all men were free and equal at birth, like the equal stars and planets of the infinite universe, human rule is not based on inherent, inborn superiority of some over others. Rather, the basis of all rule—both secular *and* ecclesiastical—must be the consent of the governed, and rulers and their laws must be selected by the people, even though their rule is sanctioned by God. Not surprisingly, these practices were to be found among the urban guild-governments.

While conservative in form, Nicholas of Cusa's ideas undercut the basic notions of hierarchy—social and cosmic—entrenched since the days of Plato.

THE NEW COSMOLOGY

Once again, though, Nicholas of Cusa's work, however radical in its implications, remained abstract philosophy. If it was to prevail over the orthodox universe it had to be made concrete. His influence spread in a practical way, in part, through his collaboration with the astronomer, geographer, and mathematician Paolo Toscanelli. Toscanelli disseminated Nicholas's new cosmology, which he had helped to develop, and linked it to the emerging observational sciences. One stream of this current led to Toscanelli's extraordinary student Leonardo da Vinci and to the development of the modern scientific method. With Leonardo, the new philosophical ideas were at last shorn of their scholastic trappings and married to the crafts and technical innovation. Having little formal education, Leonardo enthusiastically accepted Nicholas's new worldview as a justification for rejecting the outmoded authority of the "pharisees—the holy friars" and of his "adversaries" Plato and Aristotle.

For the first time since the Ionians, he put forward a conception of science that was *wholly* secular, in no way based on religious doctrines or philosophy. The gap between spirit and matter, thought and action, theory and practice, was finally bridged in reality. While philosophers from John Philloponus to Nicholas of Cusa had recognized the unity of the world, they had remained abstract thinkers. In Leonardo the craftsman, scientist, and inventor are merged into one. Liberated philosophically by the new infinite cosmology, and liberated economically by widespread social change that had weakened the authoritarian hierarchy, he went far beyond his predecessors—he observed *the whole world.*

Leonardo put into practice Nicholas's idea that knowledge must derive from observation, and linked it with the necessity of mathematical description. He emphasizes that "there is no certainty in science where one of the mathematical sciences cannot be applied." But he emphatically rejected the Platonic idea of mathematics as the master of science. He laid out his method explicitly:

> In dealing with a scientific problem, I first arrange several experiments, and then show with reasons why such an experiment must necessarily operate in this and in no other way. This is the method which must be followed in all research upon the phenomenon of nature. We must consult experience in the variety of cases and circumstances until we can draw from them a general rule that is contained in them. And for what purposes are these rules good? They lead us to further investigations of nature and to creations of art. They prevent us from deceiving ourselves and others by promising results which are not obtainable.

Thus, in Leonardo's method, experiment leads to the hypothesis of "rules of nature," mathematical rules whose utility is as an aid to human beings in their art (which includes inventions and mechanical devices) and in their lives, and to predict the results of other actions. He applied this method on a scale never before or since equaled, as his notebooks attest—a host of scientific discoveries in optics, anatomy, mechanics, and hydraulics, among many other fields.

While the majority of his discoveries and inventions were buried for two centuries, since his notebooks were not published following his death, his impact on Italian science and technology was profound. Leonardo was no hermit, writing secretly in a cloister, but an engineer, artist, and thinker, employed by the most important princes of Italy, and a close acquaintance of nearly all the leading minds of his time.

While one stream of Nicholas of Cusa's influence led through Leonardo to the modern experimental method, the other led to the Copernican cosmology. For while Nicholas worked, the voyages of discovery provided a sharp incentive for a new astronomy—moreover, a *practical* astronomy.* If the motions of the moon

* Toscanelli, as the leading geographer of Italy, prepared new charts and maps for the Medicis' sea captains. He recalculated the earth's diameter, as had the ancients, with the same goal of finding the length of a degree of latitude. Based on his calculations (which were wrong) he encouraged Columbus in the idea that China and India could be reached by sailing across the Atlantic.

and planets could be accurately known, they could act as a celestial clock, enabling sailors to gauge their course precisely in crossing the Atlantic. For this task the Ptolemaic system with its epicycles and deferents was far too cumbersome and inaccurate.

In the centuries-long effort to conform the geocentric worldview to the observations of planetary motions, complexity after complexity had been added. It was well known that the geocentric view accounted approximately for the position of the planets and moon. Yet the obvious changes in their brightness (a direct consequence of their changing distance from earth) was inexplicable. For the moon, whose distance is actually nearly constant, the epicycles introduced a variation in the distance—thus in its apparent size—that was not observed. It was so absurd that King Alfonso of Spain remarked, "If I had been present at the creation, I could have rendered profound advice."

It was at this time that Nicholas Copernicus came to study in Italy. There he learned of Nicholas of Cusa's idea that the earth moves. It is also possible that he learned of Leonardo's conception that the sun is immobile, a concept found in Leonardo's notebooks. In any case, Copernicus knew of Aristarchus' writings. But Nicholas's idea of an infinite universe was the most significant one in clearing the way for Copernicus's geocentric universe. For if the earth moved, why couldn't that be observed in the motion of the fixed stars? (In the small, finite universe of Ptolemy and Aristotle, their apparent immobility was considered strong evidence.) But if the universe was infinite, as Nicholas said, the stars could be so immensely distant that their apparent motion would be too small to see. Equally important, in the Ptolemaic system, the outermost sphere of the fixed stars was thought to rotate once a day. If the universe was infinite, such a breakneck rotation rate would be absurd.

By the time Copernicus left Italy in 1506, he had developed the basis of the heliocentric system: the earth, rotating on its own axis, orbits the sun, as do all the planets. Again, like Nicholas, he cautiously compromised with the old system, retaining Ptolemy's perfect circles and epicycles (although he needed far fewer); as a result, the accuracy of his predictions was not substantially improved. He also waffled Cusa's radical insistence of an infinite universe, asserting that the universe was immense but might or might not be infinite. Nonetheless, Copernicus's system was a clear alternative to the geocentric, hierarchical cosmos. If the earth moved, if it was a planet, the whole structure of celestial and sublunar regions would collapse—as would the ideology of a necessary cosmic and social hierarchy, the gulf between spirit and matter, and, above all, the invincibility of authority and pure reason.

Copernicus was well aware of the radical implications of his own hypothesis, no matter how conservatively it was dressed up. He did not publish it for nearly thirty years, until 1543.

By this time, there was not only an ideological alternative to the medieval worldview, there was a political, religious, and social alternative—the Protestant Reformation against the hierarchical society in both Church and State. Beginning as an attack on the Church hierarchy and its claim to being the sole religious authority, within a few years the Reformation became tied up with the political and social struggle of the commons—merchants, artisans, and peasants—against the feudal nobility. By the mid-1530s, the Reformation had sparked peasant revolts throughout Germany, and Henry VIII had led England out of the Catholic Church. In the 1540s all of Germany was embroiled in wars between Protestant princes and those aligned with the papacy and Catholicism.

It was in this epoch, in 1540, that Copernicus's colleague Rhäticus first published Copernicus's heliocentric theory—in the Protestant stronghold of Wittenberg, where the Reformation had started thirty years earlier. This publication in turn impelled Copernicus to his own publication in 1543. Rhäticus brought out in his description the clear challenge Copernicus posed to the hierarchical worldview, which the Reformation was effectively rejecting. While the medieval universe had each sphere driven by the one above it, the higher controlling the lower, Copernicus's heliocentric system assumes all motions are natural processes. "The sphere of each planet," Rhäticus writes, "advanced uniformly with the motion assigned to it by nature and completed its period without being forced into any inequality by the power of the higher sphere." The equality under the law the commons fought for and the equality before God the Protestants asserted is reflected in the equality under natural law of the heavens.

So radical were the implications of Copernicus's view that the leaders of the Reformation rejected it in horror, even as their followers in the universities turned to it with interest. Martin Luther, in particular, denounced the idea as fantastic and a contradiction of the Bible. But in England, where the power of the old Church had been uprooted by Henry VIII's decrees, the new ideas found fertile soil.

THE COSMOS AND THE COMMON MAN

It was, in fact, in England that the two streams of Nicholas of Cusa's influence—scientific method and the new infinite cosmology—first merged. England had nurtured its own scientific tradition from the time of Bacon, and English scholars and politicians kept abreast of the latest developments in Italian philosophy. The practical impetus for astronomical and general scientific research was stronger in England than anywhere else. After the feudal nobility had killed themselves off in the War of the Roses, a collateral royal line, previously involved in trade rather than landholding, came to power with Henry VII. By the time Elizabeth became Queen in 1558, English navigation was in a state of fevered expansion, attempting to wrest control of trade from Catholic Spain.

Elizabethan England, recently freed from the intolerance of Mary's rule, welcomed that antihierarchical and anti-authoritarian teaching of the Copernican system. Thomas Digges, a leading English astronomer, became the first to popularize Copernicus's ideas to a broad audience, writing a book about it in English, not scholarly Latin, in 1576. Already in 1572, Digges and other astronomers had studied the supernovas of that year, showing that the heavens do in fact change, contrary to tradition—a sight visible to all. Now Copernicus's ideas, backed by Digges's prestige as a leading scientist, became the property of the common man. Digges synthesized Copernicus's and Nicholas of Cusa's work, proclaiming the universe to be infinite, populated with innumerable suns and worlds. But above all he explicitly criticized the ancients' method: "I have perceived that the ancients progressed in reverse order from theories, to seek after true observations, when they ought rather to have proceeded from observations and then to have examined theories."

In a country where free labor was increasingly drawn into manufacture, and the need for both technological advances and an educated work force became acute, Digges championed the idea that scientific and technological advances are welded together, and that scientific knowledge must become common to all. With the help of scientific education, "how many a common artificer is there in these realms who, by their own skill and experience, will be able to find out, and devise new work, strange engines for sundry purposes in the Commonwealth, for private pleasure and for the better maintaining of their own estate?" Since technological advance would be most rapid when the common workers had combined scientific knowledge with practical experience, Digges vowed to write all his work in English. Digges and others began a series of practical scientific manuals aimed at the widest audience. By 1589 publicly sponsored scientific lectures drew crowds of artisans, soldiers, and sailors eager for knowledge.

TECHNOLOGY AND COSMOLOGY

The conflict between the old and the new cosmologies was not settled by scholarly argument, but by the battles of the old and new societies—embodied in the struggles of nations. Protestants, in manufacturing Holland, revolted against its Catholic imperial ruler, Spain; and in 1584 the main Protestant power, England, allied with Holland. The Spanish empire was based on forced labor—serfs at home and serfs and slaves in the huge empire of the New World. The English and Dutch relied mainly on free labor.

The Copernican scientific worldview gave not only ideological justification to the Protestant side, but also decisive technological advantage. By synthesizing theoretical science with craft skill, English industry moved ahead of Spain in critical areas, such as the casting of naval artillery, producing lighter guns with greater range and accuracy.

The Copernican revolution had also meant throwing out Aristotelian physics—based on the idea that moving objects sought their "proper" place in the hierarchy. This had significant application in the science of ballistics. Aristotle had taught, and the medieval scholars accepted, that a projectile flew upward in a straight line, then fell vertically to earth. Leonardo and his successor in engineering, Tartaglia, showed by experiment that the trajectory is a curve, and compiled a gunnery table linking the elevation of the gun to the range of the shot.

Digges and other English scientists systematized their results, producing widely read manuals of naval gunnery. English ships, manned by draftees drawn from the artisan and working classes, had by 1588 both seamen and officers on board trained in the basics of the new ballistics. Spain, by contrast, had no use or interest in the new sciences. Nor could their uneducated sailors use them.

The related differences in social structure, technology, and training proved decisive when the Spanish Armada sailed to invade England. The English ships mounted mostly small guns, called culvetines, whose effective range was one thousand yards. The Spaniards had crude cannons, effective only at point-blank range—that is, before the shot began to fall significantly, perhaps three hundred yards. With this and other advantages the English battered the Spaniards at long range, while the Spaniards' ammunition fell far short of the targets. For one hundred thousand cannonballs fired, the Spaniards killed one English officer and two dozen seamen, sinking no vessels. The English, with about half as many shots and lighter guns, sank or disabled seventeen Spanish ships and inflicted thousands of casualties. When the Spanish ran out of ammunition, the English chased the shattered Armada out of the channel.

Thus, in a very practical way, the superiority of the empirical worldview was demonstrated—with cannon, not with debate. In fact, the defeat of the Armada determined which worldview would triumph, since it determined which society would survive.

THE PRICE OF HERESY

The Catholic hierarchy recognized, as the Protestants had earlier, that the new cosmology was subversive—incompatible with the traditional, authoritarian society. One of the first victims of the Counter-Reformation was Giordano Bruno, a former monk. Bruno traveled to England and befriended its leading political and scientific figures; and when he returned, he popularized Copernican theory on the continent. Bruno took Digges's version of the infinite, Copernican universe and purged it of remaining Ptolemaic elements, such as the perfect spheres that carried the planets' orbits. He made this infinite universe, with its infinite inhabited worlds, the basis of his philosophy, integrating Nicholas of Cusa's thinking, even going beyond it. Bruno explicitly challenged the idea of creation *ex nihilo,* arguing that the universe must be unlimited in both space *and time,* without beginning or end.

Bruno was a philosopher, not a scientist, and he used the tradition of logical argument to support the Copernican worldview. Above all, though, he considered himself a loyal Catholic bent on reforming, not rejecting, the Church. Yet on his return to Catholic territory in 1592, he was promptly arrested. Robert Cardinal Bellarmine, a prominent leader of the Counter-Reformation and the pope's own theologian, saw in Bruno's writing an effort to subvert the Church from within. The idea of an infinite number of worlds not only undermined the primacy of the Church hierarchy, it contradicted as well all sources of authority—the idea was found neither in the Bible nor in Aristotle or Plato. Moreover, it very obviously destroyed the Catholic vision of a material, subterranean hell and an ethereal heaven beyond the cosmic spheres: it portrayed a cosmos in which these threats and enticements would have no place, and would be comprehensible to only a few—but not to the ill-educated peasants, as the simple picture of a heaven above and a hell below certainly was.

For seven years of imprisonment Bellarmine labored to get Bruno to recant the doctrine of the infinite plurality of worlds. Bruno refused, and in 1600 he was burned at the stake.

Since the charges against Bruno were never made public, other Catholic scientists, including Galileo, did not take his execution as a sign of Catholic hostility to Copernicus. But this hostility was confirmed even as the new theory triumphed.

Despite its widespread acceptance in England, there was still relatively little observational evidence for the Copernican model. Tycho Brahe, the most accurate observer of his day, formulated a compromise alternative in which the planets revolve around the sun, which in turn revolves around an immobile earth. Mathematically, this system was identical to Copernicus's, so neither seemed clearly superior.

In 1609 this situation suddenly changed. After Brahe's death Johannes Kepler used his observations, which were 150 times more precise than Ptolemy's, to find an accurate description of the solar system. Starting with the traditional conception of perfect circles, Kepler labored for years. After enormous struggle he broke with this last remnant of the ancient cosmology. By trial and error, he discovered in 1609 that the planets moved in ellipses, not circles, and not at constant speeds, but at such a rate that the areas swept within their elliptical orbit in a given time remained constant throughout their orbit. (As a planet approaches the sun in an elliptical orbit, the gravitational attraction increases, and it speeds up; when it has passed the sun, its trajectory carries it farther away from the sun, and the force of gravity slows it down.) The immensely complex system of epicycles, deferents, and eccentric spheres was replaced by simple ellipses (Fig. 3.1).

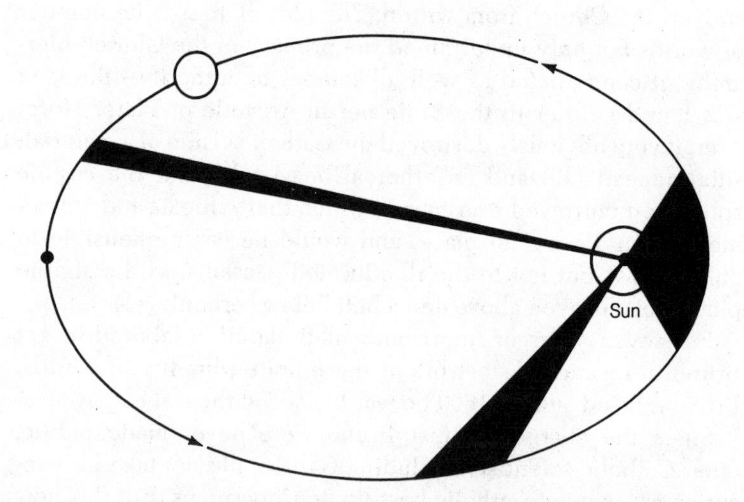

Fig. 3.1. Kepler's solar system simplified the dozens of epicycles and deferents to seven ellipses. While the ellipse was not as "perfect" as the circles that had dominated astronomy for two millennia, they are the correct orbits of the planets and the moon. Kepler's work showed unequivocally that the earth moves around the sun and that each planet sweeps out equal areas in equal times, moving faster closer to the sun and more slowly farther away.

Kepler's system was far more accurate than any other. It could not be translated to Tycho Brahe's, since then the paths of the planets would not be simple ellipses but complex compound motions.

That same year, Hans Lippershey patented the telescope in Holland. Within a year, Galileo in Italy and other astronomers had turned the new instrument to the heavens. Galileo discovered the existence of the moons of Jupiter, the phases of Venus, and the mountains of the moon. The changeless, perfect heaven so crucial to Aristotelian cosmology was shattered by observation.

Armed with his new observations, Galileo immediately became a propagandist for the Copernican worldview, actively trying to win over the Catholic hierarchy. Cardinal Bellarmine, warned by the case of Bruno, moved to quash Galileo's effort. No conflict with the literal interpretation of scripture is possible, he informed Galileo: the sun is described in the Bible as moving, rising, and setting—anything else is heretical. In 1616 Copernicus's work was added to the index of prohibited works and the new doctrine officially condemned.

Galileo, however, continued his efforts, which culminated with the publication in 1632 of his great defense of Copernicus, the *Dialogue on Two World Systems*. The response came swiftly: he was forced, with the example of Bruno before him, to recant and was placed under house arrest. The new science remained forbidden in Catholic countries for over a century.

It was only in those countries where the new society was victorious that the new science became self-sustaining—above all, in England. The English revolution of 1642 led to the decisive defeat of the landowning classes and their absorption into the new mercantile and manufacturing regime. During the period of the Commonwealth the revolutionaries proudly identified their movement with scientific rationalism and the rejection of myths and superstitions. English scientists rapidly synthesized Kepler's laws and Galileo's investigations in mechanics, which had been smuggled abroad and published in 1638. Together, these led Robert Hooke to formulate a universal law of gravitation, which Isaac Newton then proved would verify Kepler's laws.

While many of the social gains at the height of the revolution were subsequently rolled back, the fundamental outlook and goals of society had been irreversibly transformed. The English government's sponsorship of scientific research put English science far ahead of that of any other country; this, together with England's swift economic development, propelled it a century later into the industrial revolution.

The scientific revolution was thus not an inevitable process, a natural outgrowth of human intellectual development. It was the result of a fierce social conflict, in which cosmological questions were matters of life or death for individuals and whole societies.

Certainly the people of the time did not think that the defeat of Spain, the victory of England and Holland, and later the victory of the English revolution were at all inevitable. Yet without those victories, the scientific revolution would certainly have not occurred. Only the open society born in the sixteenth and seventeenth centuries could have nurtured the infinite unlimited cosmos of modern science. And only such a worldview could have given the new society the moral and material strength to prevail.

THE SCIENTIFIC ALTERNATIVE TO CREATION

The scientific revolution of the sixteenth and seventeenth centuries had, at least in England, displaced the hierarchical, finite universe with an infinite one, the appeal to authority and reason with the observational method. But, unlike the Ionians, seventeenth-century scientists had not developed a naturalistic theory of the origins of the world, an alternative to the creation from nothing of the medieval cosmology. Philosophers such as Nicholas of Cusa and Giordano Bruno had advocated the idea of a universe unlimited in time and space, eternal and without beginning. But no scientist had justified these notions with hard data.

For many scientists, it was in this realm of origins that religion still intersected with science. Isaac Newton, for example, argued that God is needed to form the solar system and to maintain it.

In the period after the English revolution, the Restoration, and the ensuing Glorious Revolution, English society settled into a conservative phase. The idea of change, implicit in any concept of evolution in nature, lost its popularity. The universe, like the unwritten English constitution, was a finished product brought into being by events (such as the revolution) that could not recur.

It was not until the middle of the eighteenth century, when the winds of change started to blow in Europe and America, that the problem of origins was again attacked. In 1755 the philosopher Immanuel Kant formulated a naturalistic explanation for the origin of the earth, in many ways strikingly similar to Anaxagoras'. Kant, who was familiar with the latest astronomical research, argued that observation showed that stars are not randomly scattered throughout the universe, but appear to be grouped into a huge disk, the Milky Way. He speculated, correctly, that the distant fuzzy nebulas astronomers were then studying are similar vast agglomerations of stars, what we now term galaxies. By analogy he reasoned that these, too, probably formed still larger systems or clusters—again a guess later confirmed by observation.

Starting with this concept of an infinite universe, arranged into a chain of larger and larger spinning agglomerations of matter, Kant proposed the idea that far in the past the universe was a nearly homogeneous, infinite gas. Certain regions, which by accident happened to be denser than others, started to attract matter by gravitation. The random motions of the gas gave to each agglomeration a slight spin, creating huge vortices, within which galaxies, then stars, then planets coalesced. Since Kant assumed that this process started in one place in the universe, and spread outward, he believed that creation was and remains a continuous process, which will spread through the infinite universe.

In the years following Kant's "Theory of the Heavens" Europe and America were convulsed by sweeping revolutions that sought to complete the overthrow of the old hierarchical societies and to replace them with democracies. By the end of the century, the spectacular changes of government and society brought about by these revolutions led their supporters to conceive of a general and continual process of human social change—the idea of progress. Both to the Founding Fathers in the United States, and to the French revolutionaries, their revolutions were part of the inevitable advance of society, perfecting its institutions and improving without limit the material well-being of mankind. Jefferson concluded that to prevent the resurgence of tyranny and to advance progress, periodic revolutions might be needed. The

American Constitution embodies in its amendment process the idea that it is not the final product, but one that can accommodate indefinite revision.

In England the chemist Joseph Priestley propounded the general theory of human progress: through the growth of scientific knowledge, he wrote in 1771, "human powers will be enlarged, nature, including both its materials and its laws, will be more at our command, men will make their situation more comfortable, they will probably prolong their existence in it and will daily grow more happy."

The revolutionary concept that society is not a fixed entity, that it continuously evolves through effort and struggle, through science and technology, toward higher forms of organization and material well-being, was swiftly taken up in the field of science. In late-eighteenth-century England geological knowledge advanced rapidly as coal became central to the steam-powered industry of the industrial revolution. Geological observation led James Hutton, an amateur scientist, to develop a theory of the continuous evolution of the earth itself.

By observing such processes as the compaction of clay into sedimentary rock, Hutton concluded in his 1795 work, *Theory of the Earth*, that mountains, rivers, oceans, and the sedimentary and igneous rocks of the world today were formed over many millions of years, not by miraculous floods or one-time cataclysms. He emphasizes that a scientific history of the world can be obtained only by examining current processes and working backward in time, not by speculating about origins and working forward. The idea of a world finite in time, with a supernatural origin, is rejected: "The result, therefore, of this physical enquiry is that we find no vestige of a beginning, no prospect of an end."

Within a decade, the French mathematician Pierre-Simon de Laplace had taken Hutton's approach a step further into the past and given a firm scientific basis to Kant's vortex theory of origin. Using Newtonian mechanics, Laplace demonstrated in 1796 that, if the sun had condensed from a spinning sphere of gas, it would have thrown off material as it contracted, since as it contracted it would have spun faster. The material thrown off would form into rings, which would, in turn, condense gravitationally into planets. The nearly circular orbits of the planets would therefore be neatly accounted for.

Hutton and his supporters rapidly accepted Laplace's nebular theory, producing an integrated approach to the history of the world since its origins. Others quickly applied the historical approach to the development of life itself. Erasmus Darwin (Charles's grandfather), in the same year as Laplace's theory, proposed that the fossils found in geological strata represent the evolution of various species of animal from one another, leading to a greater and greater perfection of life over vast stretches of time.

The revolutionary changes of the last quarter of the eighteenth century were not universally hailed, and neither were the new scientific theories. The capitalists who ruled Great Britain owed their power to the social revolutions of the seventeenth century and the industrial revolution of the eighteenth, but they had no desire to lose that power in further social upheavals. Great Britain became the major foe of all social change, fearing the development of rival industrial powers abroad and a continual evolution of social structure at home. From Britain, religious and philosophical replies were launched against the ideas of human and natural progress. Thomas Malthus, rebutting the Marquis de Condorcet, the French theorist of progress, argued that population growth will always outstrip agricultural production, condemning most people to hunger and blocking material progress. Geologist John Williams blasted Hutton's theories on theological grounds. Hutton's "wild and unnatural notion of the eternity of the earth leads first to skepticism and at last to downright infidelity and atheism. If we once entertain a firm persuasion that the world is eternal, and can go on itself in the reproduction and progressive vicissitudes of things, we may then suppose that there is no use of the interposition of a Governing Power," he wrote, concluding that "all rebellions soon end in anarchy, confusion and misery and so does our intellectual rebellion."

But these efforts proved generally unsuccessful: in the course of the first half of the nineteenth century, Europe continued to be rocked by repeated popular revolutions, and the industrial revolution transformed British society as well. By the 1840s the new geology and cosmology held wide acceptance among scientists and the public, and socialist concepts of human evolution spread throughout Europe.

In 1859 Charles Darwin systematized and popularized the theory of biological evolution, ironically seizing on Malthus's theory of limited resources to formulate a vision of continual evolution and change. By the 1860s, despite continued religious opposition, the evolutionary and historical approaches in the sciences had become dominant, as had the related idea of human progress.

The result, it should be emphasized, was not so much a victory of science over religion as the *separation* of science and religion. The Protestants (and Catholics like Galileo) who championed the scientific revolution considered themselves devout, as did the Quakers and Unitarians who led in the introduction of the ideas of evolution. Like Galileo, they believed that religious ideas spoke of morals, not of physics. Galileo's famous phrase states, "Religion teaches men how to go to heaven, not how the heavens go."

Neither religion nor philosophy could place limits on the natural universe in time or space. The triumph of the scientific revolution was the triumph of the infinite universe.

■

4 THE STRANGE CAREER OF MODERN COSMOLOGY

What makes God comprehensible is that he cannot be comprehended.

—Tertullian, c. a.d. 200

The most incomprehensible thing about the universe is that it is comprehensible.

—Albert Einstein, 1935

The more the universe seems comprehensible, the more it also seems pointless.

—Steven Weinberg, 1977

We may now be near the end of the search for the ultimate laws of nature.

—Stephen Hawking, 1988

In our century the cosmological pendulum has swung back. The universe of present-day cosmology is more like that of Ptolemy and Augustine than that of Galileo and Kepler. Like the medieval cosmos, the modern universe is finite in time—it began in the Big Bang, and will end either in a Big Crunch or in a slow decay and dissipation of all matter. Many versions, like Stephen Hawking's, are finite in space as well, a perfect self-enclosed four-dimensional sphere. There is a gap between the heavens and the earth: in space there exist strange entities, governed by the pure and ethereal mathematics of general relativity—black holes, cosmic strings, axions—which cannot, even in principle, be studied on earth.

The nineteenth-century universe evolved by laws still in action today, as did that of the Ionians, yet the universe of modern cosmology is the product of a single, unique event, qualitatively different from anything occurring today—just as the medieval cosmos was the product of the creation. While scientists of a century ago saw a universe of continuous change, evolution, and progress, today's researchers see a degenerating universe, the ashes of a primordial explosion.

To earlier scientists, and to most of today's scientists outside cosmology, mathematical laws are descriptions of nature, not the true reality that lies behind appearances. Yet today cosmologists assume, as did Plato and Ptolemy, that the universe is the embodiment of preexisting mathematical laws, that a few simple equations, a Theory of Everything, can explain the cosmos except for what "breathed fire" into these equations to make them come alive.

Big Bang cosmology does not begin with observations but with mathematical derivations from unquestionable assumptions. When further observations conflict with theory, as they have repeatedly during the past decades, new concepts are introduced to "save the phenomenon"—dark matter, WIMPs, cosmic strings—the "epicycles" of current astronomy.

Of course, just as the nineteenth-century cosmos was not merely a revival of Ionian philosophy, so the modern cosmology of the Big Bang is not a simple echo of Augustine and Ptolemy. It rests on an impressive foundation of elaborate and beautiful mathematical theory. But, like Ptolemy's theories, it provides few predictions that are confirmed by observation.

Within a century all the basic ideas rejected by scientists in their battle against medieval cosmology have now again become the accepted truth. How could this new swing of the pendulum come about? How could the cosmology of the scientific revolution be replaced by the cosmology of the Dark Ages? How could a worldview that justified a society of slavery and serfdom arise again in the twentieth century?

As in the fourth and seventeenth centuries, the battles in cosmology today cannot be separated from the evolution of society as a whole. A universe of unlimited progress from an infinite past to an infinite future makes sense when society is advancing. But when that advance halts, when the idea of progress is mocked by the century of Verdun, Auschwitz, and Hiroshima, when the prospect of human betterment is dim, we should not be surprised that the decaying cosmos again rises to dominance.

THE BEAUTY OF MATHEMATICS, THE LIMITS OF THE INFINITE

In order to see how this huge cosmological shift took place, we must first understand that the ideas of the scientific method and the infinite, evolutionary universe, which became dominant during the seventeenth, eighteenth, and nineteenth centuries, were never universally accepted, even by leading scientists. While the bulk of scientific work used the observational method pioneered by Leonardo and Galileo, the old deductive method, that of seeking final truths from indisputable mathematical principles, was never wholly abandoned.

In the seventeenth century René Descartes became the main champion of deductive, abstract logic. Descartes was a penetrating mathematician, the originator of the idea of coordinate geometry, which allows one to relate the mathematical formulas of algebra and calculus to the forms of geometry. With Cartesian coordinates any shape can be reduced to a set of numbers or an equation. But Descartes believed that scientific truth can be derived only from three basic principles: motion, extension, and God. God is essential as the originator of extension and motion, which can't then be further changed by nature. Experiment was to be used, as with the Platonists, to illustrate laws mathematically deduced from first principles.

Descartes was the first to term the mathematical rules that others had discovered "the Laws of Nature." God rules the universe through these eternal and unchangeable laws, just as the new royal governments coming into existence in Europe, and especially in Descartes's native France, ruled society by written laws (rather than by custom, as had been true in the Middle Ages). These laws were thus not mere descriptions of nature, as they were with Galileo, but the very legislation of nature, superior to it as God is superior to creation: Descartes's God was the great Lawgiver.

Descartes revived, as well, the Platonic dualism of spirit and matter. The material world is like a machine, he writes, but man has a soul that links him with a different sphere—that of the spirit, which is not governed by natural law.

The Cartesian idea of a set of universal laws which control natural occurrences exercised a powerful appeal in the succeeding centuries. Laplace, even as he developed his theory of a naturally evolving cosmos, endorsed the idea that, given the laws of gravitation, Newtonian mechanics, and the "initial conditions" of the universe, every subsequent event not only can be accurately predicted, but is *predetermined.* The whole history of the universe, and of earth, is the inevitable operation of a set of eternal laws. In modern terms, Laplace believed that mathematical physics constitutes a Theory of Everything.

Why is Laplace wrong? There are several reasons, but the simplest is that he ignores everything not governed by his basic mechanical laws. For example, he ignores the electrical and magnetic interactions of bodies, their chemical reactions, their nuclear transformations, the process by which they are heated and cooled—in short, all the phenomena now known to science but unknown to him. As long as the scientist realizes that in hypothesizing an exact mathematical law he is abstracting a single aspect of nature, no problem arises. Laplace's error was to assume that a single mathematical law can describe *all* of nature. By contrast, the Galilean approach assumes that science is open-ended, and that new phenomena, previously neglected or, like radioactivity, unknown, will come to light and require mathematical description.

The concept of an infinite universe was also questioned even after the scientific revolution. Newton was undecided on whether his laws of gravitation preclude an infinite collection of matter. He thought that only a divinely precise positioning of all the stars could prevent such an infinite collection of matter from collapsing into a series of heaps. Much later, in 1823, the astronomer Heinrich Olbers pointed out that an infinite universe seemed to imply a paradox. If there were an infinite number of stars, if one went far enough in any direction from earth, one would hit a star. This implied that the sky should be uniformly bright, as bright as the surface of the sun, which it obviously is not.

The idea that the universe had a finite lifetime also existed in the mid-nineteenth century, although only on the popular fringes of science. The first suggestion that the universe originated in a creative explosion—the first Big Bang—actually came from the pen of Edgar Allan Poe in 1849. Poe was not only a well-known poet and writer, he was also a scientific popularizer who kept himself up-to-date on the latest in astronomical research. In the book-length essay *Eureka* Poe rejected the idea of an infinite universe, citing Olbers's objections. He reasoned that a universe governed by gravitation would collapse in a heap if not kept apart by some form of repulsion. He postulated that God had, in an enormous explosion at the creation, thrust all the stars apart. Like a rocket racing into the sky, the stars and galaxies would first expand, and then contract into a final catastrophe, the end of the world.

Both Cartesian deductive methods and questions about the infinity of the cosmos remained marginal to the mainstream of science through the mid-nineteenth century. The swift advance of technological progress and the equally swift transformation of society convinced most scientists that the basic methods of science correctly yield results proved in practice, and that the thesis of an unlimited, evolutionary universe is valid. It was not until social and economic progress slowed that the corresponding scientific assumptions came under serious attack.

THE "HEAT DEATH" OF THE UNIVERSE

In the latter third of the nineteenth century, from around 1870 on, the nature of the rapid social and economic evolution of western society began to change. By this time, the last institutional vestiges of compulsory labor had been wiped out by social revolutions in Europe, the Civil War in the United States, and the liberation of the serfs in Russia. After the defeat of the Paris

Commune—the 1871 attempt to establish a workers' rule—Europe entered a period of relative political stability. The earlier ideas of revolutionary progress, progress through active participation in history, rapidly gave way to new concepts of progress as an automatic, smooth process. According to these new ideas, political rights and standards of living would gradually rise in tandem with the advance of technology, eliminating the need for the violent upheavals of the past.

But even as this automatic progress became a complacent assumption of, at least, middle-class and upper-class Europe and America, the exuberant expansion of society was slowing. The sixty years from 1820 to 1880 had witnessed the fastest economic growth in history. But by 1880, the limits of capitalist markets were being reached: European and American goods were penetrating virtually every corner of the globe, as Britain, France, and Germany rushed to carve up the only remaining land—Africa. While the actual need for goods remained immense, the market for goods that could be sold at a profit was nearing the end of its growth. For centuries, millions of new farmers and peasants had been drawn into the developing capitalist market system as feudal regimes fell apart and as new colonies were conquered and absorbed. When this expansion lost its frontiers with the formation of a global market at the end of the nineteenth century, the industrial economies could no longer continue their vigorous expansion.

After 1880 the production of iron and steel and the laying of new rail lines practically ceased their growth. Real wages continued to increase, but more slowly, peaking in Europe by 1900. Manufacturers turned to the European states for new markets, leading to the growth of a gigantic arms industry. Manufacturers found that, to paraphrase Lord Keynes, two battleships are always better than one, unlike two railroads from Liverpool to London. These arms, in turn, were used to maneuver for a greater share of the precious world markets and the resources of the colonies.

It was in this era of slowing growth that the first real scientific challenge to the unlimited universe appeared. Steam power had developed throughout the nineteenth century, as did the study of heat and its transformation, thermodynamics. In the early part of the century, scientists had discovered that energy can be transformed in various ways, but never created or destroyed, a fundamental principle that came to be known as the first law of thermodynamics. In 1850, Rudolf Clausius discovered another fundamental principle, the second law of thermodynamics. A body's ratio of its energy to its temperature, a quantity Clausius dubbed "entropy," always increases in any transformation of energy—for example, in a steam engine.

In 1877, Ludwig Boltzmann attempted to derive the second law from the newly emerging atomic theory of matter. He redefined entropy as a function of the probability of a given state of matter: if the state is more probable, it has a higher entropy. Thus, if a million atoms of oxygen mixed with a million atoms of nitrogen, it would be far more probable to find them evenly mixed than segregated. The well-mixed state has a higher entropy, and left to itself, a container with oxygen on one side and nitrogen on the other will rapidly go to the higher entropy state of an even mixture.

Boltzmann, using his new definition of entropy, went on to demonstrate, so he claimed, that all systems tend toward a state of equilibrium—defined as the state in which there is no net flow of energy. Thus a hot object and a cold object placed in contact are not in equilibrium, since heat will flow from one to another, until they're the same temperature, which is a state of equilibrium.

From this proof, Boltzmann propounded a new concept with profound cosmological implications. The universe as a whole, must, like any closed system, tend toward an equilibrious state of entropy: it will be completely homogeneous, the same temperature everywhere, the stars will cool, their life-giving energy flow will cease. The universe will suffer a "heat death." Any closed system must thus go from an ordered to a less ordered state—the opposite of progress.

Boltzmann was aware that his ideas contradicted the notion, then widely accepted, of a universe without beginning or end. The present-day universe is far from a state of equilibrium, comprising as it does hot stars and cold space. If all natural systems "run down" to disorder, the present state of order must have been created by some process that violates the second law at a finite time in the past. Conversely, at a finite time in the future, the world will cease to exist, becoming a lifeless homogeneous mass: human progress is but an ephemeral and inconsequential episode in a universal decay.

Boltzmann found his results disturbing. Since he rejected a supernatural origin of the universe, he tried to argue that, in an infinite amount of time, extremely improbable events do occur, such as the spontaneous organization of a universe, or a large section of it, from a prior state of equilibrium. The second law is, after all, a statistical one stating what is likely to happen, not what *must* happen. Just as there is an incredibly small chance that all the air in a room will rush to one side, there is a smaller chance that all the atoms in a homogeneous part of an infinite universe suddenly rushed together into one spot of low entropy. Boltzmann's argument did not much impress fellow scientists, since by his own theories the probability of these occurrences was, in fact, so tiny that it was equivalent to impossibility.

But scientists had other reasons for not accepting the second law's implication that the universe necessarily had a beginning from which it was now running down. The predictions of thermodynamics appeared to contradict what was known of geological and biological evolution. In the 1890s a debate broke out between thermodynamicists and geologists over the age of the earth. The physicist Lord Kelvin argued that, from the cooling rate of the earth as estimated from measurement of heat in mines, the earth must have been nearly molten as recently as twenty million years ago. Geologists countered that the formation of certain rock deposits must have taken at least twenty times as long, four hundred million years. Backed up not by theory but by a vast accumulation of observation, geologists doubted the physicists' theories.

In addition, some thermodynamicists pointed out that Boltzmann had proved far less than he claimed. He assumed that gas began in a high degree of disorder, close to equilibrium, and never got far from it. Moreover, he only allowed for atomic collisions, but took no long-range forces, such as electromagnetism or gravity, into account. In most real physical situations, though, these restrictions aren't valid, so Boltzmann's proof is not applicable. A century later scientists were to demonstrate that, in the general case, Boltzmann's law of increasing disorder simply isn't true.

Beyond these scientific objections, though, were cultural ones. At any moment, scientists must decide which problems or apparent paradoxes are worthwhile and which should simply be dismissed—it is here that the ideology of the age, of society as a whole, affects what scientists feel "makes sense." And Boltzmann's concept of a world running down simply didn't make sense to most nineteenth-century scientists.

In the late nineteenth century, while material advances had slowed and the ominous trends leading toward the crises of the twentieth century were beginning to emerge, progress remained the overwhelmingly dominant idea of the epoch. Standards of living continued to rise, albeit more slowly, until 1900. Technological progress was more rapid than at any other time in human history: someone born in 1870 would have grown up in a world of gaslight and horse-drawn carriages, but by age forty he or she would live in a world of electricity, telephones, phonographs, movies, radio-telegraphs, automobiles, and airplanes.

Science, too, advanced dramatically in the same period. Biology and medicine were transformed in the 1880s by the germ theory of disease leading to the widespread use of antiseptics in surgery, and the general use of vaccination. Physics saw the blossoming of the study of electromagnetism, put on a firm foundation in 1865 by James Clerk Maxwell, and later radioactivity, X-rays, Einstein's special theory of relativity, and the beginning of quantum theory.

The reality of progress in science and society was so apparent to the average scientist that Boltzmann's vision of a universe in continual decay seemed too bizarre. In practice, Boltzmann's laws were very useful in dealing with steam engines and simple gaseous systems, and were widely applied. But his broad generalizations about cosmology, which implied that the universe must have had a beginning, must have been "wound up," had no significant impact for more than a generation.

THE RETURN OF A FINITE UNIVERSE

The European and American confidence in progress was shattered in August of 1914. In the following four years the vast economic power and technological achievements of the prior century were thrown into the barbaric enterprise of slaughtering twenty million human beings. In the wake of war came revolution and counterrevolution: working-class living standards had plummeted during World War I, and workers' movements had seized power in Russia and tried to do so in Germany. Throughout Europe and America, employers and governments battled strikers.

On November 9, 1919, the front page of the *New York Times* was filled with news of turmoil. COURT ORDERS STRIKE CALL REVOKED was the lead headline as the government forbade a national coal strike, which the judge warned "would undermine the foundations of the Republic." Another headline read, 73 RED CENTERS RAIDED HERE BY LUSK COMMITTEE, the article telling of hundreds of immigrants seized as dangerous subversives in the Palmer raids and then summarily deported. OUTBREAKS IN ITALIAN CITIES ON BOLSHEVIST ANNIVERSARY, read a third.

Page six had a quite different story: ECLIPSE SHOWED GRAVITY VARIATION, and below, DIVERSION OF A LIGHT RAY ACCEPTED AS AFFECTING NEWTON'S PRINCIPLE, HAILED AS EPOCH MAKING. BRITISH SCIENTIST CALLS THE DISCOVERY ONE OF THE GREATEST OF HUMAN ACHIEVEMENTS. An observation of the May 29, 1919, solar eclipse had confirmed Einstein's prediction of the bending of light from a distant star by the sun's gravity. This vindication of his general theory of relativity was announced at a meeting of the Royal Astronomical Society.

Why was Einstein's theory, not even briefly described in this first article, so outstanding? One scientist noted that the effect on practical astronomy of the small differences from Newton's laws would not be very great. But "it was chiefly in the field of philosophical thought that the change would be felt." The *Times* reported, "Space would no longer be looked on as extending indefinitely in all directions. Straight lines would not exist in Einstein's space. They would all be curved and if they travelled far enough they would return to their starting point."

Thus the first public announcement of Einstein's theory suddenly proclaimed the falsity of a basic cosmological tenet, that the universe is infinite. More surprises came the next day when a *Times* headline declared, LIGHTS ALL ASKEW IN THE HEAVENS, MEN OF SCIENCE MORE OR LESS AGOG. Not only was the new theory shocking in its implication, but it was incomprehensible as well:

J. J. Thomson stated that it was useless to detail the theory to the man in the street, for it could only be expressed in strictly scientific terms, being "purely mathematical." In fact, the *Times* went on, Einstein himself had warned his publishers that there were not more than twelve people in the whole world who could understand his theory. But, another scientist commented, this was of no concern, since "the discoveries, while very important, did not, however, affect anything on this earth," only the heavens.

This idea too was a tremendous break with the past. For hundreds of years it had been a common belief that conveying the latest scientific discoveries to the broadest possible audience was not only possible but essential. Today's science would be tomorrow's technology, and those who manned an increasingly technological industry would have to understand the new machines. Nineteenth-century public lectures on science by leading authorities normally drew a cross section of working people and the middle class, eager to keep up with progress. And experience bore out the necessity for this process. After all, within fifteen years of Maxwell's discovery of the laws of electromagnetism, Thomas Edison made them the basis of a technological revolution.

Now scientists were saying that this new theory was not, even in principle, comprehensible, and that it would have no impact on earthly technology—only on the heavens.

The very means by which this discovery was conveyed was just as unprecedented as the discovery itself. The preceding two decades had seen their share of startling phenomena—X-rays, radioactivity, Einstein's own special theory of relativity (which concerns the effects of high velocities), and quantum theory. Most of these involve difficult ideas, and unlike the general theory of relativity, they all had immediate implications for technology. Yet none had been reported as news in the popular press; instead they were disseminated through journals and popular lectures. Now Einstein's discovery was being reported alongside the latest roundup of Reds and the dispatches from the Civil War in Russia.

The public reception of Einstein's general theory of relativity is a striking development in the history of science. Virtually overnight many of the trappings of the ancient cosmology were reintroduced. We have a finite universe—calculated to a radius of eighteen million light-years. It is again a cosmos, in which knowledge of the heavens is the privilege of an elite, without practical applications on earth, disengaged from the promise of egalitarian progress. How is it that these striking notions became so rapidly and widely disseminated by the leading newspapers of the world? Why were they viewed as major news, rivals of strikes, wars, and revolts?

Einstein's new theory appealed to scientists, reporters, and editors because it brought a vision of the universe as a whole, a vision that appeared as a solace to a tormented society. The cosmology Einstein developed in 1917, two years after formulating his general theory, had, for many scientists, a terrific aesthetic and philosophical attraction. In part, this was based on the appeal of general relativity itself. As Alfvén has written, "No one can study General Relativity without being immensely impressed by its unquestionable mathematical beauty." And, moreover, it was demonstrated not only in its prediction that light near the sun would be bent by gravity, but by subtle variations in the orbit of Mercury which Newtonian gravitation couldn't explain. Newton and other scientists had always been bothered that gravity appeared to act "at a distance," a magical influence in empty space. General relativity eliminates this problem, showing that mass curves the space around it like a weight resting on a sheet pulled taut at the edges. It is this curvature of the space that results in gravity, not the direct action of one object on another.

But beautiful as it was, this change in gravitational theory was *not* what captured the imagination of scientists and the press. It was instead Einstein's cosmological speculations of a closed, finite universe. Gravity, Einstein argued, would curve the entire cosmos around into a four-dimensional sphere, finite, yet without boundaries. Einstein's spherical universe is static, eternally unchanging, ruled by his elegant equations.

To a society shattered by World War I, this vision of a calm, ordered universe must have been tremendously reassuring. When mankind is progressing, the dynamic changing infinite universe, the "restless universe," as Sir James Jeans called it, seems exciting and challenging. But when human affairs are in shambles, and change no longer means progress but can mean upheaval and death, a finite and static universe like Einstein's can appear a balm to tortured souls, just as Augustine's hierarchical cosmos seemed to offer refuge from the confusion and misery of the fourth century.

As one of Einstein's biographers, physicist Abraham Pais, wrote, "Einstein's discovery appealed to deep mythic themes. A new man appears abruptly, the suddenly famous Dr. Einstein. He carries a message of a new order in the universe. . . . His mathematical language is sacred, . . . the fourth dimension, light has weight, space is warped. He fulfills two profound needs in man, the need to know and the need not to know but to believe."[1] In a time of death and uncertainty, "he represents order and power. He became the divine man of the twentieth century."

The great crisis caused by the war probably had much to do as well with leading scientists' promotion of the idea that General Relativity was incomprehensible. There was a grain of truth in this. The mathematics that described the curvature of space was complex. Indeed, only a dozen or so researchers had mastered it —presumably the origin of Einstein's remarks.

But this was still only a grain of truth. In 1919, as today, the vast majority of the public lacked the mathematical knowledge to understand the equations describing practically any physical theory, even Newton's. However, the story was entirely different when it concerned the physical concepts involved. As subsequent popularizations have shown, the basic idea of gravity altering the paths of objects through the curvature of space is quite simple to visualize and understand using two-dimensional models.

For the scientists of 1919, though, the temptation to emphasize the theory's incomprehensibility, rather than its conceptual elegance and simplicity, must have been very great. Like most of the middle-class in Europe, scientists felt themselves assaulted on all sides in the years of the world war and its aftermath. They lacked any of the power of the wealthy and were like any other citizens, the helpless pawns of warring states. Now, as revolution rocked Europe, the middle class felt itself threatened by working-class movements that seemed to threaten hard-won status and privileges.

For scientists like J. J. Thomson to now turn around and say to the mighty and the masses alike, "we alone can understand the secrets of the universe—they are beyond you" might well have seemed sweet revenge for the indignities of the recent past. In addition, while general relativity itself really could be explained in commonsense terms, the cosmology many scientists found attractive could not be. To the average man a finite universe seemed absurd. If this new cosmology was to be accepted, it was convenient to appeal to uncritical belief, contending that the reasons for the new view were beyond the ken of ordinary mortals and so could not be questioned.

Initially, the scientists' claims, while widely reported, were met with some skepticism. "What they say," the *New York Times* noted editorially, "hints less at the impossibility of making understandable Einstein's discovery, than at an inclination to keep a particularly interesting thing to themselves . . . If it was explained and we gave it up, no harm would be done, for we are used to that. But to have the giving up done for us is—well—just a little irritating."

Nor did everyone readily buy the idea that the universe was finite. "Critical laymen have already objected that scientists who proclaim that space comes to an end are under some obligation to tell us what lies beyond it," the *Times* reported.

Yet within a few weeks, the newspaper's line on relativity had made a complete turnaround. Under the headline "Nobody Need Be Offended," the *Times* proclaimed that for those who do not understand mathematics, "nothing is left except to accept the expert's conclusion, on the authority of its maker, supported by the acceptance of the few others like him."

It's impossible to be sure exactly what motivated this remarkable change of heart, which was echoed by the London *Times* and many other papers, who soon made Einstein's incomprehensibility proverbial. We can guess. In 1919, the newspaper owners and the other powers that be had a few motives of their own to exhort their readers to unquestioning acceptance of authority. The papers were full of modern-day fairy tales about the bomb-throwing Bolsheviks who were behind every strike and union organizing drive, the untrustworthy aliens who needed to be hounded out of the country before they sank it in anarchy and bloodshed, and the patriotic captains of industry who had only the country's interest at heart as their thugs lynched unruly workers. Perhaps the editors thought that, with all the revolutionary challenges to authority, a little bolstering of scientific authority was not such a bad idea after all.

THE REBIRTH OF MYTH

Whatever the complex motives that produced the myth of Einstein and the general theory of relativity, it has had a profound impact on twentieth-century science. Nineteen nineteen became a fault line in the history of science, and in that year the main trends that were to lead to the acceptance of the Big Bang began.

As Alfvén points out, it is quite ironic that a triumph of science led to the resurgence of myth. The most unfortunate effect of the Einstein myth is the enshrinement of the belief, rejected for four hundred years, that science is incomprehensible, that only an initiated priesthood can fathom its mysteries. Alfvén wrote sixty years later, "The people were told that the true nature of the physical world could not be understood except by Einstein and a few other geniuses who were able to think in four dimensions. Science was something to believe in, not something which should be understood. Soon the best-sellers among the popular science books became those that presented scientific results as insults to common sense. One of the consequences was that the limit between science and pseudo-science began to be erased. To most people it was increasingly difficult to find any difference between science and science fiction."[2] Worse still, the constant reiteration of science's incomprehensibility could not fail to turn many against science and encourage anti-intellectualism.

The distorted triumph of general relativity has also contributed to the revival of purely deductive methods of Descartes and Plato. Most theories, such as Newton's laws or Maxwell's theories of electromagnetism, are swiftly confirmed by hundreds or thousands of independent observations. Einstein's theories were at first accepted on the basis of only two—the deflection of starlight by the sun's gravitational field and the subtle shift of Mercury's orbit. Especially in the past few decades, they have received many more confirmations, so that the faith of Einstein and his supporters has been retrospectively justified. General relativity, we now know by observation, is almost unquestionably an accurate theory of gravitation.

But since Einstein's theories were initially "proved" on the basis of only two observations, this exceptional situation has come to be taken as a precedent in cosmology. Again and again elaborate theories have been accepted as verified on the basis of one or two apparently confirming observations. The result has been a loss of rigor in comparing theory and observation.

This tendency might not have been so important had it not been for Einstein's explicit endorsement of the deductive method. He came to believe that his theories developed as deductions from certain fundamental, primarily mathematical principles, and that this deductive method is the one that science must increasingly pursue. In 1934 he wrote, "The theory of relativity is a fine example of the fundamental character of the modern development of theoretical science. The hypotheses with which it starts are becoming steadily more abstract and remote from experience. The theoretical scientist is compelled in an increasing degree to be guided by purely mathematical, formal considerations in his search for a theory, because the physical experience of the experimenter cannot lift him into the regions of highest abstraction. The predominantly inductive methods appropriate to the youth of science are giving place to tentative deduction."[3]

However, the actual development of relativity theory, as described by Einstein himself, does not at all bear out his generalizations about method. (As Einstein commented, great scientists tend to make poor philosophers of science.) In the essay just quoted from, Einstein sought to justify the purely deductive methods he came to use in the thirties, in his efforts to create a unified field theory that would explain both gravity and electromagnetism. He then briefly stated that general relativity emerged directly from an abstract mathematical problem—how to define physical laws in such a way that they appear the same to all observers, whether or not the observers accelerated relative to one another.

But in the next essay in the same little book, *Essays in Science*, Einstein directly addresses the historical question of how relativity theory came about. He there tells a different story, which makes it clear that his theories were *not* simply deduced from pure reason. The special theory of relativity—which describes how the measurement of time, length, and mass change with velocity, and which led to the famous equivalence of matter and energy—was derived from Maxwell's equations, which describe electricity and magnetism. These relationships were, in turn, based on thousands and thousands of observations over decades of work. The key that led Einstein to relativity was the use of the speed of light in Maxwell's equations. Einstein reasoned that this speed must be constant to all observers, regardless of how fast they moved, since for all observers, Maxwell's equations were true. From this simple but fundamental observation, Einstein developed the relations of relativity.

The same method of generalization based on simple but vital observations of nature led to the general theory of relativity as well. In thinking about how gravitation would fit into his theory of relativity, Einstein was struck, as he later explained, by a news story of a man who survived a fall from a roof and remarked that he had felt no weight during the fall—that is, no pull of gravity. Einstein knew that this is a direct consequence of the well-known fact that all objects, regardless of their weight, fall at the same speed in a gravitational field. A falling man who releases a coin will see it floating next to him, since it falls with the same velocity. Einstein reasoned that for a falling man there would be no gravitational field, because he is constantly accelerating. So a gravitational field and acceleration must be equivalent. This was the basis of his gradual evolution of a new theory of gravity.

In this ten-year-long effort, Einstein continually introduced basic mathematical assumptions which seemed to him simple and necessary. But he knew that any law of gravity must agree with Newton's laws in most cases, since those laws had been confirmed by millions of observations. Again and again he rejected his assumptions when their consequences failed to agree with observation. "That fellow Einstein," he commented ironically at one point, "every year he retracts what he wrote the year before."

In this he followed exactly the method Kepler had used in the early days of the scientific revolution, when he applied various mathematical theories to the known motions of Mars. Einstein finally arrived at an answer not through deductions from first principles, but through a flash of insight—gravity can be described as a curvature of space. If objects travel by the shortest path in curved space, then their paths will curve, regardless of their mass.

It was only when Einstein turned, after 1915, to the fields of cosmology and unification theories that he actually applied the deductive methods he espoused in the abstract. And here the results were radically different from his earlier breakthroughs: his quest for a unified theory of electromagnetism and gravitation, which occupied him for the last thirty years of his life, was unquestionably a failure, even in his own eyes.

In his cosmology he again departed from standard method by adopting a fundamental premise that was actually contradicted by observation—a hypothesis that would become basic to all subsequent relativistic cosmology. Einstein assumed that the universe as a whole is *homogeneous,* that matter is, on the largest scale, spread evenly throughout space. Given this, Einstein used his general theory of relativity to prove that space would be finite. Simply put, the larger a mass of a given density is, the more it curves space. If it is big enough it will curve space entirely around onto itself. So if the universe is homogeneous, with the same density everywhere, it must be finite.

But by 1919 there was enormous evidence that the universe is *not* homogeneous. Back in Newton's time, scientists knew that almost all matter is concentrated into stars, separated from each other by vast, nearly empty spaces. Subsequent observation (prior to Einstein) showed that nearby stars form an aggregate galaxy, the Milky Way. Even by the 1850s astronomers had noted that the spiral nebulas, which many rightly believed to be other galaxies, are themselves concentrated in a broad band across the sky, a formation much more recently called a galactic supercluster.

So Einstein knew that observation indicates the universe at all scales was inhomogeneous. Yet purely for philosophical and aesthetic reasons he proposed a homogeneous cosmos, thus laying the basis for a revival of a finite universe. But for an inhomogeneous universe, when the density of a large section of space is less than that for smaller regions, the universe need not be closed over into a sphere.

Einstein's assumption of homogeneity had three profound effects on cosmology. First, it introduced the idea of a finite universe, which resuscitated the medieval cosmos—previously considered obsolete and antithetical to science itself. Second, the aesthetic simplicity of the assumption of homogeneity, combined with Einstein's prestige, embedded this assumption in all future relativistic cosmology. Third, and perhaps most significant, it set a precedent by allowing the introduction of assumptions contrary to observation, in the hope that further observation will justify the assumption. In the case of Einstein's cosmology it was the hope that, on scales larger than clusters and superclusters of galaxies, the universe would become smooth.

THE BIRTH OF THE BIG BANG

Einstein had first formulated his conception of a static, finite universe in 1917, two years after developing the general theory of relativity. But he soon saw its flaws. A static, closed universe could not remain static, because its own gravitation would cause it to collapse. This was a problem not only of his theory, but of any theory of gravity, including Newton's. As Poe had noted seventy years earlier, unless a body of matter rotates, it will collapse under its own gravity—only rotation stabilizes bodies such as the galaxy and the solar system. But Einstein ruled out a rotating universe on philosophical grounds. First, he believed that rotation itself is relative, like all other motion, and the universe could not rotate relative to anything else. Second, rotation implies a central axis, but such an axis would be a distinct direction in space, different from all others—this contradicted his belief that space is the same everywhere and in every direction. Third, Einstein believed his equations dictated a closed universe. A universe with such a powerful gravitational field would not be stabilized by rotation, even if it were rotating at the speed of light—and any faster rotation is prohibited by the special theory of relativity.

Clearly, Einstein reasoned, something prevents the collapse of the universe, something like the centrifugal force of rotation, but not rotation itself. This force must somehow increase with distance: it had never been observed on earth or in the solar system, but it must be strong enough at cosmological distances to overcome gravity. He introduced a new term into his equations of gravity, "the cosmological constant," a repulsive force whose strength increases proportionally to the distance between two objects, just as the centrifugal force of a rigidly rotating body increases proportionally to its radius. But this force, he thought, acts in all directions equally, like gravity, so it does not disturb the symmetry of the universe.

To preserve his conception of a static universe, Einstein set the cosmological constant to a level that would balance gravity exactly, so that its repulsive force neutralized the tendency of the universe to collapse.

This was the delicately balanced cosmology that captivated leading scientists like Arthur Eddington and J. J. Thomson in 1919. Matters rested there for five years. In this time, despite the excitement stirred by the general theory of relativity, only about a dozen scientists in the whole world *did* master its intricacies well enough to continue research in the area.

In 1924 new observations changed the picture radically. For a decade, astronomers had been measuring the spectra of stars in nearby galaxies. In nearly all cases, the spectra shifted slightly toward the red. Scientists had long known the simplest explanation for these redshifts is that the galaxies are moving away, shifting the frequency of light to the red (an analogous phenomenon makes the pitch of a train whistle rise as it approaches and fall as it recedes). It seemed strange that, rather than moving randomly, the galaxies all seemed to be moving away from each other and from us.

Carl Wirtz, a German astronomer, put all the forty-odd observations together in 1924 and noted a correlation—the fainter the galaxy the higher its redshift, thus the faster it is receding. Assuming that fainter galaxies are more distant, then velocity increases with distance. The conclusion was tentative, since the distances to the galaxies were uncertain. But the American astronomer Edwin Hubble and his assistant Milton Humason soon began to examine Wirtz's findings. Hubble had developed a new way of measuring the distance to a galaxy, based on the known brightness of certain peculiar stars called Cepheid variables. Soon word filtered through the astronomical community that Hubble's data seemed to confirm the relation between redshift and distance.

This news was of immense interest to a young Belgian priest and budding relativist, Georges-Henri Lemaître. Born in 1894, Lemaître received his doctorate in physics in 1920, and shortly thereafter entered a seminary to study for the priesthood. While at the Seminary of Maline, he became fascinated with the new field of general relativity, and after being ordained in 1923, went to England to study under Eddington. He then spent the winter of 1924–1925 at Harvard Observatory, where he heard Hubble lecture, and learned of the growing evidence for the redshift-distance relation.

In the next two years Lemaître developed a new cosmological theory. Studying Einstein's equations, he found, as others had before him, that the solution Einstein proposed was unstable; a slight expansion would cause the repulsive force to increase and gravity to weaken, leading to unlimited expansion, or a slight contraction would, vice versa, lead to collapse. Lemaître, independently reaching conclusions achieved five years earlier by the Russian mathematician Alexander Friedmann, showed that Einstein's universe is only one special solution among infinite possible cosmologies—some expanding, some contracting, depending on the value of the cosmological constant and the "initial conditions" of the universe.

Lemaître synthesized this purely mathematical result with Wirtz's and Hubble's tentative observations, and concluded that the universe as a whole must be expanding, driving the galaxies apart. And if the universe is expanding, then any of the cosmological scenarios that led to expansion could be a valid description of the universe. But cosmic repulsion and gravity are not delicately balanced—repulsion predominates in an expanding universe.

Lemaître put forward his hypothesis of an expanding universe in a little-known publication in 1927, and within two years his work and Friedmann's had become widely known and accepted in the tiny cosmology fraternity. By this time, 1929, Hubble had published the first results showing the redshift relation, apparently confirming Lemaître's idea of an expanding universe.

This was not yet the Big Bang, though. The equations of general relativity derived by Friedmann and later Lemaître showed only that many solutions led to universal expansion. Some solutions did indeed produce a singularity, a collapse into, or an expansion from, a universe of zero radius. If gravity was strong, the universe was dense, and repulsion was weak, the universe would collapse; if the reverse was true, the universe would expand outward from a point. But if both forces were strong, there would be no singular state: the universe could be diverging from a state near Einstein's balance, moving away faster and faster with the passage of time; or it could have contracted from an indefinitely large radius in the infinite past to a minimal radius, still perhaps very large, but now again expanding. These nonsingular solutions would assume a universe of infinite age. Indeed not all possible solutions are spatially finite, closed spheres, as Einstein envisioned—some are infinite in spatial extent. Every possible solution, though, is limited in some way, either by an origin in time, or by being closed in space, or both (Fig. 4.1).

In general when equations describing physical reality produce singularities—solutions involving either zero or infinity—it is a sign that something is wrong, since scientists assume that only measurable, finite quantities should be predicted. So initially the solutions without singularities attracted the most attention.

This is as far as general relativity alone could take the cosmological problem. A repulsive force, of unknown origin, is counterbalancing gravity and causing the universe to expand, as Hubble's data confirmed.

In 1928, Sir James Jeans, one of the most prominent astronomers of the time, revived Boltzmann's old arguments about the fate of the universe. The second law of thermodynamics, Jeans reasoned, shows that the universe must have begun from a finite time in the past, and must move from a minimal to a maximal entropy. Incorporating Einstein's equivalence of matter and energy, Jeans argued that entropy increases when matter is converted to energy, because energy is more chaotically dissipated. Thus the end state of the universe must be the complete conversion of matter to energy. "The second law of thermodynamics compels the materials in the universe to move ever in the same direction along the same road, a road which ends only in death and annihilation," he wrote gloomily.

At the same time Eddington was reaching a similar conclusion. Curiously enough he begins his book *The Nature of the Physical World* with philosophical premises similar to those used by Bruno's enemies three centuries earlier. Like Bruno's persecutors, Eddington was viscerally repelled by an infinite universe: "The difficulty of an infinite past is appalling," he writes. "It is inconceivable that we are the heirs of an infinite time of preparation." He too concludes that the second law implies a beginning in time. He isn't pleased by this idea either, but feels that it follows naturally from Boltzmann's laws.

Lemaître, hearing his former teacher's views in March of 1931,

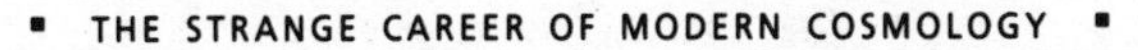

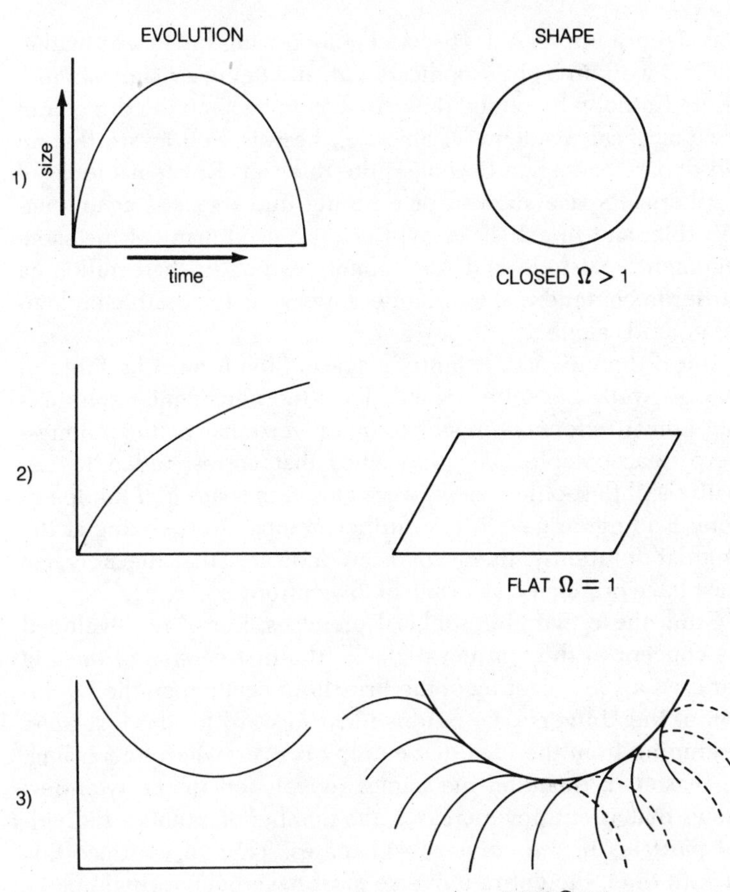

Fig. 4.1. An infinite number of different universes are allowed by Einstein's equations—all of which are based on the assumption that the universe is homogeneous. The real universe can have one of three different types of evolution, depending on its density and the strength of the universal repulsion that Einstein hypothesized, the cosmological constant. It can expand from a point or singularity and contract back to it (1); it can expand indefinitely from a point (2); or it can contract from infinite size to a minimum diameter and reexpand (3). Three different shapes in space are also allowed. The universe can be closed, like a sphere (1), flat, like a plane (2), or open, like a saddle (3)—again depending on density and the repulsion field. A closed universe can have any of the three types of evolution, but a flat or open universe can evolve only by the first two routes—both go through a singularity, and so are finite in time. Thus all the Einsteinian universes are finite in either *space or time or both.*

was deeply impressed. He had been viewing his recent mathematical work in a philosophical light, and being a rising member of the Catholic hierarchy (he was soon to become the director of the Pontifical Academy of Science), he was well aware that an old debate had again become quite relevant. Einstein's ideas of a spherical space showed that a finite universe was again conceivable, and in a 1929 essay, Lemaître used many of the same arguments Aristotle had used almost two and a half millennia earlier to contend that an infinite universe is impossible on logical grounds alone.

But if the universe is finite in space, then it must be finite in time as well, Lemaître argued. Thus the nonsingular solutions that Lemaître found—in which the universe has no beginning—were unacceptable. The only ones that corresponded to Lemaître's philosophical views were closed in space *and* limited in time. Eddington gave him a further rationale for looking at the singular solutions—the second law indicates that the universe must have originated at a state of low entropy.

From these two philosophical premises, Lemaître developed his concept of the "primeval atom," the first version of the Big Bang. At a 1931 meeting of the British Association on the Evolution of the Universe, he put his ideas forward for the first time. Beginning from the idea that entropy is everywhere increasing, he reasoned, quantum mechanics (developed in the twenties) shows that as entropy increases, the number of quanta—individual particles in the universe—increases. Thus, if we trace this back in time, the entire universe must have been a single particle, a vast primeval atom with zero radius. He identified this instant with the singularity of some relativistic solutions. Just as uranium and radium atoms decay into subatomic particles, so this giant nucleus, as the universe expanded, explosively split up into smaller and smaller units, atoms of the size of galaxies decaying into atoms the size of suns and so on down to our present-day atoms.

In defense of his fireworks theory of cosmology, as he sometimes called it, Lemaître cited one phenomenon—cosmic rays. Since the early years of the twentieth century scientists had known that the earth is bombarded by extremely high-energy radiation, either photons or other particles. They also knew that most of this radiation is absorbed by the earth's atmosphere, since it is much more intense at high altitudes reached by aircraft and balloons. Lemaître argued that cosmic rays could not have originated on any body with an atmosphere, such as a star, since they would not then escape into space. Cosmic rays, then, can only be the results of the primordial decay of the primitive star-atoms, before they broke up and formed a gaseous atmosphere.

Rather than thoroughly developing his theory in order to determine an appropriate test, though, Lemaître justified it with a single piece of evidence—the existence of cosmic rays—a phenomenon, he said, that could not be explained in terms of anything happening in the universe today, and so must be ascribed to the entirely different conditions of creation. It was also known that the cosmic rays come from every direction uniformly—they are isotropic. From this Lemaître concluded they cannot derive from any current source, like the stars and galaxies, which are distributed irregularly, but can have been scattered evenly through space only by a primordial explosion and the subsequent eruption of star-atoms.

EARLY TROUBLES

This earliest version of the Big Bang was not warmly welcomed in the scientific world. In fact, several sharp criticisms were leveled against it, all of which were confirmed by later research. After Lemaître propounded his theory Robert Millikan, a leading experimental scientist, criticized his cosmic ray theory, dismissing as "scientifically unacceptable the hypothesis that in bygone ages these rays were created by processes no longer existing and have since been wandering around like lost souls." He pointed out that Lemaître's theory that cosmic rays had penetrated no atmosphere implies not that they originated on the surface of an incredibly dense star-atom or in an equally dense primordial explosion, but rather in interstellar space, where matter is sparse. The lower-energy cosmic rays which he had actually observed and measured *could* be produced by known processes—in particular, the conversion of hydrogen into helium, with a release of energy, is a possible source.

In retrospect, it is clear that Lemaître was radically wrong and Millikan fundamentally right. By the fifties it was generally accepted that cosmic rays are mainly produced by electromagnetic processes occurring in the present universe. Just as the cyclotron atom smashers of the thirties and forties could impart great energy to electrons and protons, so magnetic and electric forces in space can accelerate cosmic rays.

Furthermore, while some cosmic rays appear to be truly isotropic, originating from all directions equally, most cosmic rays come from the Milky Way but their trajectory is scrambled by the galaxy's magnetic fields—a hypothesis proposed by Alfvén in 1939 and now completely accepted. Since most cosmic rays are charged particles, their paths are bent by magnetic fields. Thus their observed direction bears no relation to the actual location of their source, either in interstellar space or the outer reaches of a star's atmosphere.

Nor was Lemaître's theory of stellar evolution at all acceptable. He believed that a star's energy comes from some unknown process of direct conversion of matter to energy—annihilation. In the course of the thirties scientists were realizing that the process driving the stars is thermonuclear fusion—the fusing together of four hydrogen nuclei (single protons) to form a helium nucleus. Such a process was known to be a possible source of energy, since four hydrogen nuclei weigh considerably more than one helium nucleus. According to Einstein's formula, this mass difference must be transformed into a large quantity of energy. In 1938 Hans Bethe showed that the temperatures thought to exist in the center of the sun are sufficient to drive the series of nuclear reactions converting hydrogen to helium, thus providing a star's energy. At the same time S. Chandrasekhar and Robert Oppenheimer were demonstrating that stars, once they have exhausted their nuclear fuel, must end their lives in a collapsed dense state—but could not *begin* their lives in this state, as Lemaître said. Some of his colleagues in general relativity also rejected his primeval-atom theory. Willem de Sitter pointed out that it isn't necessary to assume that the solutions involving a singularity, an origin, are the right ones. Even if they are, de Sitter contended, the simplified equations used by Lemaître would no longer apply in a dense universe, and no real singularity need occur.

Finally, other scientists attacked Eddington's and Lemaître's underlying justification of the origin of the universe—the second law of thermodynamics. In a reply to an Eddington article in

Nature on "The End of the World," H. T. Poggio commented on his gloomy predictions. "Prophecy, we are told, is the most gratuitous of all forms of error and long distance forecasts have a way of going wrong, even when apparently firmly based upon all the available knowledge of the time. . . . At one time such speculations had a theological basis, and often predicted a very unequal distribution of temperature, which in some regions would be excessively high."[4] Poggio goes on to point out that it is a gross overextrapolation of the second law to assume that because it works in certain simple situations on earth, it would work everywhere in the universe. He too points out that fusion is an example of a building up, not a decay, of the universe. "Let us not be too sure that the universe is like a watch that is always running down," he warns, "there may be a rewinding. The process of creation may not yet be finished."

By the end of the thirties, the primeval-atom hypothesis was actively supported only by Lemaître himself and taken seriously by only a few others, such as Eddington. Its scientific bases, incorrect theories of cosmic rays and stellar evolution, and incorrect extrapolations of general relativity and thermodynamics, had been refuted. The philosophical pessimism of which it was born, that of Eddington and Jeans, had been widespread in the disillusionment with progress following World War I, yet it was far from universal. Many considered the war and the ensuing Depression an aberration in the general trend of human progress. Indeed, science was advancing rapidly, with the formulation of quantum mechanics in the twenties and the discovery of nuclear fusion and nuclear fission in the thirties. Few scientists were willing to accept Lemaître's pessimism, and fewer still could accept his premises. So the first version of the Big Bang was stillborn.

THE ATOMIC BOMB AND THE RETURN OF THE BIG BANG

During World War II cosmological research was suspended along with other peacetime pursuits, as scientists were drawn into the war effort. By the war's end, though, it was transformed. Prior to the war the creation of the elements that compose the universe had been a speculative theoretical subject—too little had been known of nuclear reactions. Now, with the successful production of atomic bombs, the creation of the elements was no longer a hypothesis, but a technological fact. The fuel for bombs tested in New Mexico and unleashed on Japan was itself a created element—plutonium—generated from uranium. The A-bombs had transformed common elements into new and exotic elements and isotopes, which scientists found in analyzing the fallout from the bombs, especially that of the Trinity test. And the vast expansion of nuclear research coming out of the Manhattan Project continued to yield data about nuclear reactions.

To one of the Manhattan Project scientists, George Gamow, the detonation of an A-bomb constituted an analogy for the origin of the universe: if an A-bomb can, in a hundred-millionth of a second, create elements still detected in the desert years later, why can't a universal explosion lasting a few seconds have produced the elements we see today, billions of years later? In a paper in the fall of 1946, Gamow put forward his idea, a second version of the Big Bang. Unlike Lemaître, he took as observational proof of his hypothesis the abundance of the elements, not cosmic rays; but like him, Gamow assumed that this abundance could not have been produced by any process continuing in the present-day universe.

Gamow knew that attempts in the thirties to explain the origin of the elements had failed because the theories predicted that as the atomic weight of the elements increased, their abundance would drop exponentially. That is, carbon might be trillions of times less common than hydrogen, and heavy elements like lead would be virtually nonexistent, perhaps one atom per galaxy. This was in violent conflict with observation. By the mid-forties scientists knew from the spectra of distant stars and clouds of gas that the universe is overwhelmingly composed of hydrogen and helium, with about three-quarters of the mass being hydrogen. Intermediate-mass elements, mainly carbon, nitrogen, and oxygen—the elements essential for life—constitute about one percent of the total, far more than the one part in a trillion predicted. The abundance of elements heavier than nitrogen and lighter than iron fluctuates widely around one part in a hundred thousand, while heavier elements are generally found in about one in a billion atoms—again, far more than predicted.

Gamow attributed the extreme discrepancy to earlier estimates' failure to account for the magnitude of the initial explosion. If the universe truly came from a point, the equations of general relativity indicate that within seconds the temperature of the universe would drop so far that the nuclear reactions building up and breaking apart the elements would cease. There would be time to build up the heavy elements but not enough time to break them down. The universe would begin, like an A-bomb, as a hot neutron gas; as the neutrons bombarded one another they would fuse, forming light elements, then increasingly heavy ones. By adjusting a parameter that determines the density of the universe at any given moment, Gamow was able to produce heavy elements in amounts close to those actually observed.

Unlike Lemaître, Gamow had a tremendous flair for publicizing and popularizing his own theories, a flair that, within a few years, would establish his element theory—soon to be dubbed the Big Bang, ironically, by its detractors—as the dominant cosmology. His propagandist talents are demonstrated in the first sentence of the article proposing his views—"It is generally agreed at present that the relative abundances of the various chemical elements were determined by physical conditions existing in the universe during the earlier stages of its expansion" —which was not at all the case: only a handful of scientists had accepted Lemaître's primeval atom and perhaps only two or three believed that this could explain the origin of the elements.

But if it hadn't been true before, Gamow changed that: in 1947 he published the immensely popular and well-written book, *One, Two, Three, Infinity,* which gave a lively and sweeping overview of modern physical science and astronomy. The last chapter presents the Big Bang as accepted fact.

Gamow's persuasive writing and his use of the analogy to the A-bomb, so vivid to the entire postwar population, made his theory plausible to the lay world of science writers and readers. I grew up in the fifties, and remember how exciting I found his books, which were among those that turned me toward physics and astronomy. Gamow's idea had an immediate appeal to his colleagues in nuclear science as well. With the war against fascism over and the Cold War developing, many of the Manhattan Project scientists abandoned defense work, appalled by the destruction to which their work had been directed. They were eager to turn the wartime scientific gains to peacetime research that would be equally challenging. Gamow's new cosmology was just the bridge they needed. And in the late forties and early fifties, when the field started to grow, fueled by an influx of nuclear scientists like Gamow, the Big Bang became the hottest concept in astronomy.

Yet the rapid and widespread acceptance of Gamow's theory of a temporally finite universe was as sharp a break with past scientific thinking as Einstein's spatially finite universe had been. The Big Bang completed the swing of the cosmological pendulum, to the medieval universe—finite in extent, having a definite origin in an instant in time, and created by a process no longer at work in the universe. Gamow's Big Bang was a rejection of nearly all the premises that had evolved over the course of the past few hundred years of scientific development—the infinite nature of the universe, and the assumption that its evolution could be described in terms of processes observable here and now.

To the average layman the theory was certainly a shocking and fascinating one. Yet it seemed another insult to common sense, as Einstein's had been. If the universe had an origin in time, what came before it? What started it? The Big Bang seemed, on the surface, an invitation to hypothesize some supernatural power as the initiator of this titanic explosion.

In fact, the question of what caused the Big Bang has been a weak point of the theory from the start. Gamow speculated that the Big Bang was preceded by a period, perhaps infinite in length, in which the universe contracted to a point and then "bounced" out of that singularity into the current expansion. But by the fifties, observations indicated that the expansion velocity was sufficiently high that it would overcome the gravitational force of all the matter in the universe. Thus gravity alone could not have led the universe to contract with such energy that it was "bouncing" apart at this speed. Some additional, unknown force must at some point have given it an additional push—either at the Big Bang itself, or in the distant past during Gamow's hypothetical contraction.

The situation is something like watching a ball bounce: the faster the ball is seen to rise, the higher the height it must have dropped from. But for any gravitating body, there is a speed—the escape velocity—at which an object will overcome the force of gravity and neither fall back again nor go into orbit. No object can *bounce* away from a gravitating body at escape velocity—that would be like a ball bouncing higher than the point from which it fell, it requires *more* energy than gravity initially conferred to it. So, since the universe is expanding at a rate greater than its escape velocity, gravity alone can't account for its expansion. But there was no other source of energy big enough.

Moreover, even before it was proposed, Gamow's theory of the origin of the elements had been undercut. Gamow had argued that the stars' temperatures are too low to create elements heavier than helium. From nuclear experiments it was known that hydrogen would fuse to form helium at temperatures as low as ten million degrees, which are known to exist at a star's core. But fusing helium to carbon requires much greater temperatures—more than a billion degrees—because the more protons there are in a nucleus the more they repel other nuclei, so far more energy is needed to overcome this repulsion and fuse.

Gamow contended that because these high temperatures couldn't be achieved by stars, the heavier elements must have been formed in the more intense heat of the Big Bang. But in April of 1946, several months before the publication of Gamow's theory, British astronomer Fred Hoyle had put forward an alternative hypothesis involving stars that have exhausted their hydrogen fuel. In a normal star, hydrogen is converted to helium in the dense hot core of the star. The tremendous pressure generated by the radiation pushing outward from this core supports the rest of the star, preventing it from collapsing under its own gravity. As the core of the star is depleted of hydrogen, it contracts, increasing its temperature, and burning the remaining fuel faster—thus preventing the overall collapse of the star.

Once the core is entirely converted to helium, no more fusion of hydrogen can take place; there is nothing to support the weight of the star, so it rapidly contracts, and as it does, the temperature swiftly increases at the core. Hoyle calculated that the temperature would soon reach the billion or so degrees needed to start the fusion of helium to carbon. Once again, the energy pouring out of the core would support the weight of the star, stopping its contraction, until the helium is consumed. This process would continue, producing oxygen from carbon, and so on, eventually building up all the elements, either by fusion or by the same neutron-capture process Gamow used in the Big Bang. And with each contraction the star would spin more rapidly, eventually spewing much of its mass into space.

Hoyle accounted for the production of heavy elements by a process that continues into the present-day universe, and thus can—unlike the Big Bang—be verified. Moreover, he calculated that this process would produce the elements in roughly the observed proportions. Had the Big Bang occurred, the two processes together would have produced more heavy elements than are actually observed.

▪ THE STEADY STATE

Within two years, Hoyle and two collaborators, Thomas Gold and H. Bondi, had formulated a general alternative to Big Bang cosmology—the Steady State theory. Curiously enough, this alternative developed, as the Big Bang had twenty years earlier, from philosophical premises—not scientific ones. In their 1948 paper, Bondi and Gold noted that all current cosmology was based on "the cosmological principle," the idea that the universe, on a large enough scale, looks the same to observers in any spot, in other words, is homogeneous and isotropic—the assumptions introduced by Einstein. But what about how the universe looked in different times, different epochs? If the universe is expanding, as the Big Bang proposes, it would look entirely different to observers at different times. Instead, the Steady State proposes a perfect cosmological principle—that the universe must look the same to all observers, at all places *and* times.

How could this be, if the universe is expanding, as the Hubble relation seem to show? As the universe expands, its density will drop, and, obviously, its appearance will change. To avoid this consequence, Bondi and Gold hypothesized a wholly new phenomenon, the spontaneous and continuous creation of matter: in every block of space about a hundred meters on a side, there comes into being about one atom per year. This tiny amount of matter accumulating through vast regions of space over the aeons, they figured, would maintain a constant density in an expanding universe.

The new matter, they assumed, appears in the form of hydrogen atoms. These gradually condense by their own gravity into huge clouds, then into galaxies, and finally into stars, which process them into the various heavier elements, spewing them back into space. As the fuel in a galaxy is exhausted over billions of years, the galaxy will die, its stars becoming indivisible dark embers. In the meantime, however, new galaxies come into being from newly created matter.

In this way the Steady State countered one of the strongest arguments Gamow had brought forward for the Big Bang—an apparent agreement between the age of the earth and the age of stars. Measurements of the Hubble expansion velocity had indicated that the Big Bang occurred a few billion years ago. By 1950 geologists, comparing the relative amounts of radioactive substances, like uranium, which decay at a known rate, had determined the age of the earth to be about five billion years. Astronomers had in the meantime estimated that a star like the sun would burn its fuel in about ten billion years, and that the sun is also about five billion years old. Since the Hubble constant, which determines the elapsed time since the Big Bang, is uncertain, all three figures could be said to be in rough agreement. Gamow thus argued that the sun and the earth were formed soon after the Big Bang, evidence that something extraordinary did happen back then.

According to the Steady State theory, however, this merely indicates that our galaxy and solar system were formed relatively recently, as they must have been to harbor life. Very old galaxies, having burned up their fuel over tens or hundreds of billions of years, would have stars too feeble to warm fertile planets.

The Steady State theory was, in many ways, just as different from traditional scientific ideas as its rival, the Big Bang. Although the Steady State assumes that processes in the present can account for the universe, it hypothesizes a process that cannot be observed on earth, or even observed at all. Even in the depths of interstellar space—let alone anywhere near earth—a cube of space a hundred meters on a side would hold trillions of atoms, so it would be utterly impossible to observe the creation of a single new atom in a year.

The nonevolutionary universe of Steady State theory also departs from the tradition that, since the late eighteenth century, saw science as explaining the history and evolution of nature. A universe that, for all intents and purposes, never changes is a universe in which real progress is utterly impossible, just as it is in the fading universe of the Big Bang.

Thus by the middle of the twentieth century there was no widely supported view of cosmology compatible with the open-ended, progressive worldview of the preceding century. As in Augustine's time, the halting or reversing of progress on earth undercut the idea of progress in the cosmos. To many who lived through the first half of this century, a belief in progress required either blind faith or a rather long view of history. The slaughter of World War I, the misery of the Depression, the rise of fascism, another world war, the Holocaust, the atomic bombing of Japan —all had exhausted civilization. In 1948 Europe lay shattered, with millions starving and homeless, facing a winter without warmth or shelter. Famine and epidemics stalked the colonial nations, with mortality rates rising to grotesque levels. China was in the grip of a brutal civil war, while Hindus and Muslims slaughtered each other in communal strife following Indian independence the preceding year. The allies of World War II had already squared off in preparation for a new war, and both sides were building and amassing ever more deadly weapons.

In a pessimistic science fiction novel, *October the First Is Too Late*, Hoyle explicitly linked his cosmology and a view of humankind as incapable of real progress, condemned to an endless cycle of overpopulation and war. From the standpoint of the late forties such pessimism, and the two cosmologies spawned then, is certainly comprehensible.

Yet to many others the victory over fascism and the economic recovery that began in 1948 vindicated human progress. The early fifties were, for most working people, a period of increasing optimism and confidence. For this reason, although many read Gamow's popularizations, the scientists' cosmos doomed to stagnation or decay remained distant.

▪ THE BIG BANG IN ECLIPSE

For a decade, until 1957, the Big Bang and the Steady State theories both had their proponents, although—thanks to Gamow's literary skills—the Big Bang got far more publicity. Neither side was able to make sufficiently precise predictions about the key problem, the abundance of heavy elements, to score a clear victory.

In 1957, after years of steady work—aided by advances in nuclear physics and stellar observations—Margaret and Gregory Burbridge, William Fowler, and Hoyle published a comprehensive and detailed theory showing how stellar systems could produce all the known elements in proportions very close to those observed to exist. In addition, the theory accounted for the growing evidence that the elementary composition varies from star to star, something that would not be possible if the elements were produced by the Big Bang. The new theory was rapidly accepted as substantially correct.

The researchers showed that the most common elements—helium, carbon, oxygen, nitrogen, and all the other elements lighter than iron—are built up by fusion processes in stars. The more massive the star, the farther the fusion process can proceed, until it develops iron; at that point no more energy can be derived from fusion, since the iron nucleus is the most stable of all. Thus, when a star exhausts its fuel, it collapses, and the unburned outer layers of the star suddenly mix as they fall into the intensely high temperatures of the core. The star explodes as a supernova, a "little bang" that outshines an entire galaxy for a year. In this explosion, the heavier nuclei absorb still more neutrons, thereby building up the heaviest elements, including radioactive ones like uranium. This explosion scatters the new elements into space, where they later condense into new stars and planets. The earth and the entire solar system was, five billion years ago, formed from the debris not of the Big Bang but of a supernova.

The theory was not perfect, though. Given the present brightness of the stars in most galaxies—an indication of their nuclear activity—it did not seem that enough helium, nearly a quarter of all matter, would be produced; and it was hard to see how certain light elements—deuterium, lithium, beryllium, and boron—which were burned in all stars as soon as they were created, could survive at all. But just as Lemaître's Big Bang failed when cosmic rays were shown to be produced in the present-day universe rather than the distant past, so Gamow's failed when the chemical elements were shown to be produced by present-day stars.

While the Big Bang continued to be prominent in popular accounts of cosmology, its support among scientists rapidly ebbed. For a few years following 1957 cosmology as a whole went into eclipse, since the most interesting questions seemed to be in such rapidly developing fields as stellar astrophysics. And with the first space launches, more efforts went into solar system studies as well—including the infant field of space plasma science. As a result, the number of cosmology papers published annually dropped from forty or fifty a year in the mid-fifties to a dozen or so per year from 1958 to 1960. And of these only a handful developed Big Bang theory. A second effort to develop a Big Bang cosmology had failed.

In 1961, though, new observations brought some comfort to the Big Bang theorists. Since the early fifties, radio telescopes had detected sources ever more distant in space, and with advances in radar technology (brought about in part by military research) fainter objects were turning up. If, as the Steady State theory supposed, the universe is homogeneous in space and time, the density of radio-emitting objects should have been constant, because looking farther out in space means looking farther back in time. Observations, however, showed otherwise. As one looked outward in space and backward in time, there were more and more radio sources: the universe was, in fact, changing and evolving with time—so the Steady State must be wrong.

With *both* leading theories in trouble cosmologists continued to back their favorite with a clear conscience, since there was no better alternative on the horizon. More researchers returned to the field and research papers emerged again at a rate of forty or fifty per year. This was a brief period of ferment as Big Bang diehards and Steady Staters sought to rescue their theories, and other, less conventional ideas, such as those that would later give birth to plasma cosmology, were given a hearing as well.

Three years later, help came to the Big Bang from another quarter. Some radio sources appeared to be tiny, starlike points of light. In 1964 redshifts from these "quasi-stellar objects," or "quasars," were measured and turned out to be extremely high, higher than any measured for galaxies. If, as most scientists immediately assumed, these were Hubble redshifts, the quasars must be immensely far away. But at such a distance their brightness meant that they were radiating huge amounts of energy, in some cases a hundred thousand times more than an entire galaxy. Yet the quasars' light varied noticeably over a period as little as a year, so they could be no more than a light-year across—far tinier than a galaxy, which is typically a hundred thousand light-years in diameter. Thermonuclear fusion, even a supernova, could not pack so much power into so little space.

Ever the one with a bold hypothesis, Fred Hoyle proposed that the only possible source of such power is the energy generated by the gravitational collapse of an enormously massive object, one with millions of times the sun's mass. Robert Oppenheimer had calculated in the thirties that an object of sufficient mass could not be prevented from collapsing entirely, right down to a singularity, a point. Hoyle speculated that if a *really* huge agglomeration of gas similarly collapsed—whether or not it reached a singularity (Hoyle thought it would not)—it would release a tremendous amount of energy. While Hoyle was vague about the mechanism of the energy's release, and consequently the type of energy, it was nevertheless a possibility.

Since such a massive collapsed object would have an enormously strong gravitational field, it could only be studied theoretically by using Einstein's general relativity equations. Suddenly the struggling relativists working on the Big Bang got an infusion of new blood as researchers turned Einstein's physics to the quasars. General relativity reemerged from its cosmological backwater and the Big Bang again seemed reasonable. If objects could collapse into singularity—shortly dubbed a black hole—why couldn't the universe itself have been born from a singularity?

The glamour of the mysterious quasars quickly attracted young researchers to the arcane calculations of general relativity and thus to cosmological problems, especially those of a mathematical nature. After 1964 the number of papers published in cosmology leapt upward, but the growth was almost wholly in purely theoretical pieces—mathematical examinations of some problem in general relativity, which made no effort to compare results with observations. Already, in 1964, perhaps four out of five cosmology papers were theoretical, where only a third had been so a decade earlier.

THE THIRD BIG BANG; MICROWAVES TO THE RESCUE

Despite the quasars' apparent confirmation of certain aspects of Big Bang theory, and the definite upsurge in theoretical cosmology, specific problems remained—notably the as yet inexplicable energy that initially spurred the Big Bang. If the universe was "open," infinite and expanding, as Gamow's figures indicated, there was still no explanation for the expansion rate observed. Robert Dicke and others figured that if they could return to Einstein's closed universe things would be simpler—it would expand for a while and then contract back to, or near to, a singularity. If it could be assumed that something prevented it from reaching a mathematical point, a singularity, it would "bounce" back into expansion. Such an oscillating universe would, therefore, in some sense exist forever; but the only universe we would or could have any knowledge of is this one cycle, finite in both space and time.

Gamow had calculated both the energy density and the matter density of the universe for all time, including the present. The predicted matter density for the present universe was about two atoms per cubic meter of space, and the energy density, expressed as the temperature that radiation coming from the great fireball would appear to have today, after billions of years of cooling, was 20° K—twenty degrees above absolute zero.

These figures are just shy of the amounts needed to close the universe. Dicke knew that Gamow had selected these values to make the heavy-element production come out right, but since no one now believed that the Big Bang had created the heavy elements, Dicke could dispense with them. Instead he assumed that the universe is closed, which requires a stronger gravitational field and thus a higher density of three to twelve atoms per cubic meter, depending on the observed expansion speed (something that still remains uncertain by about a factor of two).

Dicke also knew that the stellar synthesis theory (of the Burbridges, Fowler, and Hoyle) left open the source of the universe's 25 percent helium, because stars like those burning today could not have produced enough helium in the time since the galaxy seems to have formed—although they could have produced the right amounts of the heavier elements. He put the problem to a graduate student, P. J. E. Peebles: Given a closed universe, could the Big Bang at least produce the amount of helium observed?

Peebles found that it could. As with Gamow's calculations the key variable was the ratio of energy to matter: as the universe expands, radiated energy decreases because each photon would be stretched by the expansion, and the longer the wavelength the lower the energy. However, the number of photons would not change, nor would the number of protons and electrons that make up matter. So the ratio of photons to protons is a constant—an unknown one. By varying this constant in his calculation, Peebles found that as the number of photons per proton decreases, the production of helium increases. If there were just about one hundred billion photons per nucleus, then the Big Bang would have produced the proper quantity of helium. Peebles was then able to predict that the universe would now be filled with radiation, mostly radio waves with an apparent temperature of 30° K, somewhat more than Gamow's prediction.

In 1965 with Dicke's encouragement Peebles set out to test this prediction observationally by building a radio telescope to search for this primordial radiation. However, Arno Penzias and Robert Wilson, researchers at Bell Labs, had already discovered the radiation he was seeking. They had found this isotropic radiation, at least at the frequency they observed. The primordial radiation predicted by Peebles, and much earlier by Gamow, really existed. The Big Bang must have happened—or so elated cosmologists immediately concluded. And when they conveyed the news to excited science reporters, cosmology was once again headline news.

The *New York Times*, in a front-page article, described the Bell Labs discovery as clear evidence not only that the Big Bang occurred, but that Dicke's oscillating universe (which avoided the sticky question of what happened before the Big Bang) was the valid model. "SIGNALS IMPLY A BIG BANG UNIVERSE," read the top-of-the-page headline.

But the reporters had overlooked the fact that Penzias and Wilson had measured a temperature not of 30° K but 3.5° K. This was considerably worse than it looked: the amount of energy in a radiation field is proportional to its temperature to the fourth power. The observed radiation had *several thousand times less* energy than Peebles or Gamow had predicted. Even by astronomers' standards, where factors of two are often chalked up to observational uncertainty, a disagreement of thousands of times bodes ill.

Dicke told the *New York Times* that his group had predicted 10° K, which he considered acceptably close to the observations. (This figure is nowhere given in his published papers, so it's unclear where it came from.) And even 10° K yields a hundredfold difference between the energy predicted and that observed.

While the science writers ignored this problem, Peebles did not. As he pointed out in his theoretical paper that accompanied Penzias and Wilson's report of their observations, the low temperature observed implied a much less dense universe, nearly a thousand times too diffuse to close the cosmos—what he and Dicke initially wanted. Instead of making Gamow's universe denser, thus oscillating, the new observations showed that it is more diffuse, with less gravity, greatly aggravating the original problem—where the energy for the expansion came from.

Alternatively, Peebles wrote, if the universe is really dense enough to oscillate, then the low temperature and his new equations require that nearly all matter would have been converted to helium—a clear contradiction of reality.

Far from confirming the Peebles-Dicke model, the Penzias-Wilson discovery clearly ruled out the closed oscillating model. Yet Peebles initially hung on to his theoretical assumptions and introduced additional hypotheses to bridge the gap between fact and theory. (This was to become typical of Big Bang cosmology.) Since, according to general relativity, a closed universe with as little energy as was observed would produce far more helium than was observed, Peebles simply introduced a new modification of the gravity equations. The modification was unjustified except that it, like Ptolemy's epicycles, "saved the phenomenon," preserving both the finite universe and an agreement with the microwave temperature.

Within a year, a new development forced the abandonment of the closed-universe Big Bang—again, it came from the Big Bang's erstwhile foe, Fred Hoyle. Intrigued by all that helium, Hoyle, in a more elaborate version of Peebles's work, carefully calculated that a Big Bang would produce only very light elements—helium, deuterium, and lithium. The amounts, he found, depend sensitively on the density of the universe: if there were about one atom per eight cubic meters, the resulting amounts of helium, lithium, and deuterium (the latter two quite rare) would come very close to those observed.

Here then was a second major support for the Big Bang. From a single parameter (the ratio of photons to protons, which Hoyle estimated at twelve billion) and a single observation (the temperature of the microwave background), Big Bang theorists were able to account for the abundance of three elements and to predict the density of matter in the universe. The resulting density was actually quite close to the most recent estimates of the density of matter in the galaxies and stars observable from earth.

The golden age of the Big Bang, and its unquestioned dominance in cosmology, began on this basis. The evidence was no longer Lemaître's cosmic rays or Gamow's heavy elements, but the microwave background and three light elements. Again, cosmologists argued that these phenomena could not be explained by any current sources. As with the cosmic rays, the microwave background is isotropic, and this, cosmologists contended, shows that it cannot derive from current sources, which are unevenly distributed.

The Big Bang that triumphed was, to be sure, quite different from the one cosmologists had been used to. It was, in fact, a third version, far less dense—an open universe, expanding indefinitely. The vexing problem of what could have propelled this vast explosion, a hundred or more times greater than gravity could contain, was quietly swept under the rug. The new Big Bang became the standard model.

THE END OF THE GOLDEN AGE

The annual number of cosmology papers published skyrocketed from sixty in 1965 to over five hundred in 1980, yet this growth was almost solely in *purely* theoretical work: by 1980 roughly 95 percent of these papers were devoted to various mathematical models, such as the "Bianchi type XI universe." By the mid-seventies, cosmologists' confidence was such that they felt able to describe in intimate detail events of the first one-hundredth second of time, several billion years ago. Theory increasingly took on the characteristics of myth—absolute, exact knowledge about events in the distant past but an increasingly hazy understanding of how they led to the cosmos we now see, and an increasing rejection of observation.

In a decade the field of cosmology was transformed from a small group of squabbling theorists trying to develop theories that would match observation, to a huge phalanx of hundreds of researchers, virtually all united in their basic assumptions, mainly preoccupied with the mathematical nuances of the underlying theory.

This tremendous expansion of theoretical cosmology was encouraged by powerful economic incentives, for both the researchers and their institutions. In no other field of science, excluding mathematics itself, could research be accomplished as inexpensively as in cosmology. The seventies saw a rapid contraction in the research money available for physical sciences, especially in the U.S., with the end of the defense spending boom of the Vietnam War and of the Apollo Project. In most fields of science, advance was based on experimentation which required expensive equipment and an arduous search for money to build it. In these fields theoreticians were a minority in need of experimenters' data to inspire or to test a new theory. In astrophysics too theoreticians relied on extensive data from nuclear scientists and their accelerators, or on observers' giant radio and optical telescopes—or on even more expensive satellites. By contrast, theoretical cosmologists seemingly need no data at all. A few, especially in the later seventies, started using computers for simulations; but most of their time-consuming calculations needed nothing more than paper and pencil. Cosmology was scientific research on the cheap!

The tremendous growth of the theoretical side inevitably biased the entire field against observation, which became secondary to the "real" work of manipulating equations. Cosmologists came to look down on the observing astronomer who spent long nights at the telescope but could not fathom (or did not care to fathom) the complexities of a Bianchi universe.

At the same time, the social investment in Big Bang theory greatly increased. For an experimental scientist, the bulk of the working scientific world, the discrediting of a theory can redirect his or her work, but it can't make it useless. Good data, competently obtained and analyzed, is of scientific value even if the theory that inspired it is wrong. Other theorists will find uses for it that were little imagined when it was first gathered. Even in theoretical work, honest efforts to compare a theory to observation almost always prove useful regardless of the theory's truth: a theoretician is bound to be upset if his or her pet idea is wrong, but time won't have been wasted in ruling it out.

But with hundreds of researchers engaged in examining theoretical, mathematical, hypothetical universes, the case is different. It took no great insight to realize that if the Big Bang theory was basically wrong, as had been thought as recently as the early sixties, then these researchers were simply wasting time and talent. A challenge to Big Bang theory would threaten the careers of several hundred researchers. It could hardly be surprising that by the end of the seventies virtually no papers challenging the Big Bang in any way were accepted for presentation at major conventions or for publication in major journals. It became simply inconceivable that the Big Bang could be wrong—it was a matter of faith.

Yet in the course of this golden age, not a single new confirmation of the theory had emerged. No new phenomena predicted by theoreticians had been observed, or any additional feature of the universe explained. In fact, serious conflicts between theory and observation were developing.

The first and most serious was the problem of the origin of the galaxies and other large-scale inhomogeneities in the universe.

The extreme smoothness of the microwave background posed another, more theoretical problem. According to Big Bang theory, points in the universe separated by more than the distance light can have traversed since the universe began (about ten or twenty billion light-years) can have no effect on one another. As a result, parts of the sky separated by more than a few degrees would lie beyond each other's sphere of influence. So how did the microwave background achieve such a uniform temperature?

This simple question demonstrates that one of the basic parameters of the theory, the number of photons per proton, is wholly arbitrary. Why should there be twelve billion photons for every proton, rather than twelve thousand or thirty-six? Why is the temperature of the microwave background 2.7° K rather than some other temperature?

As described in Chapter One, this isotropic microwave background created other problems as well. The anisotropies, or irregularities, in the background were supposed to reflect tiny clumps in the matter of the early universe, which eventually grew to become galaxies. But the observed anisotropy was so small that these fluctuations would not have had time enough to grow into galaxies unless there was far more matter—and thus much more gravity—than there appears to be. The microwave background was simply too smooth to fit into the Big Bang theory.

And then there was the "flatness problem"—why omega, the ratio of the universe's density to that needed to "close" it, was so near to, but not equal to, 1. If omega were *exactly* 1, it would remain constant as the universe expands, creating a perfect universe, a four-dimensionally flat universe neither positively curved like a sphere nor negatively curved like a saddle—hence the "flatness problem." But if omega were less than 1, as it seemed to be, the disparity would increase as the universe expands and its relative density decreases. Conversely, as we go back in time toward the Big Bang, omega would get closer and closer to 1. If, for example, we know that omega is .01 now, in a universe twenty billion years old, omega would have been about .95 at two hundred million years, .99995 at twenty thousand years, and so on. Cosmologists had calculated that at 10^{-43} seconds of age omega would vary from 1 by one part in 10^{58}—and even to theoretical cosmologists a *crucial* number fine-tuned to fifty-eight decimal places seemed suspiciously convenient. A discrepancy of only one part in 10^{40} would have caused the universe to collapse or disperse in less than a second, which it evidently hasn't done. So why was omega "in the beginning" equal to .999999999 . . . ?

All these problems derive from the basic premise of the Big Bang, that the universe originated as a "perfect" world, an Eden of symmetry whose characteristics conform to pure reason. Cosmologists had to explain how such perfection—isotropy, a perfect omega of 1—came to be. Yet they also had to explain how their perfect world gave birth to the present clumpy and "imperfect" one. On both sides there were difficulties, and success on one side tended to lead to defeat on the other.

THE FOURTH BIG BANG: INFLATION

Thus, despite its unquestioned dominance, the third version of the Big Bang was internally implausible and, in at least one respect, its predictions about microwave smoothness clearly contradicted observation. But abandoning the theory was by now out of the question, so a new generation of theorists set about to overhaul it once again.

At this point, cosmologists appealed to their colleagues in particle physics, who were probing the fine structure of matter. The cosmologists knew that an omega of 1 would solve at least the flatness problem and probably the problem of anisotropy. Yet all the known matter added up to a few percent of that density—there just wasn't enough. If the Big Bang was to be saved, there had to be far more than we can see, so cosmologists decided that most of the universe was dark, or "missing." Like a worried pet owner searching for a lost dog, cosmologists asked particle physicists if they could help find a missing universe.

The particle physicists were only too willing, since an alliance with cosmology would aid them with their own quandaries. In the late seventies theoretical work in particle physics had sought a theory that would unify the three forces of nature that are important on the small scale—electromagnetism and the weak and strong nuclear forces. (The weak force causes radioactive decay, while the strong force holds the nuclei together and is responsible for the release of nuclear energy.) Such a Grand Unified Theory, or "GUT," was to explain these forces as aspects of a more fundamental principle, much as Maxwell had united electricity and magnetism a century earlier.

Like cosmologists, the particle physicists approached their theory mainly on the basis of certain a priori, "perfect" mathematical assumptions (described in more detail in Chapter Eight). As a result, the postulated GUTs made few testable predictions. For the most part they predicted new particles and phenomena that could be detected only at extremely high energies, around one hundred million trillion electron volts (eV) or higher. (An electron volt is the energy acquired by an electron falling through an electrical potential of one volt.) The largest accelerators conceivable on earth could accelerate to less than a millionth of that gigantic energy—the Big Bang, however, allegedly released such fantastic energies in the first fraction of a second of the universe's existence. Perhaps, some particle theorists thought, we can find in some aspect of cosmology a confirmation for GUTs, and in the process find the cosmologists' missing universe.

One such theorist, Alan Guth, succeeded after a fashion. Guth knew that all GUTs assume a hypothetical, omnipresent force field called the Higgs field. In 1980 he realized that it could provide energy not just for a Big Bang, but for a far faster expansion, an exponential explosion he dubbed the "inflation." The inflationary universe would double in size every 10^{-35} seconds, attaining to a fantastic size in an instant, a trick the old Big Bang took far longer to accomplish. Once inflation ended, after 10^{-33} seconds or so, the ordinarily sedate Big Bang expansion at the speed of light could begin.

Inflation solved the flatness problem, because the universe blew up to such a huge size, far bigger than the part we can observe, that it *must* appear flat (omega equal to 1), just as the earth appears flat because we see only a minute part of it, up to an apparent horizon. Moreover, inflation explains the smooth microwave background: because inflation proceeds far faster than the speed of light, regions at one time in contact with each other, and thus at the same temperature, are blown farther away from each other than the distance light can have traveled in the duration of the universe. All the observable universe had once been contained in such a small region, so it should all have the same temperature.

Finally, since inflation dictated that omega is 1, cosmologists could happily use this value to calculate how the galaxies formed from the tiny anisotropies in the microwave background.

But that's not all. Since Gamow, the source of all the matter and energy of the universe, and the impulse driving the Big Bang itself, had remained a mystery. In the laboratory, matter and energy can be transformed into each other but never created or destroyed. In Guth's theory, the Higgs field, which exists in a vacuum, generates all the needed energy from nothing—*ex nihilo*. The universe, as he put it, is one big "free lunch," courtesy of the Higgs field.

Guth's theory wasn't *perfect*, though. It did not say what that missing 99 percent of the universe is, but only gave theoretical justification to the cosmologists' desire for it. And the theory had, it turned out, internal inconsistencies. But both these problems were of minor importance in light of its major result—the link between particle theory and cosmology had been made.

A period of enormous theoretical ferment now began. Every year, or even twice a year, theorists from around the world would replace existing inflationary theories with newer versions—inflation was followed in 1983 by New Inflation, and then by Newer Inflation. At the same time, new GUTs were formulated by particle theorists at a similarly frantic pace, generating new ideas like superstrings and supersymmetry. Reputations were made and unmade in a twinkling as some of the young theorists like Guth and Edward Witten at Princeton became media figures, subjects of features in national newsmagazines.

As a philosopher wrote of a similar period in the nineteenth century, "Principles ousted one another, heroes of the mind overthrew each other with unheard-of rapidity and in three years more of the past was swept away . . . than at other times in three centuries. All of this is supposed to have taken place in the realm of pure thought."[5]

Indeed, in all these intense theoretical battles, duly reported in the scientific press, there was virtually no reference to observation. Every critique involved only mathematical consistency or the relation of one theory to another. Some underlying difficulties were ignored—for example, there wasn't a shred of evidence that omega equals 1, in fact evidence suggested it is around .02, as we've seen. So, despite the fact that the GUTs themselves lacked any experimental confirmation, omega *became* 1 because this was predicted by all of the GUTs through Guth's inflationary models. One hypothesis without any observational foundation was used to support other such baseless speculations.

But the GUTs did make *one* testable prediction, a dramatic one: they all predicted that protons decayed. Since protons make up the vast bulk of the observable mass in the universe, this meant that the universe is bound to decay. The lifetime of a proton, though, was enormously long—10^{30} years (one thousand billion billion billion years). A ton of water contains about 10^{30}

protons, so within that mass one proton on average should decay each year, emitting a characteristic energetic particle that should be observable.

To test this, scientists set up arrays of detectors around swimming pool–size bodies of water deep in mines, where they would be shielded from cosmic rays that could confuse the experiment. Such large amounts of water were used to increase the probability of observing a proton decay.

But nothing happened—for days, weeks, months, years. Protons do not decay. By 1987 it was clear that the GUTs were wrong. However, that didn't stop the particle physicists or the cosmologists. They went back to their blackboards and proved that the lifetime of the proton stretched to 10^{33} years, beyond the limits set by experiments, and everyone got back to his work. (If the scientific method is a way to ask questions of nature, then the particle theorists and cosmologists are people who won't take "no" for an answer!)

Cosmologists weren't perturbed, though, because particle theorists had provided an entire zoo of particles to make up the missing mass. First came heavy neutrinos. Neutrinos are real particles, observed in laboratory experiments, but they are quite hard to detect because they interact so little with matter. They appear to travel at the speed of light, so must have no mass. However, particle theorists postulated that neutrinos do have mass, and some cosmologists decided that these massive neutrinos could be the missing mass.

A supernova blew away this idea. Supernovas produce huge quantities of neutrinos when they explode. In 1987, when a supernova occurred in the Large Magellanic Cloud, a satellite galaxy of our own Milky Way, scientists were able to detect the neutrinos released, using the same arrays that had been patiently waiting for a decaying proton. The neutrinos all arrived in a single bunch, showing that they all travel at the speed of light and have either no mass or so little that they couldn't fill up the universe.

So cosmologists, except some diehards, turned to other particles, which, being wholly hypothetical, could not be eliminated as missing-mass candidates by inconvenient supernovas. Particle physicists supplied these in large numbers, equipped with whimsical names—axions (named after a detergent), WIMPs, photinos, and so on. None had ever been observed, but all came with good credentials, having been predicted by someone's GUT.

As the eighties progressed, the level of theoretical fancy rose higher. The Higgs field began to produce objects like cosmic strings; these too served to explain away such problems as galaxy formation. Finally cosmologists took off on their own, going the particle theorists one better by postulating quantum gravitational theories that bring gravity under the same theoretical framework as the GUTs' three forces. From this effort came the most bizarre theoretical innovation of the eighties—baby universes—pioneered by Stephen Hawking. At the scale of 10^{-33} cm, less than one-million-trillionth of a proton's diameter, space itself is, according to this idea, a sort of quantum foam, randomly shaping and unshaping itself; from this, tiny bubbles of space-time form, connected to the rest by narrow umbilical cords called wormholes. These bubbles, once formed, then undergo their own Big Bangs, producing complete universes, connected to our own only by wormholes 10^{-33} cm across. Thus from every cubic centimeter of *our* space, some 10^{143} or so universes come into existence every second, all connected to ours by tiny wormholes, and all in their turn giving birth to myriad new universes—as our own universe itself emerged from a parent universe. It is a vision that seems to beg for some form of cosmic birth control.

This theory was an attempt to eliminate an embarrassing problem, which had always beset the Big Bang: what happened before that? While some cosmologists were perfectly content to make the link between the Big Bang and the biblical creation, others, including Hawking, were not, and sought to avoid a beginning to time. The many-universe idea is one "solution" since it assumes that each universe is part of an infinite chain of universes. Yet because all of these universes are, in principle, unobservable from our own, it leaves our own universe finite in time.

Earlier, in his book *A Brief History of Time*, Hawking had attempted to solve the same problem with a mathematical analogy comparing the universe in four dimensions to the surface of the earth in two dimensions. Time would be, he explained, like latitudes on earth: "before the Big Bang" is as meaningless as "south of the South Pole." Time, therefore, has neither beginning nor end, like a circle, yet is still finite in extent. This analogy caused no end of confusion, since many reading his book concluded that he had abandoned the Big Bang and was advocating a universe infinite in duration—which he was not. In many parts of the book, Hawking himself refers to the beginning and end of the universe. His analogy with the lines of latitude is just a word game to minimize the theological implications of a beginning and end to time.

During this entire period, none of the cosmologists' speculations received observational confirmation—in fact, the foundations of this theoretical structure were being undercut. Even with dark matter, the Big Bang still could not account for the low level of microwave anisotropy, or the formation of galaxies and stars. Nor could it accommodate Tully's large-scale supercluster complexes (described in Chapter One). And the dark matter itself was ruled out by new observation and analysis. The Big Bang in all its versions has flunked every test, yet it remains the dominant cosmology; and the tower of theoretical entities and hypotheses climbs steadily higher. The cosmological pendulum has swung fully again. Today's cosmologists have, as Alfvén puts it, "taken Plato's advice to concentrate on the theoretical side and pay no attention to observational detail." They are creating a perfect edifice of pure thought incapable of being refuted by mere appearances.

They have thus returned to a form of mathematical myth. A myth, after all, is just a story of origins, which is based on belief alone, and as such cannot be refuted by logic or evidence. Neither can the Big Bang. Entire careers in cosmology have now been built on theories which have never been subjected to observational test, or have failed such tests and been retained nonetheless. The basic assumptions of the medieval cosmos—a universe created from nothing, doomed to final destruction, governed by perfect mathematical laws that can be found by reason alone—are now the assumptions of modern cosmology.

Certainly this development is due in part to the growing legitimacy within cosmology of a purely deductive method, justified by Einstein himself. In 1933 he said, "It is my conviction that pure mathematical construction enables us to discover the concepts and the laws connecting them, which gives us the key to the understanding of nature. . . . In a certain sense, therefore, I hold it true that pure thought can grasp reality, as the ancients dreamed."[6] Today's cosmologists, with the support of this lofty authority, proudly proclaim that they have abandoned experimental method and instead derive new laws from mathematical reasoning. As George Field says, "I believe the best method is to start with exact theories, like Einstein's, and derive results from them."

As we have seen, Einstein himself did not use this deductive method in making his great breakthroughs. More important, I think, he would have been horrified to see what his words have been used to justify: even in his unsuccessful later work he ruthlessly rejected theories clearly contradicted by observation. Yet today's cosmologists take the deductive method as a rationalization for clinging to long-disproven theories, modifying them into bizarre towers of ad hoc hypotheses and complexities—something Einstein, the lover of simplicity and beauty in both nature and mathematics, would never have tolerated.

COSMOLOGY AND IDEOLOGY

There has always been an intimate relation between the ideas dominant in cosmology and the ideas dominant in society. It would be astonishing if that relationship had come to an end in our present enlightened times. Not that cosmologists directly derive their theories from social or political ideas—far from it. But what sounds reasonable to them cannot but be influenced by events in the world around them and what they and others think about it.

So it is certainly no coincidence that the period during which the Big Bang was in eclipse, from around 1957 to 1964, corresponds to the time of the most vigorous expansion of postwar recovery and a resurgence of confidence in progress. The Big Bang's golden age in the seventies, on the other hand, corresponds to the end of the postwar boom and a new decade of growing pessimism. In fact, the links between cosmological and social ideas were made explicit by both cosmologists and political writers of the period.

In the late sixties and the early seventies the postwar recovery ended in all the market economies. Real wages peaked in the United States and Western Europe, and somewhat later in Japan. In the Third World, per capita grain production, the best overall indicator of food supply and living standards, reached 340 kilograms per year and stopped rising, only to remain there for the next twenty years. While the food supply had increased by 50 percent from the depths of the late forties, it only recovered the levels of 1913.

As the seventies wore on, the economic problems facing the entire world, east as well as west, became more obvious. As had happened before, new markets were being saturated and increasing pressure was put on wages and living standards worldwide, as industrialists strove to maintain and increase their companies' profitability.

Almost as soon as this cessation of growth began to manifest itself, social ideas that justified the situation as inevitable started to circulate. In 1968 the Club of Rome, bringing together industrialists and academics, championed the idea of zero growth: the earth is finite, the universe is running down, it is impossible to continue the increase in living standards. The two oil crises were interpreted as warnings of the exhaustibility of finite resources—a logic that must appear quaint to oil producers who now go to war in a struggle against a persistent glut.

Many writers used the Big Bang cosmology and the idea of universal decay to buttress the argument that consumption has to be restrained. In his 1976 book *The Poverty of Power* Barry Commoner begins from the cosmological premise that "the universe is constantly, irretrievably becoming less ordered than it was," and concludes that, given this overall tendency, Americans must make do with less in order to postpone the inevitable day when total disorder reigns on earth. The faltering universe of the Big Bang became a metaphor for the faltering economy—both equally inevitable processes, beyond the control of mere mortals.

Nor were cosmologists and physicists immune from the influence of such analogies. In the popular 1977 account of the Big Bang, *The First Three Minutes,* Nobel Prize winner Steven Weinberg concludes by contemplating the philosophical lessons of this universe, which will end either in the icy cold of final decay and infinite expansion, or in the fiery collapse to a new singularity:

> It is almost irresistible for humans to believe that we have some special relation to the universe, that human life is not just a more

> or less farcical outcome of a chain of accidents reaching back to the first three minutes, but that we were somehow built in from the beginning. As I write this I happen to be in an airplane at 30,000 feet, flying over Wyoming en route home from San Francisco to Boston. Below, the earth looks very soft and comfortable—fluffy clouds here and there, snow turning pink as the sun sets, roads stretching straight across the country from one town to another. It is very hard to realize that this all is just a tiny part of an overwhelmingly hostile universe. It is even harder to realize that this present universe has evolved from an unspeakably unfamiliar early condition, and faces a future extinction of endless cold or intolerable heat. The more the universe seems comprehensible, the more it also seems pointless.[7]

For Weinberg, as for others, the universe of the Big Bang is irreconcilable with human progress. The end may come billions of years from now, but in the end all that the human race accomplished in aeons will be nothing, of no consequence. Progress, then, is an illusion, as it was for Augustine sixteen hundred years ago. The only question is when it will stop—now, or at some point in the future. It is thus no surprise that the Big Bang flourished simultaneously with the social ideas, like zero growth, that deny the reality of progress, and with a growing economic crisis that, at least in the short term, had stalled that progress. Once again, cosmology justified the course of events on earth.

But there is probably no better example in this century of the interaction of social ideology and cosmology than the development of the inflationary universe in the eighties. Nineteen eighty, with the coming to power of conservative administrations in America and elsewhere, marked the end of a period of fashionable pessimism and the beginning of a decade of speculative boom. Alan Guth arrived at his idea of cosmic inflation just as the worst monetary inflation of the century was coming to a climax. He concluded that the universe is a "free lunch" just as the American economy began its own gigantic free lunch—a period of speculation which rewarded its wealthy participants while actual production stagnated.

Throughout the decade, the rise of financial speculation in Wall Street was shadowed by the rise of cosmologists' speculations in Princeton, Cambridge, and elsewhere. As Witten and his colleagues were acclaimed by the press as geniuses for theories that produced not a single valid prediction, so men like Michael Milken and Donald Trump earned not only far greater fame but also incomes that peaked, in Milken's case, at half a billion dollars per year for paper manipulations that added not a single penny to the nation's production.

In the realm of finance, fortunes were built on a tower of debt. A speculator would borrow four billion dollars to buy a company, sell it for five billion to another speculator, who would, in turn, break it up to sell it in pieces for six billion dollars—all on borrowed money. All involved reaped handsome profits and were hailed as geniuses of financial wizardry—until their indictments.

The result of this was an actual decline in living standards both in the U.S. and throughout the world: by the end of the eighties real family income in the U.S. had dropped by 10 percent and was at the same level as it had been twenty-five years earlier, despite the fact that most families by now had *two* incomes.

Obviously, the small-scale speculators of cosmology did not, in any conscious way, imitate the large-scale speculators of Wall Street. Yet, as in every other epoch, society's dominant ideas permeated cosmology. If the wealthiest members of society earned billions by mere manipulation of numbers, without building a single factory or mill, it didn't seem too strange that scientific reputations could be made with theories that have no more relation to reality. If a tower of financial speculation could be built on debt—the promise of future payment—then, similarly, a tower of cosmological speculation could be built on promises of future experimental confirmation.*

There was, however, a more direct relationship between the development of the economy over the past decade and the development of cosmology and science generally. The eighties saw a slashing, particularly in the U.S., of the amount of money devoted to nonmilitary research and development and a drastic slowing of technical advance.

To a large extent, this intensified a tendency evident in the seventies and even in the sixties. Since 1960 there has not been a single major qualitative breakthrough in physical technology.

* Some cosmologists themselves have noticed the similarity of the two types of speculation. In 1990, University of Chicago cosmologist Michael S. Turner commented that the "go-go junk bond days of cosmology" are over and theoretical speculation will now be checked by observation.

The thirty years *before* 1960 saw a series of fundamental developments: television in the thirties; the transistor, computer, radar, and, of course, nuclear energy in the forties; the development of space travel and the laser in the fifties. In the subsequent three decades there have been dramatic improvements in all these areas, particularly in computers, but not a single qualitatively new, functional idea. Only in biology has genetic engineering brought about a qualitative advance.

This is a profound change for modern society: not since the beginning of the industrial revolution 250 years ago has there been a similar period of three decades without major technical advances. Such technical stagnation has a deep impact on science and technology. An advancing society, which requires and thus supports fundamental work in science and technology, continually generates challenges for the pure sciences and provides the materials needed to meet those challenges. Thus the problems arising from the development of electricity and electrotechnology in the late nineteenth century led directly to the study of nuclear structure and eventually to the release of nuclear energy. When technological progress slows or ceases, that cross-fertilization of theory and experiment, thought and action, begins to wither and scientists begin to turn to sterile speculation.

The slowing of technology is, today, directly linked to the growth of financial speculation. Five billion dollars invested in buying, say, Hughes Aircraft, is five billion dollars that the buyer, General Motors, will not put into new factories or new research. To the extent that the world market appears to be saturated, as it does today, then profits are easier to make through speculation than in production. What use is new technology if new factories aren't profitable? The diversion of financial resources from technical advance has pushed thousands of scientists away from the challenges of the real world into the deserts of speculation.

Fortunately for science, even the perfection of existing technologies, such as the computer, requires a broad base of scientific research. But it is fundamental research—investigations whose findings don't seem to be immediately useful—that suffer first when technological development slows. Today those areas are clearly cosmology and particle or high-energy physics—where the link between science and technology, theory and human progress, has been broken almost completely. It is here that, as in postclassical Greece, the stagnation of society has led to the return of mathematical myths, a retreat from the problems of base matter to the serene contemplation of numbers.

Today cosmologists often pride themselves on the isolation of their work from the everyday world and from any possible application. They and their particle theorist colleagues give their hypothetical entities whimsical and comical names to flaunt their belief that their activity is, at base, an elaborate and difficult game, the "free play of the mind." In a society beset by growing crises, a world of poverty, crime, drugs, and AIDS, a world without progress, the pure realm of mathematics offers a serene cloister.

Fortunately, this tendency is not the only one that has characterized the study of the universe in the present century. While this mathematical speculation dominated the field, an entirely different development arose out of the study of electromagnetism. It is this path, which led to plasma cosmology, that we will now examine.

▪

NOTES

1. Abraham Pais, *Subtle Is the Lord*, New York, Oxford University Press, 1982, p. 311.

2. Hannes Alfvén, "How Should We Approach Cosmology?" in *Problems of Physics and Evolution of the Universe*, Academy of Sciences of Armenian SSR, Yerevan, 1978, p. 14.

3. Albert Einstein, *Essays in Science*, Philosophical Library, New York, 1934, p. 69.

4. H. T. Poggio, "Science and Prediction," Supplement to *Nature*, March 21, 1931, p. 454.

5. Karl Marx and Friedrich Engels, *The German Ideology*, New York, International Publishers, 1970, p. 39.

6. Einstein, *Essays in Science*, p. 18.

7. Steven Weinberg, *The First Three Minutes*, New York, Basic Books, 1977, p. 154.

5 THE SPEARS OF ODIN

Space is filled with electrons and flying electric ions of all kinds.

—KRISTIAN BIRKELAND, 1904

It was the question why the wanderers—the planets—moved as they did that triggered off the scientific avalanche several hundred years ago. The same objects are now again in the center of science—only the questions we ask are different. We now ask how to go there, and we also ask how these bodies were formed. And if the night sky on which we observe them is at a high latitude, outside this lecture hall—perhaps over a small island in the archipelago of Stockholm—we may also see in the sky an aurora, which is a cosmic plasma, reminding us of the time when our world was born out of plasma. Because in the beginning was the plasma.

—HANNES ALFVÉN, Nobel Lecture, 1970

Big Bang cosmology, as we have seen, is based on the very latest in physical theories—theories so new and abstract that they have no confirmation in the real world. By contrast, plasma cosmology relies on basic physics that was developed well over a hundred years ago—the physics of electromagnetism. These underlying concepts may not be new, but they have been confirmed in practice—not only by millions of experiments, but by the entire structure of modern technology. Without electromagnetism, we would have no electricity, nothing that requires electricity for its operation or even for its *production*: we would be back to the technical level of Andrew Jackson's era. The concepts of electromagnetism are also, unlike the arcane ideas of the new physics, easily understandable.

Probably the most important single discovery about electricity and magnetism is that they are closely related phenomena. In 1751 Benjamin Franklin pointed out that relation for the first time, proving by experiment that electrical discharges could magnetize and demagnetize iron. But it was not until 1820 that the relation was systematically studied, and the key concepts of electromagnetism began to be formulated.

That year, Hans Christian Ørsted demonstrated that an electrical current moving through a wire creates a magnetic field around it. That is, a magnet suspended near the wire will be pulled around the wire in a circle, transforming the electrical energy of the current into motion. Ørsted had discovered the basic principle of the electrical motor.

Eleven years later Michael Faraday proved the converse, that a moving magnetic field can generate an electrical current. If a conductor of any sort—a coil of wire or a metal disk—passes through a magnetic field, a current is generated in it. Mechanical motion can thus be converted into electrical currents—the principle of the dynamo or electrical generator.

Together these two principles obviously had profound technological implications. Not only can one translate motion into electricity; one can then retranslate that electricity back into motion again, even at a distant location. Within a decade, this led to the invention of the telegraph, enabling messages to fly across a continent in an instant, rather than creeping across it by foot or horseback.

Faraday developed laws that relate magnetism and electricity. He based them on the notion that magnetic fields pervade space, and can be imagined as bundles of curved lines (Fig. 5.1). This was a radical innovation in a time when scientists regarded the space between particles of matter as completely empty. But it wasn't until 1862 that James Clerk Maxwell unified all the elec-

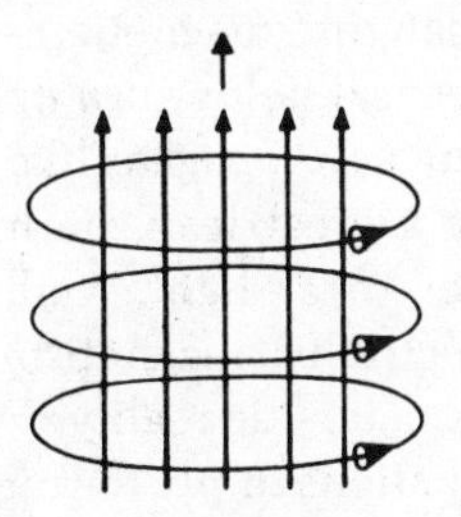

Fig. 5.1a. Ørsted discovered that an electrical current, a flow of electrical charges, creates a magnetic field around it. The field pulls one pole of a magnet around in a circle.

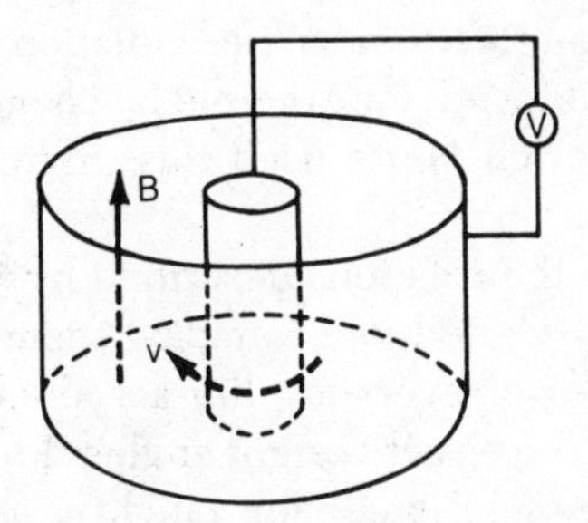

Fig. 5.1b. Faraday later discovered that a conductor moving through a magnetic field will produce a current between the outside and axis of the disk.

trical and magnetic phenomena then known into a single set of equations—Maxwell's laws.

The laws describe, with mathematical precision, four basic principles of electromagnetism: (1) a changing magnetic field generates an electrical field at right angles to the direction of change; (2) similarly, a changing electrical field generates a magnetic field at right angles; (3) an isolated, motionless electrical charge attracts opposite charges and repels like charges with a force that decreases as the square of the distance; (4) there are no isolated magnetic poles—north poles and south poles always come in pairs.

These laws, simple though they are, have enormous consequences. For one thing, the existence of electrical and magnetic fields doesn't require a physical medium—a changing electrical field can produce a changing magnetic field, which in turn will produce changing electrical fields *even in empty space*. Maxwell realized that such changes propagate like waves, moving at the speed of light. Taking a great leap, he hypothesized that such electromagnetic waves *are* light. Subsequent experiments proved him right—electricity, magnetism, and light are all aspects of a single electromagnetic reality.

What's more, the equations imply that any accelerated charged particle will emit electromagnetic radiation. If a charged particle moves at a *constant* velocity in a straight line, it will produce a steadily changing electrical field and an unchanging magnetic field. But if it is accelerated, it will produce a changing magnetic field, which will in turn produce a changing electrical field, and so on—a wave of electromagnetic radiation will be emitted. So waves can be produced, for example by changing currents. A few years later, Heinrich Hertz used this discovery to produce the first radio waves.

At first glance the relations described by Maxwell's laws, and first observed by Ørsted and Faraday, seem a bit peculiar. The mechanical forces of everyday life act along straight lines, yet electromagnetic forces act at right angles. Freshmen physics students have for generations been taught a useful mnemonic for remembering how these forces operate. For example, if your thumb represents the direction of an electrical current, the fingers of your right hand will curl around in the direction of the magnetic field created by the current. If you stretch your right thumb, forefinger, and middle finger out at right angles to each other, your thumb indicates the direction of the magnetic field, your middle finger the direction of a charged particle's motion, and your forefinger the direction of the magnetic field's force on the particle. A freely moving charged particle, like an electron, will therefore move in a circle around a magnetic field line (Fig. 5.2).

While all this may seem confusing, there is a familiar analog—the motion of fluids. Late-nineteenth-century scientists studying fluid dynamics found that the equations they came up with to describe fluid motion and, in particular, the action of fluid vortices, exactly match Maxwell's equations in some situations. Specifically, if there are no unbalanced charges (no excess positive or negative charges), the magnetic lines of force around an elec-

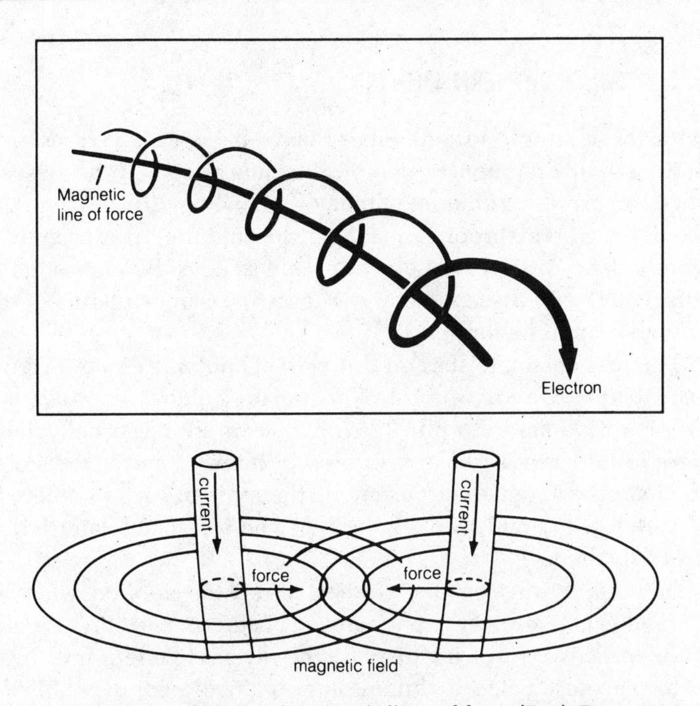

Fig. 5.2. Electrons circle around magnetic lines of force (Top). Two currents moving along parallel lines of force will create magnetic fields that pull the two currents together (Bottom). This is similar to the behavior of fluid vortices—vortices moving in the same direction attract, those moving in opposite directions repel.

trical current are identical to the motion of fluid around a vortex. However, when electrical charges are introduced, such as the static charge that builds up in a clothes dryer or indoors during a dry winter's day, the analogy breaks down. As a result, the fluid dynamicists' physical models were of limited use.

In the 1890s J. J. Thomson performed a series of experiments that led to the discovery that negative charges are carried by the electron. The electron is the light, highly mobile particle that carries currents, while the proton—discovered in 1911 by Ernest Rutherford—is nearly two thousand times heavier, and is the relatively immobile core that makes up the nucleus of every atom. (Later work in the thirties led to the discovery that the neutron, a neutral, uncharged particle, is another part of the nucleus.)

THE NORTHERN LIGHTS

With these simple ingredients—Maxwell's laws, electrons, and protons—an enormously complex technical society has developed. A surprising amount of today's technology still rests on the theoretical breakthroughs made in the mid-nineteenth century and the later discovery of electrons and protons. Perhaps equally surprising, with these same ingredients an accurate picture of the universe could be formed.

The first person to start on this path of applying electromagnetism to the cosmos was the Norwegian scientist and inventor Kristian Birkeland. Born in 1867, five years after Maxwell's laws were formulated, Birkeland studied in Bonn, Geneva, and Leipzig, learning about electromagnetic theory from such pioneers as Heinrich Hertz and Henri Poincaré (whose research later led to relativity theory).

In 1895, fresh from his studies, Birkeland applied what he learned to explaining the phenomena of aurora. The aurora is one of the most awe-inspiring displays the sky has to offer. At its best, in the far north, it is a shimmering, multicolored, ever-shifting curtain of light, everywhere broken up into spikes and streamers that move across the sky—apparitions the Vikings called "the Spears of Odin" (Fig. 5.3).

Fig. 5.3. The Spears of Odin—the Aurora.

Birkeland had seen something like this phosphorescent glow in his laboratory. At the time, the frontier of electromagnetic research was in cathode rays. When an electrical field is applied to a partially evacuated tube that has been coated with a fluorescent material, mysterious rays light up one end of the tube, creating an opalescent glow. (By studying the rays, Wilhelm Röntgen discovered X-rays, which are emitted from the tube.)

Birkeland reasoned that if an electrical current moving in the laboratory through a near vacuum can produce this glow when it hits the fluorescent material, perhaps electrons moving through space can similarly light up the northern skies when they hit the earth's atmosphere—the aurora were giant natural cathode ray tubes. (Today, cathode ray tubes are found in every American home as TV screens. The aurora could be considered nature's television show.)

Birkeland hypothesized that the current originates on the sun, where he thought sunspots emit streams of charged particles. But why, he wondered, does the aurora occur only near the poles of the earth, and why does it take the startling form it does?

In his own experiments with beams of electrons, Birkeland noticed that the electrons are guided toward a nearby magnet. The earth has a powerful magnetic field. Birkeland reasoned that it too would guide the currents: when the currents in space encounter this field they will be forced to spiral around the magnetic field lines. It would be easy, therefore, for currents to move along the lines of magnetic force, but nearly impossible for them to move across them. If the currents follow the field lines, then they will enter the atmosphere only where the earth's magnetic field is nearly vertical—and that is near the poles.

There was a further conclusion. The currents could not remain evenly distributed as they moved along the field lines. Neighboring currents, each producing magnetic fields around itself, will be pulled together. The currents will thus be forced to merge into large filaments, with very little current in between. The streamers that make up the delicate curtains of the aurora are formed this way.

To test his theory, Birkeland built an experimental device to model the aurora in the lab. For the earth he substituted a magnetized metal sphere, and for the glowing atmosphere, he painted a phosphor on the sphere which glowed when hit by electrons. When he fired an electron beam at the sphere the phosphor glowed in the same latitudes as the real aurora (Fig. 5.4). Further detailed experiments confirmed the correlation between model and reality.

Fig. 5.4. Birkeland (left) with his model.

Not satisfied with this lab data alone, Birkeland sought measurements of the magnetic field that he hypothesized is created by the currents flowing during intense auroral "storms." But to obtain these measurements, he would have to organize an expensive expedition to the north of Norway, where he intended to set up a network of magnetic-field detectors. To fund this expedition he relied on the other side of his scientific work—his inventions.

Birkeland had been involved in developing high-voltage equipment for hydroelectric power stations. While working on a new design for a circuit breaker, Birkeland noticed, like others before him, that loose pieces of iron are sucked into an electromagnetic coil, termed a solenoid, with such force that they fly like projectiles. Immediately he saw a practical application—an electromagnetic cannon. In 1901 he took out a patent and organized Birkeland's Firearms to develop the device. Swiftly gathering a core of investors, Birkeland built a test model that fired large shells at a speed of one hundred meters per second. Since this was too slow for a cannon, Birkeland decided that his invention was better at launching torpedoes over short range.

Fig. 5.5. Birkeland's largest electromagnetic cannon, now at the Norwegian Technical Museum.

In 1903, in order to raise funds for his planned expedition, he organized a public demonstration of his biggest gun (Fig. 5.5). In a large hall he pointed the gun at a three-inch-thick plank of wood and explained its operation to the assembled crowd, including representatives from Krupp and Armstrong, Europe's leading arms manufacturers. He later described what followed: " 'Ladies and Gentlemen,' I said, 'you may be seated. When I pull that switch, you will not see or hear anything except that slam of the projectile against the target.' With this I pulled the switch. There was a flash, a deafening and hissing noise, a bright arc of light due to three thousand amperes being short-circuited and a flame shot out of the cannon. Some of the ladies shrieked and a moment later there was panic. It was the most dramatic moment of my life. With this shot, I shot my stock from 300 to zero. But the projectile hit the bull's eye."[1]

The spectacular failure of the gun didn't faze Birkeland (nor did it invalidate the concept: eighty years later, his electromagnetic gun is the subject of intense research for purposes ranging from lifting space payloads to shooting down missiles). Birkeland was fascinated by the jolt of lightning his short-circuited gun had accidentally produced.

Within a week, he found an application for artificial lightning. Sam Eyde, who sought to produce nitrogen fertilizer directly from the nitrogen in the air, avoiding the use of guano or other imported materials, told Birkeland he needed "the largest lightning which could be produced on earth." Within a year the two men collaborated, using inexpensive hydroelectric power and Birkeland's giant sparking machines to produce fertilizer by the ton—and created what remains one of Norway's largest industries.

With the funds he generated from his various inventions, Birkeland mounted a series of expeditions to study the aurora. He set up his network of detectors in the face of dreadful weather. "In high winds, it was impossible to go out," he later wrote, "and more than once it took three men with a great effort to close our little door. Temperatures of −20° C accompanied by winds of 20–30 m/sec [up to 70 MPH] were pretty frequent. No one who has not tried it can imagine what it is to be out in such weather!"[2] Birkeland's measurements showed that the magnetic fields generated by the aurora are so localized on the ground that they can only have been produced by nearly vertical currents—aligned along the magnetic field of the earth. By studying the correlation of magnetic storms on earth with the rotation of sunspot groups on the sun, he also confirmed the source of the charged particles, and estimated quite accurately their speed—around 1,000 km/sec.

In the succeeding decade Birkeland generalized his theory of the aurora to other astronomical phenomena, asserting that sunspots, Saturn's rings, and even the formation of galaxies can be explained by electrical currents and magnetic fields moving through the tenuous conducting gases of space. In 1904 he wrote that "space is filled with electrons and flying electric ions of all kinds." For the first time, he had glimpsed the plasma universe.

ALFVEN AND CHAPMAN

At the time of his death in 1917 Birkeland was the best-known scientist in Scandinavia and was under consideration for a Nobel Prize. But, despite his prestige, his ideas about currents in space were eclipsed for over sixty years. In large part this was due to the personality that came to dominate the fields of auroral study and magnetic fields in space generally—Sydney Chapman, a British scientist whose approach to science was the opposite of Birkeland's.

Around 1920, shortly after Birkeland's death, Chapman began introducing to the study of the aurora the same mathematical precision that was becoming popular in the more arcane and glamorous field of cosmology. Birkeland had been an exemplary proponent of the inductive method of science—formulating hypotheses inspired by observation and using mathematical approximations to describe his theories. Chapman employed the deductive method: he formulated his "rigorous," mathematical hypotheses and applied them as necessary to observation. Currents confined to the spherical shell of the earth's atmosphere could be treated rigorously, so Chapman decided that they must be the basis of any sound theory—auroral storms must, he believed, be caused by disturbances in the earth's atmosphere. Confusing the limitations of his deductive methods with the limits of physical reality, he ruled out the currents in space that would not fit into elegant, spherically symmetrical equations. Birkeland's vision of a universe filled with fields and currents came to be all but buried with him.

Its revival was mainly the work of another Scandinavian, whom we've already met in these pages—Hannes Alfvén. Seeing Alfvén today, one would hardly guess that this soft-spoken man is one of the most controversial figures of twentieth-century science. He appears to be a kindly, grandfatherly type (as indeed he is), and he recalls his many battles with a ready smile. But although Alfvén has always been soft-spoken, he has never softened his words on matters of science. Sixty years after he began his scientific career, twenty years after he was awarded the Nobel Prize, he remains at the center of scientific conflict—and for all his quiet demeanor, he has always enjoyed a good fight.

Alfvén was born in 1908 in Norrköping, Sweden. As a graduate student in physics at the University of Uppsala, Alfvén was initially drawn to the field of nuclear physics, which was flourishing in the early thirties. Nuclear physicists were making a rapid series of discoveries that would lead, within the decade, to the discovery of nuclear fission and fusion and, later, the release of nuclear energy.

Even before he received his doctorate in 1934, Alfvén started to drift away from the study of nuclear physics. Initially he worked with a research team studying cosmic rays, the enormously energetic particles whose origin in space seemed so mysterious. Lemaître had recently proposed that cosmic rays derive from the explosion of his primeval atom—the earliest Big Bang. Alfvén's job was to design and build large Geiger counters to record the energies of the cosmic rays. For Alfvén it was a simple step from monitoring the rays to wondering what they are and how they came to be.

The most prominent scientists of the epoch, James Jeans and Robert Millikan, had proposed that cosmic rays are nuclear in origin, and that they result from some unknown annihilation reaction or nuclear fusion. Alfvén, a mere graduate student, thought he had a better idea and published it as a brief letter in the prestigious British journal *Nature*.

The paper, "Origin of Cosmic Radiation," published in April of 1933, reveals the main themes of Alfvén's subsequent decades of research. He proposed that cosmic rays are accelerated by electrical interactions with charged grains of dust in interstellar space. The dust grains, enormously larger than the electrons and protons of the cosmic rays, impart great energy to them and accelerate them to the high velocities observed. Rather than exotic nuclear interactions or an even more exotic primeval atom, Alfvén envisioned commonplace collisions between dust and atoms as an adequate explanation of cosmic rays.

Characteristically, Alfvén assumed that cosmic phenomena are similar, if not identical to forces and processes we observe on earth—in this case electrostatic forces and collisions between particles of different mass. Equally characteristically, he contradicted the received wisdom of the day. For Alfvén, the laboratory is a far better guide to the heavens than the authority of the most prestigious scientist.

Alfvén's initial idea was wrong—collisions between dust particles and cosmic rays are far too rare to produce the number of rays observed. But it turned him from nuclear physics to electromagnetic studies. Starting in 1936 Alfvén outlined, in a series of highly original papers, the fundamentals of what he would later term cosmic electrodynamics—the science of the plasma universe. Convinced that electrical forces are involved in the generation of cosmic rays, Alfvén pursued Birkeland's method of extending laboratory models to the heavens—though on a much larger scale. He knew how high-energy particles are created in the laboratory—the cyclotron, invented six years earlier, uses electrical fields to accelerate particles and magnetic fields to guide their paths. How, Alfvén asked, would a cosmic, natural cyclotron be possible?

Powerful electrical fields can be generated by moving an electrical conductor through a magnetic field. The simplest such apparatus, developed a century earlier, was Faraday's disk generator, also called the homopolar, or unipolar, generator: a conductor, moving in a circle in a magnetic field, produces an electrical field between the axis and circumference.

Observations from earth had already proved that the sun has a large magnetic field, so it seemed likely to Alfvén that all stars have similar fields. He chose the case of a double-star system to incorporate the needed motion. Here, two stars would revolve around one another—a common enough occurrence—creating giant electrical fields as they move.

But what about the conductor? Space was supposed to be a vacuum, thus incapable of carrying electrical currents. Here, Alfvén again boldly extrapolated from the lab. On earth even extremely rarefied gases can carry a current if they have been ionized—that is, if the electrons have been stripped from the atoms. In the twenties the American chemist Irving Langmuir had initiated the systematic study of such current-carrying gases, which he termed "plasma." Alfvén reasoned that such plasma should exist in space as well. Ions and electrons in space could be accelerated by a double-star generator and could carry enormous currents, a billion amps or more. If this were so, the particle would be accelerated to a trillion electron volts (1TeV), nearly as high as the highest energies then observed in cosmic rays.

At the time, one of the most mysterious things about the cosmic rays was their isotropy. Like the microwave background discovered decades later, cosmic rays issue evenly from all parts of the sky. Most scientists assumed that the source of rays cannot lie within the Milky Way galaxy: if they originated within it, they would appear to be concentrated in a narrow band across the sky, like the Milky Way stars.

Alfvén, however, explained their isotropy proved nothing of the sort. Double-star systems produce currents of high-energy particles, which in turn produce a magnetic field. Therefore, he argued, a galaxy must be pervaded by a weak magnetic field, perhaps a few trillionths of a gauss (a gauss is roughly the strength of the magnetic field at the earth's surface). Cosmic ray particles encountering this field will be forced into a complex spiral curving around in a few light-years of space—whatever direction they came from originally would be hopelessly scrambled by the time they reached earth.

Thus, Alfvén showed that, though apparently isotropic, cosmic rays need not pervade the entire universe uniformly—just the interior of a galaxy. This, moreover, eliminates the problem of explaining the giant amount of energy needed if the rays are spread evenly throughout the universe.

In just two years of studying the cosmic ray problem, Alfvén developed five basic concepts, which would be used again in the development of cosmic electrodynamics: the electromagnetic acceleration of particles such as cosmic rays, the homopolar generation of large electrical fields in space, the existence of large-scale currents and magnetic fields, and finally, a current-carrying plasma in space.

Each of these concepts had to wait twenty years or more for general acceptance and observational confirmation. The electromagnetic acceleration of cosmic rays and the existence of a galactic magnetic field were not generally accepted until the mid-fifties, while homopolar generators and the large-scale currents they produce were not confirmed until the early seventies. At the time Alfvén formulated these ideas, virtually every other scientist assumed that space, especially the space between the stars, is basically empty, a vacuum. But it was this empty space that Chapman's elegant models fit so well—his currents and fields were all safely nestled in the earth's own atmosphere. The idea of a universe filled with plasma, currents, magnetic fields, huge cyclotrons, and other scaled-up pieces of electrical equipment seemed simply bizarre.

BACK TO THE AURORA

Alfvén's initial work on cosmic rays wasn't refuted, just totally ignored. But a collision with orthodox thinking was inevitable when in 1939 Alfvén took the concepts he had developed for cosmic rays and applied them to the problem of auroral storms, or magnetic substorms, as they were then called.

Alfvén began by reviving Birkeland's theory that the storms occur when particles emitted from sunspots create currents near the earth aligned with its magnetic field. But he knew there was a basic flaw in Birkeland's theory. Birkeland speculated that the current comes ready-made from the sun in a beam consisting only of electrons. However, Alfvén knew that if the sun were to emit *only* electrons, it would rapidly develop such a huge positive charge that the current would cease—the electrons would be attracted right back to the sun. So, he assumed, the sun emits a flow of plasma with equal amounts of protons and electrons. In this he agreed with Chapman and others—but only in this.

For Chapman nothing of note happens to this stream until it reaches the currents and fields at the top of the earth's atmosphere. Alfvén, however, thought the most important things happen in space, specifically the generation of the currents that feed the aurora. As with his earlier cosmic ray model, Alfvén found the three elements of a generator: a magnetic field (the earth's), motion (the flow of the solar particles past the earth), and a conductor (the plasma surrounding the earth). As the stream of plasma flows past the magnetic field, distorting it in the process, an electrical field is generated, pulling electrons in one direction and protons in the other. This generator, Alfvén saw, was the energy supply for the auroral currents.

As the stream of solar particles reaches the earth's magnetic field, an electrical potential results, moving protons to the west and electrons to the east (Fig. 5.6). The attraction between the opposite charges impels them to complete the circuit—the electrons "want" to flow back toward the protons—so they move along the earth's magnetic field lines, as Birkeland had pointed out, spiraling around the lines. To get from east to west, they flow down the lines to the ionosphere (the electrically conducting layer in the atmosphere), flow through the ionosphere, and then flow up another magnetic field line on the other side of the earth. When the electrons, accelerated by this vast generator, hit the atoms in the atmosphere, they excite them, creating the powerful auroral storms.

Because Alfvén's theory completely contradicted Chapman's

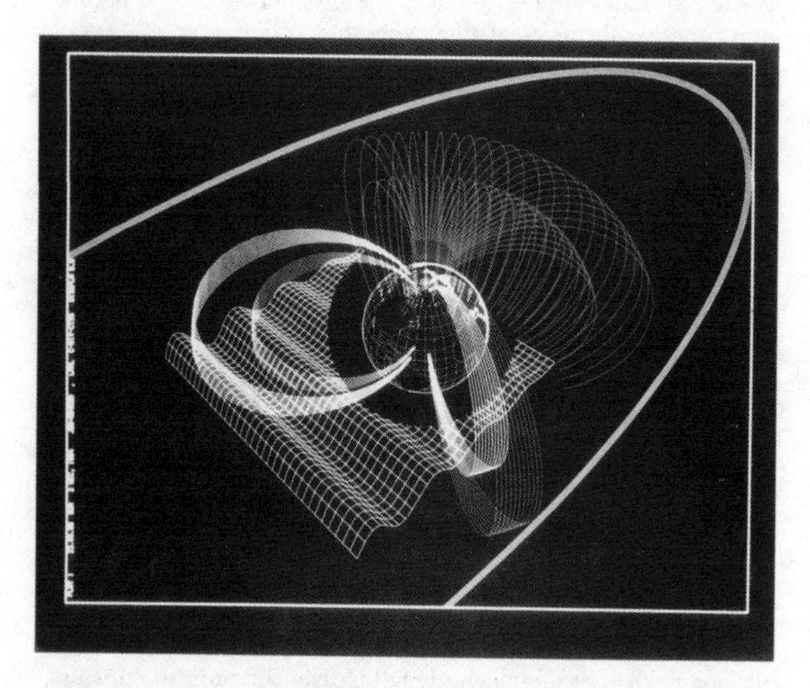

Fig. 5.6. In Alfvén's theory of the aurora, plasma from the sun (orange) moves across the earth's magnetic field (red) (see back of book jacket). At the equator, protons in the plasma are forced west and electrons east (to the right). The electrons then flow back to the west along the magnetic field lines. First they descend to the ionosphere near the poles (green and blue lines); this creates the aurora as the electrons collide with atmospheric atoms, causing them to emit light. The electrons then flow through the ionosphere to the west and then back to the equator (along the field lines to the left).

dominant ideas, he found it nearly impossible to get it published. In the end it was published in a relatively obscure Swedish journal. Worse still, Chapman refused to debate his ideas in any way. Despite Alfvén's polemical presentations at conferences Chapman would rise, say briefly that he and his colleagues disagreed, and add, "We are presently preparing a paper that will clarify these issues." Alfvén would protest, but Chapman would sit down and the matter would be closed.

For thirty years, until Chapman's death in 1970, Alfvén vainly tried to engage him in debate. Their personal relations remained friendly despite sharp scientific differences. On one occasion, Alfvén and his wife, Kersten, were Chapman's guests at Cambridge. Chapman took Alfvén, twenty years his junior, on a walking tour of the ancient campus. It proved a frustrating experience for the younger man. Alfvén recalls, "Every time I tried to raise our differences, when I started to say, 'Doesn't it seem reasonable that, in the substorms . . .' Chapman would politely interrupt and point to some quaint old tower. He would then go on for a half hour about its history. Then I'd try again to get the conversation back to science and the same thing would happen."

Another year, Chapman was Alfvén's guest in Sweden. Instead of a tour of Stockholm, Alfvén had carefully prepared a modern replica of Birkeland's magnetic sphere experiment. Perhaps Chapman, seeing the currents "in the flesh" would at least discuss why he thought they couldn't exist in space. "But he flatly refused to go down into the basement and see it," Alfvén remembers. "It was beneath his dignity as a mathematician to look at a piece of laboratory apparatus!"

Given Chapman's dominant position in the field, it was inevitable that Alfvén would have an uphill battle. As Charles Kennel, professor of physics at UCLA, notes, his scientific style added to the problem: "Alfvén's method of work attracted controversy. He imagines radically new ideas without always working out the detailed physics. Since he then announces his conclusions to the scientific community in a forceful manner, many individuals who find their pet conceptions challenged are antagonized, and initially, there may be good scientific reasons to challenge Alfvén. However, in the end, Alfvén has proven right on big issues enough times that I, for one, believe that one ignores what Alfvén thinks at one's own peril."

Like many original thinkers, Alfvén emphasizes new concepts over establishing the certainty of exact detail. As a result, the details of his initial ideas often turn out to be wrong. For example, many of the details of Alfvén's aurora theory later turned out to be wrong, although the broad outlines are completely correct. But because his concepts are based on phenomena well studied in the laboratory, not arbitrary mathematical constructions, he is confident that he can extrapolate well-tested theories to the far larger scales of astrophysical objects. The most important characteristics, both qualitative and quantitative, of the observed phenomena are then compared with the model's predictions. If they correspond well and if the model appears to be physically reasonable—in accord with what is seen in the lab—then it is probably close enough to warrant publication. Even if there are many loose ends, the model will point research in the correct direction. In most cases, the details are worked out only over decades, either by others or by Alfvén as he returns repeatedly to a problem.

The approach is thus the diametric opposite of deduction, which simplifies the physics of a problem until it can be handled in all detail with mathematical rigor and exactness. For Alfvén it is the physical process, not mathematical description, that is primary.

As a result of his approach, and because his ideas are often far ahead of conventional wisdom, decades have elapsed between the formulation and acceptance of his ideas. By the time they are confirmed, most scientists have forgotten whose ideas they were in the first place, and Alfvén himself has moved to other problems. This is the case with the galactic magnetic fields and the electromagnetic acceleration of cosmic rays. The latter was reformulated a decade after Alfvén's paper by Enrico Fermi, and has since been known as the Fermi process.

▪ INTO THE SOLAR SYSTEM

During World War II, long before arguments with Chapman had been settled, Alfvén was moving on to larger scales, applying the notions of cosmic generators and cosmic currents. His work on the aurora led back to the sun itself, the ultimate source of the northern lights. It too is prone to spectacular storms, which dwarf the entire earth in size—sunspots and prominences soaring hundreds of thousands of miles above the solar surface (Fig. 5.7). In papers published in 1940 and 1941, Alfvén hypothesized that sunspots, whose whirlpool-like rotation had long been known, can act as generators, producing powerful electrical fields as they twirl in the sun's magnetic field. Accelerated by these fields, electrical currents shoot out along the sun's magnetic field lines, heating the ions in the sun's atmosphere and making them glow brilliantly—almost a solar aurora.

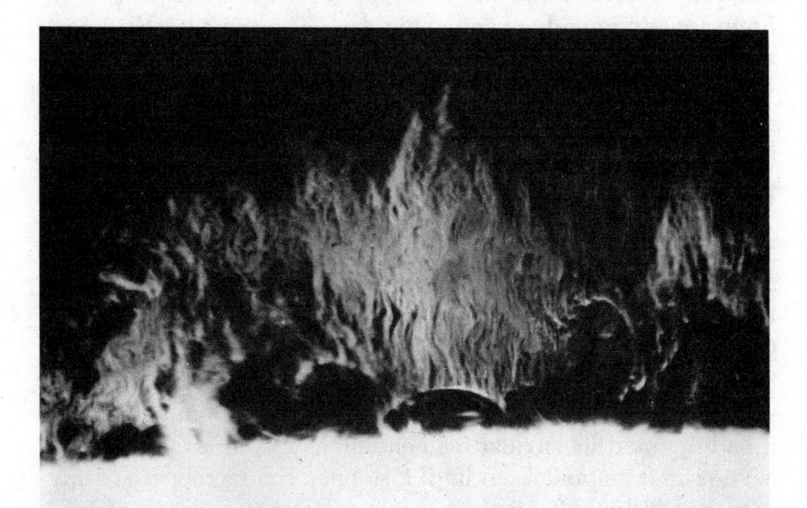

Fig. 5.7. A prominence erupting from the surface of the sun. Alfvén explained the creation of the prominence through the generation of electrical currents in sunspots: accelerated particles speed out along the sun's magnetic field lines, creating the glowing prominences.

From the sun it was but another step to the entire solar system. One of the great puzzles of the solar system's origin is why the sun has so little spin or angular momentum. Since Laplace, most scientists had thought that the sun and planets had condensed from a single nebula or gas cloud. But there was a difficulty: For any isolated object, the total angular momentum—the product of radius, velocity, and mass—is constant, so as the radius shrinks, velocity increases. So as this nebula contracted, it would have spun faster.

Given the known mass and orbit of each planet and the sun, and the sun's rotational velocity, it is simple to calculate the solar system's angular momentum and which objects in it have the most angular momentum around the center of the sun—and the answer is puzzling. If the sun had retained the angular momentum of the whole system, it would now spin around once every thirteen hours or so, as fast as a typical planet. But it takes fifty times longer, roughly twenty-eight days. The sun has only 2 percent of the solar system's total angular momentum, while Jupiter, with only one-thousandth the mass, has 70 percent; Saturn has nearly all the rest, about 27 percent. Somehow, an enormous amount of angular momentum was transferred from the sun to the planets, especially the giant planets.

Even the solar system as a whole seems to have very little angular momentum. Presumably it contracted from a cloud about a light-year in radius, roughly the distance between the stars. Even if the cloud completed a rotation at the same rate as the galaxy (about once every four hundred million years) it would have had seven hundred times the angular momentum of the solar system today. Such a cloud would not have formed a star if it had retained its angular momentum: as it contracted it would have spun faster and faster until it stopped contracting at a radius of twenty billion kilometers or so. Even if, in the process of contraction, the central star had retained only 2 percent of this angular momentum, it would have stopped contracting at about a size of ten million kilometers—more than a dozen times the size of the sun, and far too big for a star. Its gas would be too cool and diffuse to burn hydrogen to helium. In short, for the solar system to form, it must have lost about 99.9 percent of the initial angular momentum, and transferred 98 percent of the remainder to the planets. How could this happen?

Alfvén believed that the electrical currents created by a protostar's magnetic field could do the trick. Suppose a rotating magnetized body is surrounded by clouds of plasma that are not rotating as rapidly (Fig. 5.8). The magnetic field will rotate with

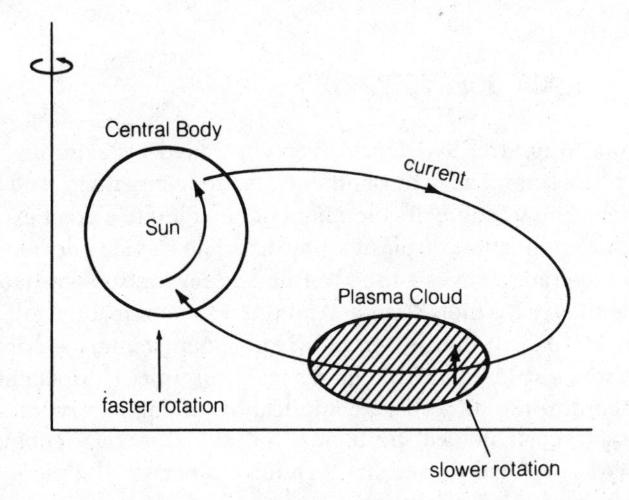

Fig. 5.8. Alfvén explained the transfer of angular momentum from the sun to the planets via a current system. Plasma clouds moving in the magnetic field of the sun generate an electrical current that flows toward the sun. The interaction of this current and the magnetic field produces a force on the cloud, which accelerates it. The current returns to the sun, where its interaction with the sun's own magnetic field slows its spinning. In this way the sun loses enough angular momentum to contract into a star.

the central body, sweeping through the clouds and generating an electrical current within them. This electrical current, because it is in a magnetic field, will exert a force which will cause the cloud to move in the direction of the rotating magnetic field. Like the blades of a gigantic fan, the magnetic field transfers angular momentum from the central body to the clouds. The resulting circuit will be a gigantic current carrying trillions of amps, flowing out along the solar magnetic field lines, through the cloud and *back* to the sun at its equator. Just as the current flowing out to the cloud will accelerate its rotation, the current flowing into the sun will slow its rotation. In this manner most of the angular momentum of the protosolar cloud can be transferred either to surrounding plasma, which will be flung outward, or to the magnetic field itself, while much of the remainder will be transferred to the planets as they form from smaller and denser plasma clouds.

PLASMA GOES TO WAR

Working in neutral Sweden, Alfvén continued his astrophysical work while scientists in most other countries concentrated on war research. The wartime development of radar led to a great expansion and application of plasma physics. The devices developed to produce radar waves—the klystron and magnetron—relied on plasma theory for their design. And the war soon made itself felt in Alfvén's lab in the person of a Norwegian scientist—Nicolai Herlofson, a student of Birkeland's main disciple. Herlofson had been continuing Birkeland's application of laboratory plasma physics to cosmological problems, but after the Nazi conquest of Norway, he had turned his technical talents to the more immediate problems of the Norwegian resistance by purchasing arms and carrying microphotographed intelligence across the Swedish-Norwegian frontier. One day, the Gestapo came to his door, and he went out a window. "Fortunately he was a fast runner, so he got away, and soon turned up at our lab in Sweden," Alfvén recalls.

Herlofson contributed substantially to the Swedish plasma work, eventually becoming director of the Plasma Physics Laboratory at the Royal Institute of Technology, where Alfvén worked. "Many of my best ideas came from discussions with Herlofson," says Alfvén. "Yet he is such a modest man that he rarely allowed his name to appear on papers he had contributed to."

Herlofson soon left for England, where he joined the Central Meteorological Establishment. But in 1946 he used plasma physics to solve a mystery left over from the war. During the V-2 attacks in 1944 and 1945, Britain relied on radar to give a few minutes' warning of the rockets' arrival. Many times the radar produced false alarms, detecting only echoes. Dr. J. S. Hey, of the Army Operations research groups, guessed that somehow meteors were causing the false alarms. Yet how could tiny meteors imitate the big German rockets?

Soon after the war ended, Herlofson learned of the mystery and developed a theory: the meteorites, too small themselves to produce a radar echo, left thin trails of plasma as they passed through the atmosphere. The radar waves, Herlofson calculated, would make the plasma resonate, much as a singer's high note makes crystal resonate. It was this plasma resonance that produced a huge echo like that of German rockets.

Herlofson's work showed in practice that radio waves indicate something about distant plasmas. In the course of the war, the enormous increase in radio technology brought about by the use of radar had provided astronomers with a new tool to study plasma in space—the radio telescope. Prior to the war the American engineers Karl Jansky and Grote Reber had found that some celestial objects, including the sun, are powerful sources of radio waves. Military radar research allowed the tremendous development of the radio telescope, a large antenna designed to study these sources. To Alfvén and Herlofson, the mysterious radio waves from space were signals from distant plasma, just like the enhanced radar reflections from the meteorite trails.

In 1950 the two scientists proposed that radio signals are produced by energetic electrons trapped in magnetic fields. Any electron, accelerated by the forces of the magnetic fields, radiates electromagnetic waves. But as energetic particles travel close to the speed of light, the frequency of the radiation shifts, increasing as the square of the energy. Thus the powerful high-frequency signals coming from sources such as the Crab nebula implied the existence of strong magnetic fields and accelerated electrons in interstellar space. Now plasma could be studied far beyond the solar system, to the far reaches of the universe.

The war had a second, profound impact on plasma studies, one that enormously strengthened the links between space and laboratory studies. The development of the atomic bomb had led scientists like Edward Teller to advocate even more powerful weapons in which the heat generated by the fission of an atomic bomb would be used to fuse hydrogen nuclei to helium, as the sun does at its core. Other scientists, revolted by such weapons of mass destruction, wondered how this same power, nuclear fusion, could be harnessed for the peaceful production of cheap and abundant energy.

Fusion uses fuels that are abundant in nature and produce far less radioactivity than the fission reactors the U.S. Navy was developing to drive its submarines. But to achieve fusion, temperatures of one hundred million degrees are needed, which no physical container can withstand. The only thing that could contain such a hot plasma was a magnetic field.

For decades scientists had known that if a high current is discharged through a plasma, the magnetic fields created pinch the current and the plasma together (a process described in Chapter One). In the early fifties scientists in the U.S., England, and the Soviet Union, working in secret, showed that indeed superhot plasma can be created and confined with such pinches. But the plasma proved extraordinarily balky. Instead of smoothly pinching to high temperatures and staying stable while fusion occurred, it bucked and bowed like a wild bronco. The fusion scientists desperately needed a way to control the unruly plasma.

Alfvén provided the theory. In 1950 he had collected much of his unpublished work of the past decade into a ground-breaking textbook, *Cosmic Electrodynamics*. Covering a broad range of problems and phenomena, it was to become extremely influential, sometimes in rather surprising ways. The book provided for the first time a detailed theoretical analysis of how electrical discharges become constricted through their own magnetic fields. He applies this analysis to two cosmic problems he had long worked on—the aurora and the solar prominences. Here Alfvén shows that the filamentary structure of both can be explained in detail by the pinch effect.

Alfvén demonstrates that the problems of fusion in the lab and the prominences in space are closely linked. From Maxwell's laws he derives rules with which a researcher can develop small-scale laboratory models of large-scale astrophysical processes. He also discusses how such processes can be used to predict plasma behavior in the lab.

He found that certain key variables do not change with scale—electrical resistance, velocity, and energy all remained the same. Other quantities do change: for example, time is scaled as size, so if a process is a million times smaller, it occurs a million times faster. Thus the stately processes of the cosmos, ranging from auroras lasting hours to prominences lasting days to galaxies lasting billions of years, can all be modeled in the lab by rapid discharges lasting millionths of a second. When densities of astronomical objects are scaled down to lab proportions, their densities become those of ordinary gases (Table 5.1).

Equally important, though, is the converse use of these scaling rules. When the magnetic fields and currents of these objects are scaled down, they become incredibly intense—millions of gauss, millions of amperes, well beyond levels achievable in the laboratory. However, by studying cosmic phenomena, Alfvén shows, scientists can learn about how fusion devices more powerful than those now in existence will operate. In fact, they might learn how to design such devices from the lessons in the heavens.

By 1956 fusion scientists, still under secrecy wraps, were gathering at the international conferences of cosmic electrodynamicists. That year, Alfvén hosted the International Astronomical Union Symposium on Electromagnetic Phenomena in Cosmic Physics in Stockholm. One researcher, Winston Bostick of Stevens Institute of Technology in Hoboken, New Jersey, reported just the sort of laboratory modeling Alfvén had described. Bostick found that tiny plasmas fired at high speed toward each other pinch and twist themselves into the graceful shapes of spiral galaxies.

THE COSMIC POWER GRID

As fusion work grew, fusion researchers increasingly turned to Alfvén's textbook for guidance in their efforts to control pinched plasma. Alfvén himself, dividing his time between technological and cosmic research, rapidly advanced in his understanding of the filamentation process. By now he was aided by colleagues like Carl Gunne Fälthammar, twenty years his junior, who were far more willing to delve into the mathematical details of the phenomena studied. Moreover, experimental results from the fusion pinch experiments were now giving him the data needed to inspire and check further theoretical work.

In the reedition of *Cosmic Electrodynamics,* written with Fälthammar in 1963, Alfvén gives filamentation a central role in producing homogeneities in plasma, on scales from laboratory up to stellar nebulae—the vast clouds of glowing gas surrounding many star clusters in a galaxy. When a current flows through a plasma, Alfvén shows, it must assume the form of a filament in order to move along magnetic field lines. The flow of electrons thus becomes force-free: because they move exactly along the lines of a magnetic field, no magnetic forces act on them. In a

TABLE 5.1
PLASMA SCALING RELATIONS

Phenomena	Size	Density		Magnetic Field		Current		Time	
		Actual (/ cm³)	Scaled (/ cm³)	Actual (G)	Scaled (KG)	Actual Billion A	Scaled (KA)	Actual	Scaled (microsec)
Typical lab plasma	10 cm		$10^{14}-10^{18}$		10–100		500–5,000		1
Aurora	30,000 km	1 trillion	3×10^{20}	.5	8	7.5	450	3 hours	30
Solar corona	million km	100 million	10^{18}	20	2,000	10,000	100,000	20 minutes	.1
Interplanetary space	10 billion km	100	10^{16}	1/100,000	.1	50	5	1 year	.3
Galaxy	50,000 light-years	1	5×10^{21}	6/1,000,000	400	1.5 billion	20,000	400 million years	2
Intergalactic space	5 million light-years	1/10,000	5×10^{19}	2/1,000,000	1500	50 billion	70,000	15 billion years	1

Alfvén discovered scaling laws that relate plasma phenomena on different scales. These make it possible to use laboratory plasma as a model for astronomical plasma and vice versa. What is most surprising is that when space plasma are scaled down to laboratory size, they are comparable in density to laboratory plasma. Their magnetic fields and currents and the duration of the processes are also comparable, although in some cases the currents and magnetic fields are ten to a hundred times larger than what has been achieved in the laboratory, and thus could serve as models for future experiments.

In this table, modified from Alfvén's original work, the characteristics of various plasma are scaled to a single size of ten centimeters. Note that taking this scaling into account, intergalactic evolution over billions of years must be viewed as transient events, lasting in the lab no more than a millionth of a second, just the duration of a typical plasma discharge.

force-free filament, the electrons, in effect, cooperate to minimize the difficulty of flowing. Those along the center of the filament flow in straight lines, producing a spiral magnetic field along which outer electrons can flow. The outer electrons, in turn, flowing in spirals around the edge of the filament, produce the straight magnetic field lines on the axis along which inner electrons flow. Together, the electrons move in a complex pattern of helical paths with increasingly steep pitch as they approach the filament's axis (Fig. 5.9).

In Alfvén's new view inhomogeneity—produced by the formation of filamentary currents—is an almost inevitable property of plasmas, and thus of the universe *as a whole*. The universe, thus, forms a gigantic power grid, with huge electrical currents flowing along filamentary "wires" stretching across the cosmos.

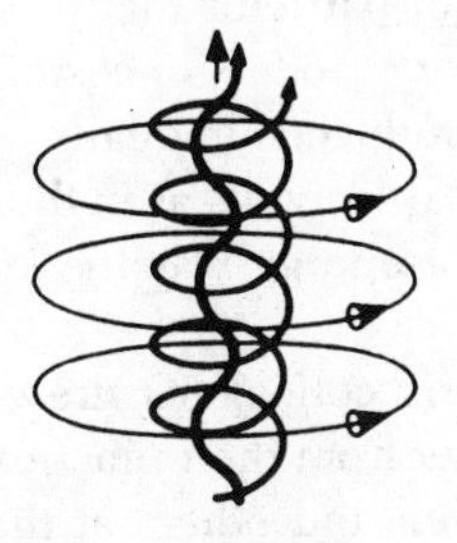

Fig. 5.9. A force-free filament. Electrical current and plasma flow along helical magnetic field lines that are straighter toward the center.

Not only are current and magnetic field thereby concentrated into the spiraling filaments, but the plasma itself is pinched together by magnetic fields, sucked into an electromagnetic tornado.

The discovery of the force-free helical form of the filaments was doubly significant. On the other hand, it yields an unmistakable method of identifying filaments if their peculiar form is observed. On the other, their helical forms show that pinch filaments are not simply "wires" carrying currents in one direction; rather, they are vortices of current, plasma whirlwinds.

Filaments do not form simply by being pulled together, but are in reality twisted together. As each small current, moving along the lines of the background field, tries to move across the field lines toward its neighbor, it is drawn into a spiral, where it contributes its energy to the larger pattern of the filament, much as wisps of cotton are spun together and twisted into a single thread.

This understanding was to prove critical in the study of the filaments' behavior and their role in the universe.

▪ SOLAR FLARES AND THE ENERGY PROBLEM IN SWEDEN

By forming the filamentary structures observed on the smallest and largest scales, matter and energy can be compressed in space. But it is clear that energy can be compressed in *time* as well—the universe is filled with sudden, explosive releases of energy. One example that Alfvén was familiar with is the solar flare, the sudden release of energy on the sun's surface, which generates the streams of particles that produce magnetic storms on earth. His "generator" models of cosmic phenomena showed how energy can be produced gradually, as in a well-behaved power station, but not explosively, as in the flares. Understanding the explosive release of energy was the key to the dynamics of the cosmos.

Again, Alfvén and his colleagues drew inspiration not from mathematical theory, but from their practical work in technology. In the late fifties, Alfvén and others at the Royal Institute had been called in by the Swedish power company, ASEA, to solve an urgent problem. Most of Sweden's electrical supply is generated by hydroelectric power in the north of the country, and is then transmitted over six hundred miles to the industrial south. ASEA found that it was cheaper to transform the alternating current to direct current for transmission with large mercury rectifiers. A rectifier allows a current to pass in only one direction, holding it back for the other half of the cycle, thus producing DC. But every so often a rectifier would explode, causing considerable damage.

Herlofson and Alfvén were consulted because the rectifier mechanism, consisting of a low-pressure mercury vapor cell, employs a current-carrying plasma. The team from the Royal Institute rapidly located the problem: the pressure of the mercury vapor in the rectifiers was too low. As a result, at high currents nearly all the electrons carried the electrical flow, creating an unstable situation in which the plasma started to slosh about within the rectifier.

At low current, this sloshing was not serious. If too many ions (the positive charges) piled up on one side, the electrons would be attracted to them, neutralizing them. But at high current something else happened. If the ions accidentally spilled out of a region, the electrons in the current would rush toward the ions with such momentum that their collision pushed the ions farther out of the region. This accelerated the electrons more, and so on. However, a few ions would break away and accelerate toward the electrons on the other side, pushing most of them back. An ever-widening tear in the plasma would open up, with electrons bunching up on one side and ions on the other. As the gap widened, fewer electrons could pass, so the current (the number of electrons passing a point per second) would drop. This is like suddenly unplugging an appliance. The drop in current produces a sudden drop in the magnetic field created by the current, and the changing magnetic field creates a powerful electrical force that further accelerates the electrons. In the case of an unplugged appliance, the voltage becomes high enough to make a spark jump across from the socket to the plug. In the case of the rectifier, the voltage builds and builds until the electrons heat the rectifier plasma so hot that an explosion ensues, and gigantic sparks jump through the air in the station. Energy is suddenly released, creating the explosive damage.

The phenomena Herlofson, Alfvén, and their colleagues explained is called an exploding double layer. Double layers—charged gaps in plasma—had been observed since the twenties, but this was the first time it had been shown that they can be the source of an explosive release of energy. Moreover, an exploding double layer permits energy generated over a period of time and through extended space to be suddenly released in a small space—a giant power compression. The rectifier explosion's energy had been generated far away in northern Sweden by a spinning dynamo and stored in the magnetic fields of several hundred miles of transmission line. In effect, this magnetic field is the stored momentum of *all* the electrons in the current, not just the few actually present in the rectifier itself. The phenomenon is inherently *nonlocal* and can be understood by only considering the rectifier as part of a huge circuit.

This gave a vital idea to two of Alfvén's colleagues, Carl Jacobsen and Per Carlqvist. Alvén had explained solar prominences as filamentary, constricted currents, generated by a vortex's motion in the sun's atmosphere. At times the prominences are preceded by, and obviously connected with, the mysterious solar flares—explosions that can last as little as a few seconds but can release as much as 10^{34} ergs in a region no bigger than the earth. (For comparison, the sun itself releases this much energy in ten seconds; it is enough energy to boil all the oceans on earth.)

Carlqvist and Jacobsen reasoned that the filamentary currents in a prominence, generated slowly and gradually, might develop exploding double layers at some point along their path. In that case, as with the rectifier, energy generated over days and spread over a vast circuit would be released in a small area in seconds. Through the action of the magnetic field, the energy could be concentrated at the speed of light and released nearly as suddenly.

Thus explosive events in the universe could be explained not only by looking at where they occurred but by viewing them as part of a global process, part of a cosmic power grid, whose circuits were defined by the filamentary currents.

▪ MATHEMATICAL BEAUTY RAISES ITS HEAD

By the mid-sixties Alfvén and his colleagues at the Royal Institute of Technology had created a picture of the solar system's plasma processes. The model was inhomogeneous, with currents concentrated into filaments, and global, the currents carrying energy generated in one place to others millions of kilometers away. As with earthly technology, energy generated over vast regions and long periods could be released explosively in a tiny space.

This solar system was a dynamic and complex place—and anything but mathematically elegant. The complexities of filaments and double layers were still understood only partially and could be quantified only approximately. Nevertheless, their model corresponded in its general development both to experiments in the laboratory and to observations of space.

Unfortunately, in 1965 this model remained a vision of a single group of collaborators. It was rejected almost unanimously by other scientists whose study of the solar system was still dominated by mathematically elegant and pure models. Ironically, by this time their beliefs derived not only from Chapman's still influential work, but also from some of Alfvén's own concepts—Alfvén was forced to fight with the ghosts of his own theories. But as with the continuing battle with Chapman, the underlying issue was the same—observational and physical accuracy versus mathematical rigor and elegance, the empirical method versus the deductive method.

In order to understand this phase of plasma theory's development we have to return to Alfvén's work of 1943. While continuing his studies of solar phenomena, especially sunspots, he had realized that plasma supports waves—not sound or light waves, but waves of changing magnetic fields, waves he called magnetohydrodynamic waves, MHD waves for short. Like his other ideas, this too contradicted conventional wisdom. Scientists had long believed that electromagnetic waves cannot penetrate a good conductor, such as a metal or plasma. (This is why you can't receive radio signals when your car goes under a metal bridge.) Conductors short out electrical fields—electrons move so freely that they cancel out the changing electrical fields of the wave: no electrical field, no electromagnetic wave.

But Alfvén had shown that waves can form in a plasma when the magnetic fields and the plasma move together. In a perfectly conducting medium, thus one with no electrical field, the imaginary magnetic lines of force would be visualized as moving exactly as the plasma did—the lines of force are, in Alfvén's terms, "frozen into" the plasma. (The lines are a description of the magnetic field's direction.) The logic is simple: any motion of the plasma across the field lines would generate an electrical field, but a perfect conductor would not allow such a field. So when the plasma moves, the magnetic field lines must "move" as well. Thus waves in the plasma would create waves of magnetic field direction.

In the fifties the reality of MHD waves, or "Alfvén waves," as they are now universally called, was brought home forcibly to plasma scientists working to tame thermonuclear fusion. As we've already mentioned, fusion scientists were hoping that pinches could squeeze, heat, and confine a plasma long enough for fusion to occur. But by the end of the fifties, things were going badly: whenever the plasma was squeezed it became unstable, pinching itself into a string of sausages, or twisting and hurling itself against the walls of its container. Fusion scientists rapidly learned that this results from Alfvén waves—they are MHD instabilities. So to tame them, the researchers turned to Alfvén's models of waving, frozen-in field lines; in doing so they found that MHD theories produce very elegant mathematical forms and apply to a wide variety of new devices, most notably the tokamak, the Soviet-invented plasma device for producing fusion.

Enthusiasm for MHD theory quickly spread as satellites probed deeper into space, for it became clear that Chapman's "empty" space—incapable of carrying current because of its infinite resistance—was no longer tenable. But a mathematics almost as elegant as Chapman's could be derived from the infinite conductivity—*zero* resistance—assumed by MHD theory. So, almost without transition, the dominant model of space went from perfect insulation to perfect conductivity. In any case, the heavens were still perfect!—too perfect.

This perfect conductivity and the frozen-in magnetic fields completely ruled out all of Alfvén's work following 1943: in such a plasma there can be no electrical fields, no current aligned with magnetic field lines, no generation of electrical power—all this needs finite resistance and a flow of plasma past magnetic fields. In the perfectly conducting MHD model, energy would dissipate instantly, no voltage could accumulate—it would be like a power station without insulators, it can't be done. The infinitely conductive MHD plasma would never develop inhomogeneities; instead, it would be perfectly smooth plasma, dominated by local effects, in which no large-scale structures could transmit power over large distances.

Once again, all Alfvén's talk about filaments, currents, and double layers was rejected as impossible in an infinitely conducting plasma—just as it had been with Chapman's perfectly nonconducting plasma. The fact that the MHD model itself is Alfvén's was gradually obscured as physicists came to rely on new textbooks by astrophysicists like Lyman Spitzer and S. Chandrasekhar.

Alfvén thus was confronted with his own ghost. He was aware of the conflict between his earlier MHD models and his later work even in 1950, when another scientist, T. G. Cowling, pointed out that space plasma should be almost perfectly conducting. Alfvén knew that this was not the case, because of the way the plasma behaved, but he was not sure why. "It gave me quite a headache," he later confessed.

Over the years, however, he began to point out how the MHD model he had developed, like any mathematical theory, is *only an approximation*, applicable to *specific* situations, and thus has very real limits—and, like every mathematical formula, it is not a universally applicable law. There are real barriers that prevent space plasma from achieving infinite conductivity—the magnetic fields created by the currents themselves.

As early as 1939 Alfvén discovered limits to the amount of current a plasma can carry. In his studies of cosmic rays he imagined what would happen if a large current of cosmic rays, traveling at high energies, were to form. As the current grew, so would the circular magnetic field it formed, eventually causing electrons or ions to circle around it. Their paths would become tangled up in it, preventing further growth of the current (Fig. 5.10).

For particles traveling near the speed of light special relativity shows that as their energy increases, so does their apparent mass. These more massive particles are bent less tightly by the magnetic field, and so sustain higher currents. Thus, as with an ordinary circuit on earth, the higher the energy or voltage of the particle, the higher the maximum current. This relation can be expressed as a resistance of about thirty ohms. (Resistance is voltage divided by current. For comparison, the resistance of a one-

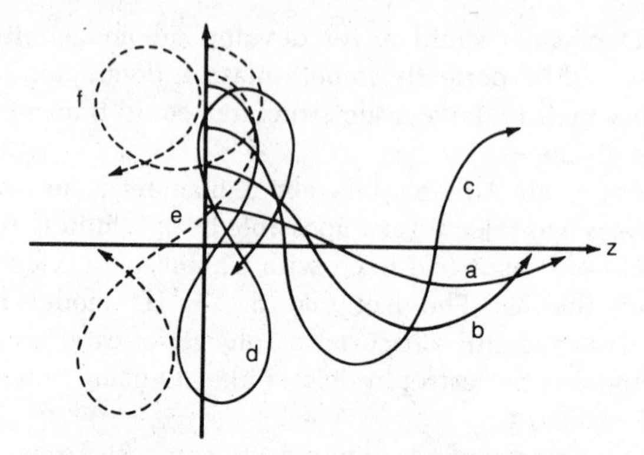

Fig. 5.10. In the thirties Alfvén calculated that currents above a certain value could not easily be carried by plasma. As the magnetic field produced by the current increases (lines going from a to e), the electrons become more tangled in the field, until they can no longer go forward. Later scientists discovered that electrons in force-free filaments are not so limited.

hundred-watt light bulb is 120 ohms.) Thus, for concentrated beams of particles in space, the cosmic plasma is neither purely conductive nor purely insulative—it has a very earthly resistance.

Later, in the fifties, Alfvén pointed to less specific restrictions on currents, due to the essentially transient nature of most celestial phenomena. Even currents well below the maximum (the "Alfvén current") require a significant amount of time to build up: energy must be poured into creating a magnetic field, just as when a current is suddenly switched off, energy flows from the magnetic field into the current itself. These inductive effects, well known in electrical engineering, are particularly important in rapid, localized electrical discharges, such as solar flares. This energy drain acts as a resistance, again impeding the free flow of electrons. Finally, Alfvén, while working with Fälthammar on the revised *Cosmic Electrodynamics* in the early sixties, realized that even for relatively long-term plasma phenomena, the plasma's resistance can be significant under certain conditions. Most plasmas in space, he found, are diffuse enough that electrons and ions spiral around magnetic field lines without colliding with one another. Consequently, it is extremely difficult for the electrons or ions to move perpendicularly to the magnetic field's direction

—thus in *that* direction, there is, in fact, a strong resistance. So, as plasma slowly moves past magnetic fields, significant electrical fields are generated and not shorted out.

Equally important, even along the lines of force, double layers can develop, introducing enormous resistance into a circuit; the increasing voltage that results can produce explosive energy releases.

Because plasma does have an effective resistance, field lines cannot in fact be "frozen in." The stronger the resistance, the faster a plasma can slip by the field lines: a resistance of about one-tenth of an ohm produces slippage velocities up to 1,000 km/sec—as high as velocities in space generally get.

As a result of these various effects, MHD theory applies only to very dense plasmas, like those within a star—where, it's worth noting, MHD theory had originally been applied. For most plasmas, Alfvén's frozen-in schema of the forties is simply not valid.

Once Alfvén was certain of the limits of MHD theory and frozen-in field lines, he promulgated his newer ideas. "Originally I thought the frozen-in concept was very good pedagogically," he explains. "But in reality, it was a dangerous pseudo-pedagogical concept that makes you believe you understand a phenomenon when you have drastically misunderstood it."

Since it was his idea in the first place, Alfvén believed he could clear up the misunderstanding and gain acceptance for the far more valid circuit and current viewpoint in a short time. But Alfvén's ghost was more powerful than he thought. The MHD approach fit in beautifully with astrophysicists' increasing use of a mathematically rigorous, deductive approach—a development that also markedly slowed fusion research.

Alfvén inveighed with increasing exasperation against the misuse of frozen-in field lines, calling it a "pseudoscience" about "pseudoplasmas." But for the remainder of the sixties he met with no success, and his model of the solar system was rejected. The ever-increasing prestige of his magnetic-hydrodynamic work overshadowed his more recent and more generally applicable work on inhomogeneous plasma.

The height of this conflict came in 1970 when Alfvén was awarded the Nobel Prize. In presenting the award, Torsten Gustafson lauded Alfvén's development of magneto-hydrodynamics and his concept of frozen lines of magnetic force—the very concepts whose limitations Alfvén had been striving to explain. For perhaps the first and last time in the history of the Nobel Prize, Alfvén criticized in his address the work for which he was awarded the prize, and decried its persistent misapplication to bolster elegant theories. "But it is only the plasma," he said, "that does not 'understand' how beautiful the theories are and absolutely refuses to obey them." Contrasting the "perfect" universe of deduction with the dynamic, filamentary universe that is actually observed, Alfvén asserted that only observation linked to laboratory experiments can lead to an understanding of the solar system and its origin. Mathematical theory, he emphasized, must always be the servant of physical understanding and close observation—never the master.

THE POLITICS OF PLASMA

Alfvén faced other battles in the second half of the sixties. He had long been politically engaged, and was particularly active in the international disarmament movement. Since Sweden had no nuclear weapons and was neutral, these activities had not led to conflict with national authorities. However, in the mid-sixties, Sweden began to consider nuclear power research and development—an issue Alfvén felt well qualified to deal with.

Alfvén became involved in an increasingly heated debate with government policymakers. He felt that the proposed new policies underestimated the contribution fusion could make to solving the energy problem, and underfunded the fusion research required. He was equally critical of the specific plans for a nuclear reactor, scorning them as technically unfeasible and misguided. He found himself at odds with local bureaucrats, and their hostility toward him was not softened when his technical critique of the reactor turned out to be well founded. (It was later converted to conventional power.)

Alfvén became an increasingly public gadfly, and his relations with policymakers deteriorated further in 1966 when, writing under the pseudonym Olaf Johannesson, he published a biting political-scientific satire, *The Great Computer*. In the novel he describes the future takeover of the planet by computers—a popular idea among science fiction writers. Alfvén, however, used it as a vehicle to ridicule the growing infatuation of government and business with the novel power of computers, and to pillory much of the Swedish establishment—greedy corporate leaders, shortsighted bureaucrats, and power-hungry politicians banding together to create a utopia for computers. In modern Sweden, a state run by an alliance of politicians, bureaucrats, and corporate leaders, Alfvén's satire didn't endear him to those already nettled by his sharp criticism of nuclear policy.

By 1967 Alfvén's relations with those running the Swedish scientific establishment had soured so much that he decided to leave Sweden. "They told me that my funding would be severely cut unless I supported the reactor," he recalls. He was instantly offered chairs at both Soviet and American universities. After a two-month stay in the Soviet Union, he moved to the University of California at San Diego.

THE TIDE TURNS

Nineteen sixty-seven was, for Alfvén, a turning point because, for the first time decisive observations proved that his concept of a dynamic, inhomogeneous cosmic plasma was right. The space probes that had recently been launched and the data they returned meant death to the idea of perfect, unchanging heavens, ruled by elegant mathematical laws. In its place was a far more interesting and restless universe.

Astronomers no longer had to guess from afar, because the space probes were able to measure a plasma's properties as they passed through it. Equally important, radiation that is absorbed by earth's atmosphere and cannot be observed from the surface—X-rays, gamma rays, infrared and ultraviolet rays—were opened up to space-based telescopes. Astronomers would now *see* much of the universal plasma.

Already in 1959 the discovery by early satellites of the Van Allen radiation belts that girdle the earth seemed to substantiate Alfvén's theory of the aurora, proposed twenty years earlier. He had hypothesized that protons and electrons are trapped in separate belts by the earth's magnetic field, interacting with the solar wind—just what the satellites found. But the belts could also be understood on the basis of MHD theory. The key question was whether electrical currents exist in space.

In 1967 the answer arrived. Satellites carrying magnetometers, magnetic field–measuring devices, reported extremely localized magnetic fields in the auroral zones. Alex Dessler, then editor of the *Journal of Geophysical Research,* had seen the data in a paper submitted by A. J. Zmuda, J. H. Martin, and F. T. Heuring. Dessler had been convinced from hearing Alfvén and Fälthammar at several scientific meetings that Alfvén's field-aligned currents—currents moving along magnetic field lines—probably existed. He realized that the localized magnetic fields were certain evidence, for only localized currents can produce them. When he could not persuade Zmuda to include this explanation in his own paper, Dessler and a graduate student, W. David Cummings, published on their own—dubbing them "Birkeland currents" in honor of their initial proponent.

Within a few years, as more sophisticated satellites probed outward into the earth's magnetosphere, the evidence became overwhelmingly convincing. The currents and filaments that Birkeland had postulated and Alfvén had elaborated into a concrete model really do exist—and dominate the earth's immediate region. In the course of the seventies Alfvén's earlier arguments about the limits of his own MHD approximation were widely accepted by geophysicists studying the aurora. Chapman, who had died in 1970, was rapidly rejected.

By the end of the seventies, confirmation of filamentary currents had spread outward as fast as space probes could travel. In 1979 the Voyager probes, equipped with sophisticated plasma instruments, cruised past Jupiter and then in the eighties went on to probe Saturn and Uranus. "As the craft went by, they completely changed the preexisting theories about the magnetic field and plasma environments of all three planets," says Dessler (Fig. 5.11). Everywhere the data showed filamentary currents, twisting plasma vortices, huge homopolar generators, double layers—the whole plasma zoo Alfvén and his colleagues had theorized. His concepts dominated the study of the solar system and were used as the basis for planning new probes to the planets and comets.

Fig. 5.11. One of the key places where Alfvén's theories were confirmed is Jupiter. Voyager's instruments showed that a sheet of electrical current (disk of dashed lines in this NASA computer simulation) is produced by the rotation of Jupiter's magnetic field (arching lines). Here, scientists could study directly for the first time the homopolar generator Alfvén had proposed forty years earlier to explain a variety of astrophysics processes.

THE GALACTIC CURRENTS

Characteristically, by the time his theories of the solar system were widely accepted, Alfvén had already moved to a new frontier of controversy—applying his plasma models to the realm of the galaxies.

Not surprisingly, his new theories were derived from new developments in the laboratory. In the sixties Alfvén had emphasized that filamentary, pinched currents are essential to the plasma universe, forming the vast power network that connects regions where energy is generated to regions where it is released. Although he had proposed a few concrete models of filamentation on a large scale, there were still some theoretical loose ends.

Alfvén knew that arbitrarily large currents can be carried by plasma along magnetic field lines, although phenomena like double layers can develop resistance that hinder them. But what happens when the current pinches itself together? If the magnetic field of a current were to increase to the point that it surpassed the background field, the electrons would no longer be able to flow along the field lines. If their current were to exceed the limit Alfvén had calculated back in 1939, their own field lines would tangle their paths and destroy the flow. This meant that, in many cases, a filament could not effectively concentrate magnetic fields, except on a small scale. So how could gigantic filaments, obviously carrying vast currents far beyond the Alfvén limit, form on such a large scale?

By the end of the sixties, researchers had developed powerful electron beams capable of carrying currents close to the Alfvén limit, and, by using external magnetic fields, were able to push past it. Heideki Yoshikawa of the Princeton Plasma Physics Laboratory wondered if it was possible to carry still larger currents without an external field; through theoretical calculations he found that it is possible—if the currents were to form a force-free filament. In this case, all the electrons would move along field lines, so their paths wouldn't get tangled up. Yet the field itself, with helical lines wrapped around the straight ones at its axis, is produced by the current, so there is no limit to either the current or the strength of the field.

Yoshikawa published his proof that force-free currents can form arbitrarily large filaments in 1970, and less than a year later his results were confirmed by laboratory experiments.

The ability of the force-free filaments to carry large currents and to concentrate them, producing strong magnetic fields, was the key that solved a number of vexing problems in Alfvén's model. One, for example, was the question of exactly how currents can transfer angular momentum in a developing solar system. Since the forties Alfvén had returned repeatedly to the problems of angular-momentum transfer, because it was obvious that the solar system could never have come to be without an efficient mechanism. At times he had proposed that the MHD waves could do the trick, but always the magnetic field as theorized seemed too weak and unstable. If the currents were evenly spread out, the magnetic field would be too weak to accelerate the gas cloud's plasma. This, in turn, would fail to slow the star's rotation down—rather, the magnetic field might itself be tangled up by the plasma.

With the powerful vortex filaments that experiments and Yoshikawa's theoretical work had shown are possible, the situation changed. Instead of trying to accelerate the protoplanetary clouds with a flimsy net of magnetic fields, the sun would hit them with "baseball bats" of self-pinched magnetic filaments (Fig. 5.12). *Their* concentrated magnetic fields would transfer the momentum efficiently, and the filaments themselves would roll through the plasma without becoming unstable, as a dispersed field would.

In 1972 Alfvén and a colleague, Gustaf Arrhenius, developed a detailed model of solar system formation which uses the filaments—"superprominences," they called them—to transfer the angular momentum. And because the filaments strongly pinch the plasma together, they vastly speed up the planets' condensation. An identical though smaller process produces satellites around each planet.

Alfvén now had, in his and Arrhenius's model of the solar system's formation, a springboard to jump to a larger scale. For if stars and planets can be formed by the action of the filamentary currents, why can't whole solar systems be similarly formed by currents in a galaxy? With Per Carlqvist, Alfvén developed in 1977 the idea that the same pinch effect leads to the formation of dense interstellar clouds—the birthplace of the stars. Again, the process is identical, but this time immensely larger: filaments sweeping through a protogalactic nebula pinch plasma into the building materials of the sun and other stars. Once the material is initially pinched, gravitation will draw some of it together, especially slower-moving dust and ice particles, which will then create a seed for the growth of a central body. Moreover, the filament's vortex motion will provide angular momentum to each of the smaller agglomerations within it, generating a new, smaller set of currents carrying filaments and a new cycle of compression that forms a solar system. (In 1989, this hypothesis, now widely accepted, was definitively confirmed when scientists observed that the rotation axes of all the stars in a given cloud are aligned with the cloud's magnetic field—clearly, a magnetic field controlled stellar formation.)

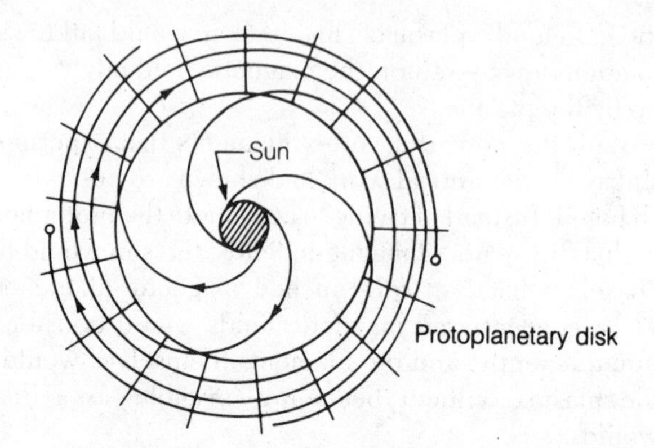

Fig. 5.12a. If the currents around the sun and the magnetic fields associated with them were evenly spread out, they would be "wound up" by the solar system plasma, eventually becoming circular, and thus incapable of transferring momentum from the sun.

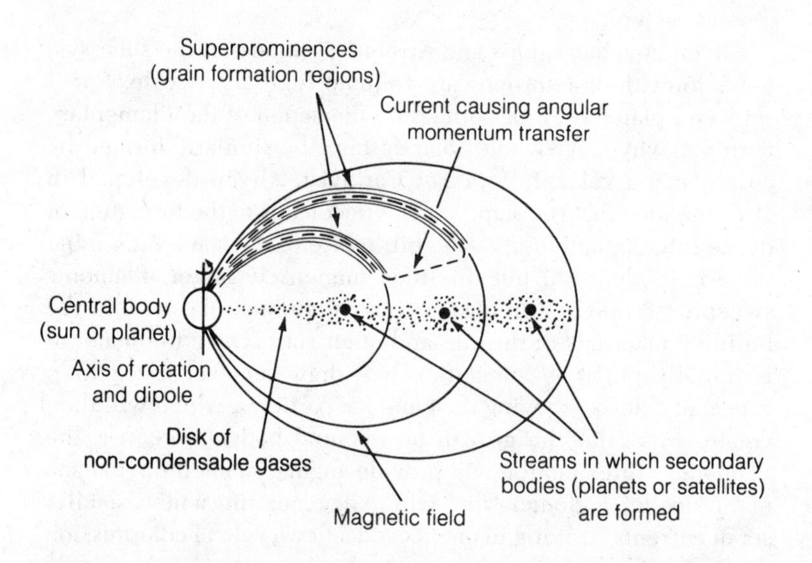

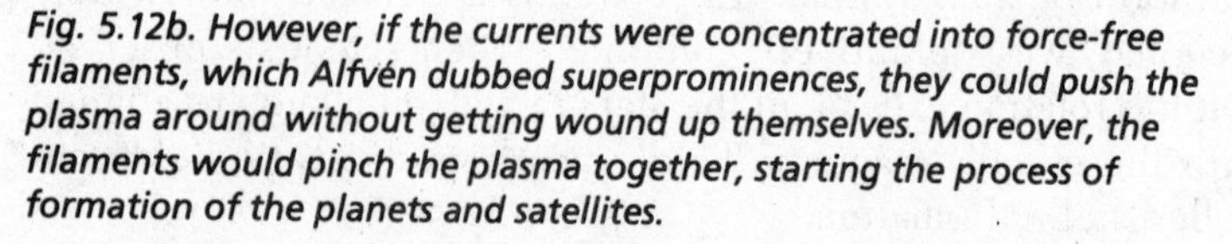

Fig. 5.12b. However, if the currents were concentrated into force-free filaments, which Alfvén dubbed superprominences, they could push the plasma around without getting wound up themselves. Moreover, the filaments would pinch the plasma together, starting the process of formation of the planets and satellites.

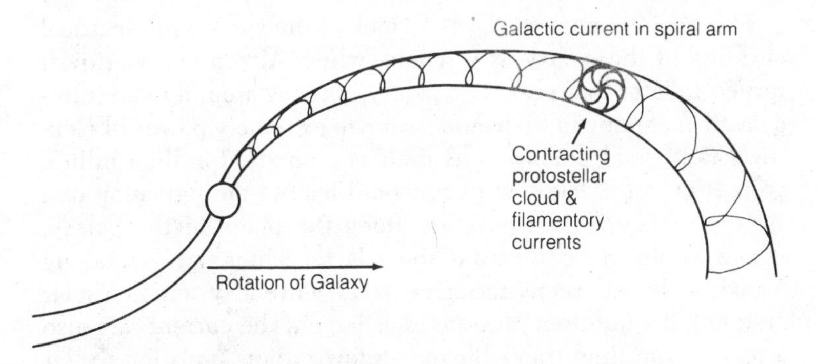

Fig. 5.13. The galactic circuit: here the entire galaxy acts as a disk generator, spinning in an intergalactic field. Currents flow inward on the plane of the galaxy, along the spiral arms, and out along the axis of rotation.

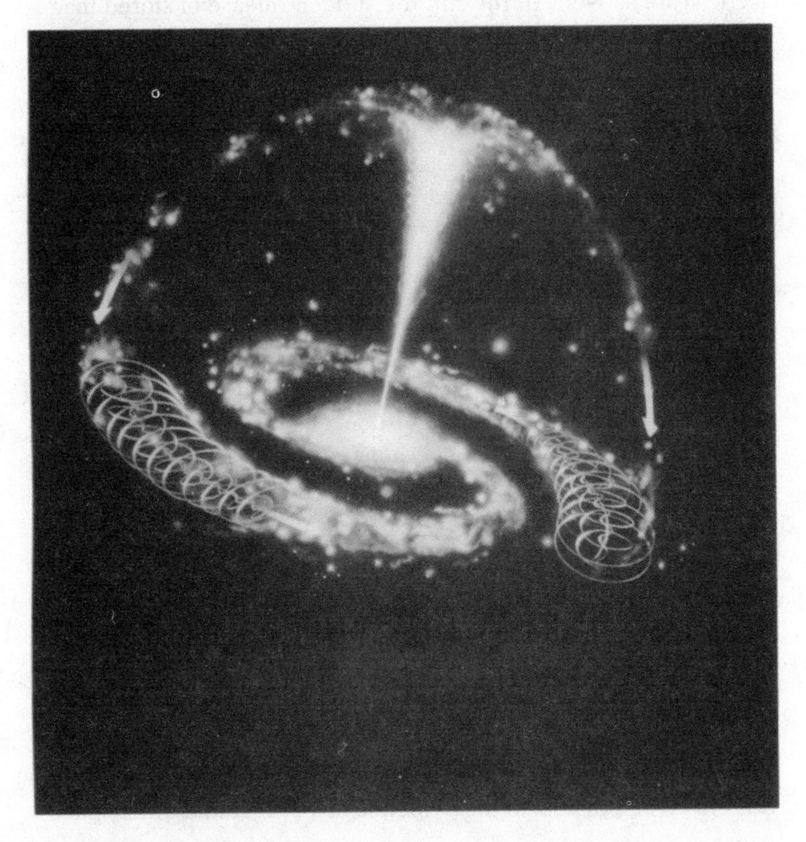

The obvious next step was to look at the system of electrical currents in the galaxy as a whole, which Alfvén and Carlqvist turned to later that year. The galaxy, rotating around in an intergalactic magnetic field, would generate extremely powerful electrical fields and potentials as high as a hundred million billion volts (Fig. 5.13). This, in turn, would lead to the formation of a system of filamentary currents along the plane of the galaxy, which would snake inward to the galactic center and "up" along its axis—almost exactly the same as the current system in a solar system but a hundred million times larger. The currents are also a hundred million times bigger—a few billion amps for a solar system and nearly a billion billion amps for a galaxy.

If double layers occur naturally in all filamentary currents as they pinch together, reasoned Carlqvist, they can occur on a galactic scale as well. In the sun the sudden release of stored magnetic energy in a double layer would produce 10^{34} ergs, but the stored energy of the galactic circuit is a whopping 10^{57} ergs—as much energy as the galaxy produces in thirty million years. The sudden release of this energy would accelerate a beam of electrons and ions along the galactic axis, radiating powerful radio waves as the electrons spiral around the field lines.

This is exactly what radio astronomers had observed for decades in the mysterious "radio galaxies," objects that shoot out single or twin beams of energy to distant clouds in intergalactic space. Since the mid-sixties most astronomers had believed that the radio galaxies' energy source is the same as the broadly similar but still more spectacular quasars—some form of monstrous, large black hole. The received wisdom claimed that as streams of matter circle toward the black hole, accelerating to nearly the speed of light, some are shot out along the rotational axis, forming beams of glowing gas—the radio jets.

Now here were Alfvén and Carlqvist with a far less exotic explanation. The galaxy as a whole acts as an ordinary electrical generator, and double layers, scaled-up versions of an explosion found in Swedish generator plants, are the accelerators of the beams. The jets themselves are just currents of high-energy particles, part of a vast current system surrounding the galaxy.

By the late seventies Alfvén and his colleagues had conquered the solar system with their theories of filaments, currents, and double layers, thanks to the evidence of Voyager and other space probes. Now, however, they were invading the astrophysicists' turf—far beyond the reach of spacecraft. Here, in these outer reaches, the ghost of Alfvén's early work still roamed. Nearly every astrophysicist had learned about plasma from old textbooks, which use the MHD frozen-in field approximation indiscriminately. Naturally, they dismissed Alfvén's currents and generators as impossible, since they contradicted everything known (to the astrophysicists) of plasma in space. Even worse, Alfvén's explanation of radio galaxies raised questions about the very existence of black holes—by 1980 a vast field of theoretical work. In either case, Alfvén was yet again challenging a "reality" constructed from neat mathematical equations—either those of MHD, or for black holes, of Einstein's general relativity.

But that, it turned out, was just the beginning. For Alfvén was now mounting a challenge as well to the very foundation of modern cosmology—he was, for the first time, starting to raise very real questions about the validity of the Big Bang.

NOTES

1. O. Devik, *Blant Fiskere*, Aschehoug, Oslo, Norway, 1971 (cited in A. Egeland and E. Leer, "Professor Kr. Birkeland," *IEEE Transactions in Plasma Science*, Vol. PS-14, p. 666).

2. A. Egeland and E. Leer, *IEEE Transactions*. Ibid.

6 THE PLASMA UNIVERSE

To try to write a grand cosmical drama leads necessarily to myth. To try to let knowledge substitute ignorance in increasingly larger regions of space and time is science.

—HANNES ALFVÉN

Alfvén had been skeptical of the Big Bang from the first time he heard of it, back in 1939. Lemaître, the theory's originator, had come to a Stockholm astrophysics conference to expound his controversial idea of the primeval atom. "I felt at the time that the motivation for his theory was Lemaître's need to reconcile his physics with the Church's doctrine of creation *ex nihilo*," Alfvén recalled years later. His skepticism was deepened by his general approach of closely linking theory to experiment. Lemaître's method of deriving a history of the universe from the predictions of general relativity resembled the elegant mathematics of Chapman: more importance was given to the equations than to the physical plausibility of the theory or to its agreement with observation.

More concretely, though, Alfvén was already deeply involved in the question of cosmic rays, so central to Lemaître's hypothesis. As we've seen,

Alfvén correctly explained the cosmic rays as the product of electromagnetic acceleration, not as the mysterious fragments of the primeval atom or its stellar descendents.

But it was not until over twenty years later, in 1961, that Alfvén himself turned to cosmology. At that time, having enjoyed great popularity in the fifties, the Big Bang theory was apparently on the ropes. Gamow's theory of the origin of the elements had been clearly refuted, and its main rival, the Steady State theory, was also in hot water because radio observations had indicated that the universe is indeed evolving.

Alfvén, elaborating work of an older colleague and teacher, Oskar Klein, entered the dispute with a third alternative. In the fifties, Klein had proposed another cause for the Hubble expansion—his culprit was antimatter, one of the strangest phenomena observed in the laboratory. First predicted theoretically by Paul Dirac in the thirties, it was shortly thereafter discovered in nature; now it is used routinely in the huge particle accelerators of high-energy physics.

Antimatter is identical to ordinary matter except for two things. First, its charge is the opposite of ordinary particles—antiprotons are negatively charged, while antielectrons (called "positrons") are positively charged. Second, and far more startling, when matter and antimatter collide, they annihilate each other, converting each other into pure energy. Conversely, matter and antimatter can be created together from pure energy. Because of this property, antimatter is created when ordinary matter particles collide in particle accelerators (the resulting antiparticles are themselves used for further experiments).

Antimatter has long been a cosmological puzzle, conspicuous in its absence. On earth, antimatter essentially does not exist in nature, because it is quickly annihilated if it does form. But it's hard to tell if it exists elsewhere in the universe, since its properties, when viewed from afar, are identical with those of ordinary matter. Entire stars and solar systems made of antimatter might easily exist without our knowing it. On the other hand, there is no direct evidence that it actually exists elsewhere. Its nonexistence, however, would be an immense puzzle, since on earth when one creates matter from energy, equal quantities of antimatter are produced. Why would this not be true of the universe?

Klein suggested that the Hubble expansion could be explained if the universe indeed consisted of equal quantities of matter and antimatter. If the mixture were sufficiently dilute, collisions between matter and antimatter particles would be rare. But if it became dense, for example through gravitational collapse, the rate of annihilation would rapidly rise, leading to the explosion that the Hubble expansion seems to indicate.

The idea was intriguing. Antimatter certainly has the power to create the huge velocities observed in the Hubble expansion—a pound of the stuff combined with a pound of matter would explode with the energy of a twenty-megaton hydrogen bomb. The matter or antimatter flung out the fastest would move the farthest, producing the Hubble relationship between distance and velocity.

The theory, as Alfvén realized, has significant problems. If matter and antimatter were evenly mixed, why wouldn't they annihilate each other completely, destroying the universe, or at least part of it? How could purely matter regions, like the solar system, be separated out? The problem *again* is how a homogeneous universe can become inhomogeneous, antimatter in one place and matter in another.

In 1961 Alfvén, collaborating with Klein on a revised antimatter theory, proposed a plasma mechanism that can separate the matter and antimatter into distinct regions. The process begins with some gravitational clustering of matter creating a gravitational field. With normal matter such a field will separate out heavier elements from lighter ones—the heavier ones, being slower moving, will "sink." (This occurs only if a sufficient number of collisions equalizes the energy of the heavier and lighter particles; otherwise they all follow the same paths in space.) Normally, protons, though far heavier, can't be separated from electrons, since the opposite charges attract each other. But if this were a mixture of matter and antimatter, the antiprotons and protons, an electrically neutral mixture, would sink while the positrons and electrons, another neutral mixture, would rise. Here, Alfvén introduced a mechanism from his early work on the aurora: if these mixtures moved through a magnetic field, a current would be produced. In both cases the antimatter would flow in one direction, the matter in the other, so a separation would occur (Fig. 6.1).

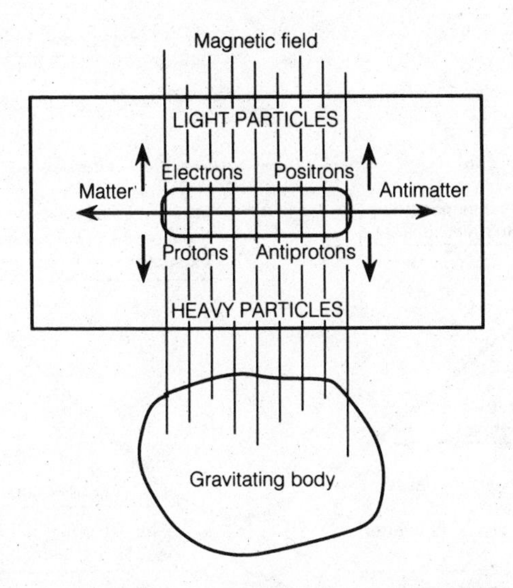

Fig. 6.1.

Alfvén and Klein calculated that such a process can separate matter or antimatter masses large enough to produce a solar system. But larger clouds will grow, they reasoned. When an antimatter cloud bumps into an ordinary-matter cloud, they will not totally annihilate each other; instead, only a thin layer will be annihilated, generating a hot, low-density plasma layer, which will push the clouds apart. (This is like the "Leidenfrost layer" of steam that suspends a droplet of water on a hot frying pan, allowing the droplet to scoot around so freely.) Clouds of opposite type will repel each other, but will combine with similar-type clouds, producing ever-larger masses of ordinary matter or antimatter.

However, according to their theory, as the observable universe contracts gravitationally its density will increase and with it the velocity of the cloud collisions. Finally, when the universe is about a hundred million light-years across (one-hundredth of its present size), the collisions will be so violent that the Leidenfrost layers will be disrupted, and some of the clouds will mix in an enormous explosion. The surviving unmixed clouds will be thrust outward, in the Hubble expansion (Fig. 6.2).

Thus, in Alfvén and Klein's scenario, only a small part of the universe—that which we see—will have first collapsed and then exploded. Instead of coming from a singular point, the explosion

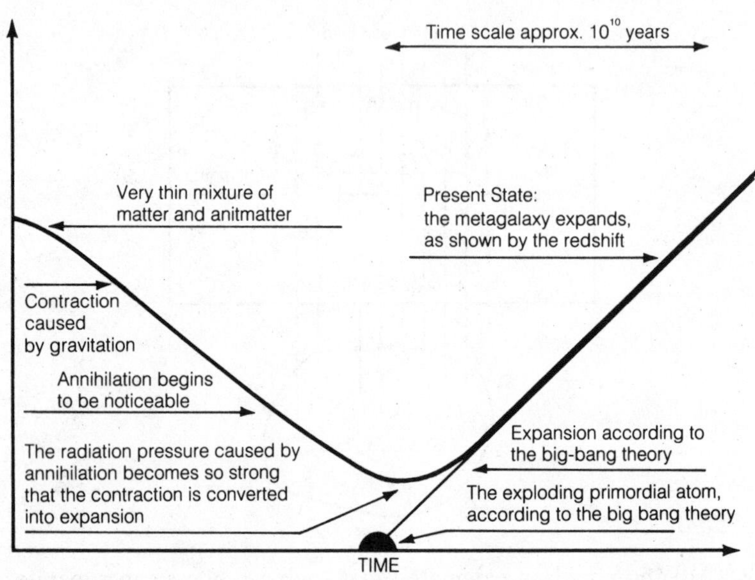

Fig. 6.2.

comes from a vast region hundreds of millions of light-years across and takes hundreds of millions of years to develop—no "origin of the universe" is required.

Alfvén and Klein had outlined a bold alternative to both the Big Bang and the Steady State—but it was only an outline, and there was an enormous leap between the plasma physics of the lab and that of the cosmos. In 1962 even Alfvén's aurora theory was not yet accepted. And while his plasma theories were being proven in fusion labs, the antimatter or "ambiplasma" theory, as he calls it, was basically without laboratory support. Even today antimatter is produced in such tiny quantities that lab tests of Alfvén's hypothesis are still impossible.

The new alternative briefly received attention because both the Big Bang and the Steady State were in eclipse. But with the resurgence of the Big Bang in the mid-sixties the Klein-Alfvén cosmology was forgotten. For a decade and a half astronomers did, in fact, have reason to pursue the Big Bang, since the microwave background and the light-element abundance did seem to confirm it. And Alfvén's theories, for once, seemed just as speculative as those of the opposition, if not more so.

THE HIERARCHICAL UNIVERSE

By the late seventies, the Big Bang was in trouble again—observations had shown that the universe is both too clumpy and too diffuse for the Big Bang to account for. The problems were related. In a diffuse cosmos, gravity could not act fast enough to bring matter together into the galaxy clusters that we see today. And in a diffuse cosmos, with omega far less than 1, the Big Bang has huge inconsistencies.

Alfvén had returned his attention to cosmology. In the intervening twenty years, he had extended his plasma theories up to the scale of the galaxies, and observations from space probes rapidly confirmed his theories, at least on the scale of the solar system.

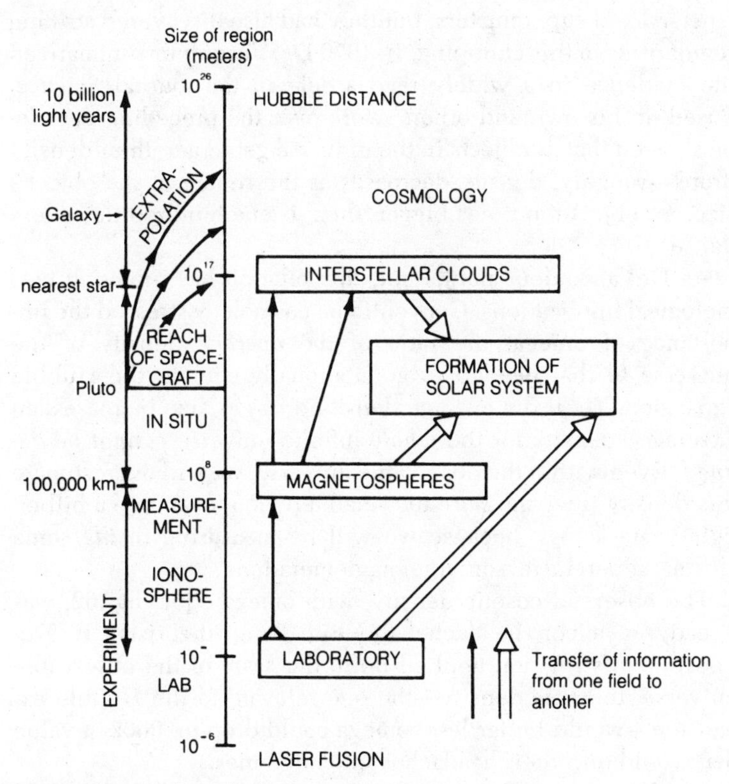

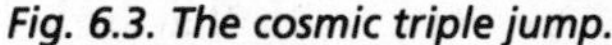

Fig. 6.3. The cosmic triple jump.

Given the increasing scale of his theories, he developed the idea of the "cosmic triple jump" (Fig. 6.3). With it, from the laboratory scale of a few centimeters one can extrapolate to the scale of the earth's magnetosphere, ten billion centimeters or a hundred thousand kilometers. Then, from observation on that scale, one can extrapolate another factor of a billion to the scale of light-years—relevant to the galaxy—and from *these* observations, one can then jump a third factor of a billion to the scale of the observable universe.

In returning to cosmology, he immediately focused his attack on the weakest point of conventional cosmology—the inhomogeneity of the universe. Here, he called attention to the work of Gerard DeVaucouleur.

By the seventies, astronomers knew not only that the universe is inhomogeneous, clumped into a hierarchy of stars, galaxies, clusters, and superclusters, but they had also discovered striking regularities in this clumping. In 1970 DeVaucouleur summarized the evidence in a widely read article in the journal *Science.* Based on his own and others' work over the preceding decade, he showed that as objects in the universe get larger, their density drops—roughly, density decreases as the square of the object's size. An object ten times bigger, then, is one hundred times less dense.

As DeVaucouleur points out, this relation has profound cosmological implications. Conventional cosmology stressed the importance of omega, the ratio of the *average* density of the universe to that needed to gravitationally contain the Hubble expansion. Yet if the average density drops as size increases, an "average" density for the *whole* infinite universe *cannot be defined.* Even within the observable universe, we will overestimate this density if we measure too small a region, even one a billion light-years across, because we will be measuring *inside* some cluster, supercluster, or larger agglomeration.

The observed cosmic density, with omega equaling .02, was already a major headache for Big Bang theorists. If DeVaucouleur's relation applied up to the scale of the observable universe, the *true* density—the one relevant to the Hubble expansion—would be far less: omega could drop to .0002, a value that would turn their headaches into migraines.

As Alfvén gleefully pointed out, in a universe with that little matter, gravitation will be so weak that the difference between Newtonian gravity and general relativity will be of little account. For all practical purposes general relativity, the foundation of conventional cosmology, can be *ignored!* In essence, the DeVaucouleur relationship places an upper limit on the escape velocity of any gravitating body. Newton's laws dictate that the square of the escape velocity is proportional to the density of an object times the square of its size or radius (or proportional to its mass divided by its radius—the equivalent thing, since density is mass divided by volume or radius cubed). But DeVaucouleur had shown that, over a range from stars to superclusters, density times radius squared remains *constant*. As a result, there is only a narrow range of escape velocities, with an upper limit of at most one or two thousand kilometers per second (Fig. 6.4). But general

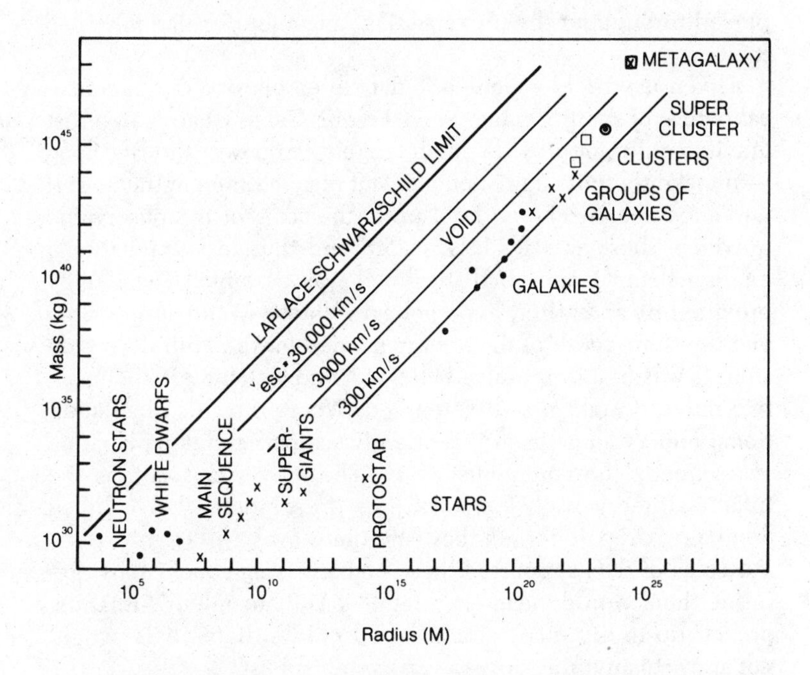

Fig. 6.4. If the density of an astronomical object is plotted against its radius, an object with a constant orbital or escape velocity falls along the diagonal lines. When actual objects are plotted (crosses and dots), virtually none falls above about 1,000 km/sec, far below the theoretical limit of the speed of light, 300,000 km/sec.

relativity makes a major difference only when the escape velocity approaches the speed of light—three *hundred* thousand kilometers per second. DeVaucouleur's discovery shows that nowhere in the universe—except perhaps near a few ultradense neutron stars—is general relativity more than a subtle correction.

As DeVaucouleur himself points out, the relation also eliminates dark matter. If dark matter existed with the density cosmologists assumed, its strong gravitation could whip objects around faster than DeVaucouleur's speed limit of a few thousand kilometers per second. The velocity of an object orbiting in a gravitational field is always a large fraction of the escape velocity. This, in turn, depends on the size and density of the gravitating body. Bodies of dark matter, the size of superclusters or larger, could produce velocities in galaxies far higher than those observed. So, unless the dark matter was somehow smoothly dispersed throughout the universe, DeVaucouleur's relations ruled it out.

Alfvén used DeVaucouleur's data to emphasize the paradoxical nature of the Big Bang. Years before, Gamow had feared that gravity could not provide enough energy to power the Big Bang—to make the universe "bounce" out of an earlier contraction. If DeVaucouleur's relation held up to the scale of the observable universe, there was far less matter, and thus less gravitational energy, than Gamow had imagined was possible. The energy provided by an earlier, hypothetical collapse would supply only one ten-thousandth of the energy needed for the Hubble expansion. It was like dropping a ball a foot and watching it "bounce" two miles straight up. Alfvén asked, Where did this huge additional energy come from? "Evidently a supernatural explanation is assumed," he comments wryly. The universe is not, as the inflation theory was soon to say, a "free lunch," but a lunch Someone had paid for ten thousand times over.

Reams of data supported these embarrassing conclusions. Refuting them would mean claiming that DeVaucouleur's relation, proven up to distances of a hundred million light-years, would not apply to anything larger—a risky bet at best.

For Alfvén, the low cosmic densities implied by DeVaucouleur's relation were a big help to his antimatter cosmology, which assumes that the observable part of the universe had been perhaps a hundredfold smaller in size than it is at present.

With the higher density estimate, the "escape velocity" of such a small universe would have been nearly the speed of light, bringing all the complications of general relativity into play. The much lower density, however, permits an escape velocity of about 20 percent of the speed of light, low enough to ignore general relativity and its curved spaces.

In 1977 Alfvén introduced a revised version of his and Klein's ambiplasma theory, incorporating DeVaucouleur's hierarchical cosmology. First, Alfvén noted that the hierarchy of entities fit neatly into his theory that only a part of the infinite universe participated in the Hubble expansion. This part is just one in the series of hierarchical objects—one larger than the observed superclusters. Alfvén dubbed this entity, coextensive with the observable part of the universe, the "metagalaxy."

As in the earlier version, this metagalaxy will contract gravitationally until matter and antimatter, previously separated, mix and explode. But rather than a single explosion, Alfvén postulated a series: as each subfragment expelled from the metagalactic explosion cools and condenses, it too reaches the critical instability limit and explodes, in turn, into smaller fragments (Fig. 6.5).

Alfvén argued that this model, which he dubbed the "fireworks model" (in ironic imitation of Lemaître's earlier nomenclature) can explain DeVaucouleur's speed limit. If an object's escape velocity stays within this speed limit, its internal motions are too gentle to disrupt the Leidenfrost layers. But as an object composed of separate clouds of matter and antimatter collapses gravitationally, its internal motions speed up, the collisions between the clouds become more violent, the Leidenfrost layers are disrupted, and explosions ensue. Thus any object existing today must obey DeVaucouleur's speed limit, and the related mass-radius or density-area relation. The fireworks explosion would naturally create a hierarchy of such objects.

The fireworks model can also account for the fact that the Hubble expansion is roughly symmetrical at all scales—that is, the velocity of all galaxies at a given distance is *roughly* the same. Alfvén noted that if the symmetry were exact—that is, if all galaxies at the same distance were to recede at *exactly* the same velocity, and if galaxies at different distances were to have the same ratio of velocity to distance—then it would seem that they

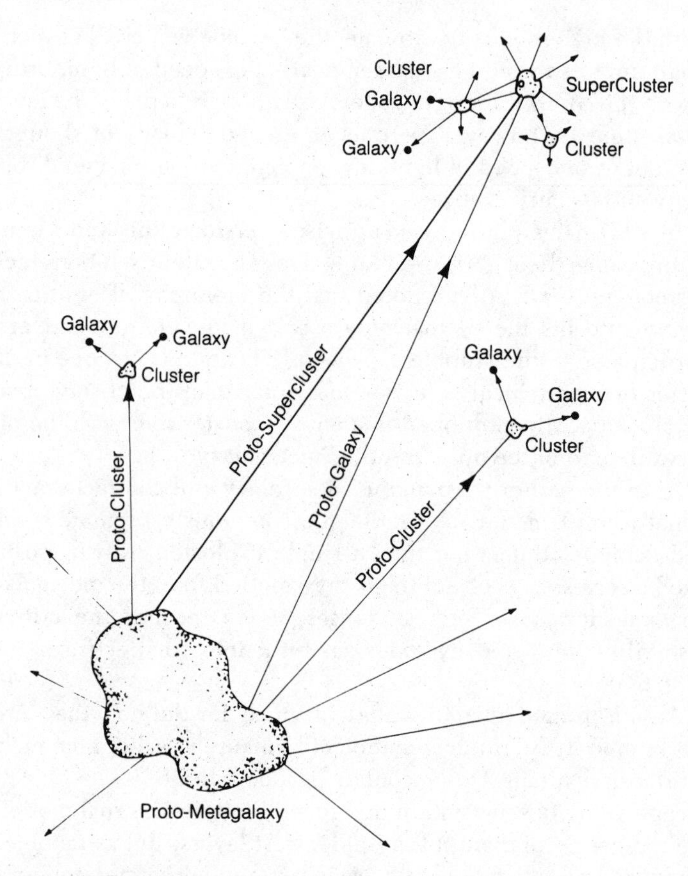

Fig. 6.5. The fireworks model.

were all receding from one point at one moment—a Big Bang. However, the various observed Hubble velocities vary by at least 10 percent, depending on distance and direction. Thus the smallest radius of the metagalaxy that could be justified on the evidence is about 10 percent of the universe's present radius, one or two billion light-years. Such a loose symmetry on the large scale can easily be explained by the explosion of a metagalaxy a few hundred million light-years across.

But a single explosion, as in Alfvén's old model, could not even produce rough symmetry on a small scale. Since some galaxies will be expelled from the nearer side of the metagalaxy and some from the center, there will be a random mix of velocities within any small volume. Faster galaxies from the center will overtake slower ones from the edge. In volumes smaller than a hundred million light-years across, the Hubble effect would be washed out.

The fireworks model would solve this problem. In each local volume, a smaller explosion would impose its own local Hubble relation. Even locally, galaxies moving faster from a local explosion will move farther away, generating the redshift-distance relation on all scales.

ANTIMATTER OR NOT?

One key aspect of Alfvén's cosmology has been a hot debate and remains an open question. Are there substantial amounts of antimatter in the universe? Conventional wisdom said no, none, and in the sixties there was no evidence that this isn't the case. But in 1976 Carlqvist and Bertil Laurent had come up with some possible signs of antimatter.

They took Alfvén's new model and asked what happens to the electrons and positrons that absorb the tremendous energy produced by annihilation. They calculated that most of the particles will have between 10 and 100 MeV (million electron volts)—equivalent to temperatures of between one hundred billion and a trillion degrees—and at such high energy will travel at very close to the speed of light. The particles would occasionally collide with photons of starlight and transfer a small amount of their energy to the photons, converting them to X-rays.

Astronomers had already observed a universal background of such X-rays, similar in its isotropy or smoothness to the microwave background. Carlqvist and Laurent compared the spectrum of the X-ray background to that which should be produced by the electrons and positrons, and found an excellent match. This spectrum requires a density of about one electron or positron per every thousand cubic meters of space—the same amount needed to produce the Hubble expansion's energy at the matter density Alfvén had extrapolated from the DeVaucouleur relation.

This evidence of antimatter's existence was published in 1976 in the prestigious journal *Nature* and brought an immediate reply from orthodox cosmologists. Gary Steigman attacked the antimatter theory, arguing that the gamma rays produced with the electrons and positrons would be millions of times more intense than had actually been observed. In addition, the electrons and positrons themselves would annihilate each other from time to time, releasing .5 MeV gamma rays in greater numbers than had been observed. Since the gamma rays observed are far fewer, concludes Steigman, antimatter must be extremely rare.

Carlqvist and Laurent answered in the same issue of *Nature*, noting that Steigman assumes a *homogeneous* universe (a critical point in the debate between Alfvén's cosmology and that of the Big Bang). With a metagalaxy explosion, occurring in a specific, limited area, ten billion years or more ago, the gamma rays from the initial annihilation would have escaped the metagalaxy aeons ago.

They argue further that the electron-positron annihilations depend on the matter density of the universe. Given the density implied in Alfvén's model and DeVaucouleur's data—one hundred times less than that used by Steigman—the gamma ray intensity agrees closely with the background observed. So both the X-ray and the gamma ray background are neatly accounted for.

The debate did not end there, for Steigman later repeated his objections on a galactic scale. If antimatter exists on a galactic scale, he contended, then collisions between matter and antimatter galaxies would release far more gamma rays than observed. Here, the key question was how effectively the Leidenfrost layers can separate matter and antimatter regions. Alfvén, though, had made no detailed calculations about how fast the layers form in a collision, a critical question, because the faster they form, the fewer gamma rays will be produced and the less annihilation will occur before the two regions of plasma bounce apart.

Although Bo Lehnert had made preliminary calculations on this problem, the main work was done by a physics professor at San Diego, William Thompson. He himself was skeptical of Alfvén's thesis but wanted to examine it, at least in theory. His work indicated that the Leidenfrost layers would form in just a few years—an instant, considering intergalactic collisions take hundreds of millions of years. When an ordinary-matter galaxy collides with an antimatter one, their stars will pass by each other without collisions, since they are so widely spaced in any case: the plasma in each will be stopped by a Leidenfrost layer and bounce off into space without releasing any gamma rays. Thompson concluded that no hard evidence contradicts the existence of antimatter, although he remained skeptical of any evidence implying that it does exist. In this, as in other cosmological questions, more observation is crucial.

THE COSMOLOGICAL PENDULUM

While Alfvén and his colleagues were developing an alternative cosmology, he opened a broad attack on the methodological and philosophical underpinnings of the Big Bang. In 1978 he formulated the broad thesis that I have elaborated here—that the Big Bang is a return to an essentially mythical cosmology. Over the millennia, Alfvén argued, cosmology has alternated between a mythical and a scientific approach—an alternation he termed the cosmological pendulum.

The mythical approach begins from certain assumptions about the "initial conditions" and proceeds forward to explain the universe from that beginning. The assumptions derive from some authority—religious, philosophical, mathematical, or aesthetic. The scientific approach, in contrast, begins from observation of the here and now, working outward and backward from this basis. "The difference between myth and science is the difference between divine inspiration of 'unaided reason' (as Bertrand Russell puts it) on one hand and theories developed in observational contact with the real world on the other," Alfvén writes. "To try to write a grand cosmical drama leads necessarily to myth. To try to let knowledge substitute ignorance in increasingly larger regions of space and time is science."

The Ptolemaic system—based on the unquestioned acceptance of the unchanging heavens, the centrality of earth, and the necessity of perfect circular motion—is a mythical cosmology. The Copernican system, as perfected by Kepler and Galileo, is an empirical one: ellipses are not more beautiful than circles, but they *are* the planets' orbits.

Current cosmology represents a return to Ptolemaic myths, Alfvén believes: "Both the Ptolemaic and Big Bang cosmology started from unquestionably correct and extremely beautiful philosophical-mathematical results. No one can study the Pythagorean science comprising the mathematical theory of music and the theory of regular polyhedra without being immensely impressed. The same holds for the theory of relativity. . . ." But neither the Ptolemaic nor the Big Bang cosmology corresponds to observation. In particular the Big Bang's use of general relativity is valid *only* if the universe is dense enough to be "closed," or near to that. Otherwise, as observation shows, general relativity and its whole mathematical approach are a mere nuance on the cosmological scale. For similar reasons—as we have seen—Big Bang theorists' initial predictions were wrong: the microwave background temperature is lower than predicted, by an order of magnitude, and the universe is in *no way* homogeneous.

Since it is without empirical support, Alfvén concludes, "the Big Bang is a myth, a wonderful myth maybe, which deserves a place of honor in the columbarium which already contains the Indian myth of a cyclic Universe, the Chinese cosmic egg, the Biblical myth of creation in six days, the Ptolemaic cosmological myth, and many others." The underlying method of relativistic cosmology is based on flawed philosophy:

> The reason why so many attempts have been made to guess what was the state several billion years ago is probably the general belief that long ago the state of the Universe must have been much simpler, much more regular than today, indeed so simple that it could be represented by a mathematical model which could be derived from some fundamental principles through very ingenious thinking. Except for some vague and unconvincing reference to the second law of thermodynamics, no reasonable scientific motivation for this belief seems to have been given. This belief probably emanates from the old myths of creation. God established a perfect order and "harmony" and it should be possible to find which principles he followed when he did so. He was certainly intelligent enough to understand the general theory of relativity, and if He did, why shouldn't He create the Universe according to its wonderful principles?

Worst of all, this approach allows theory to rule over observation, like the Ptolemaic astronomers who refused to look through Galileo's telescope. Today cosmology is in the hands of scientists who

> had never visited a laboratory or looked through a telescope, and even if they had, it was below their dignity to get their hands dirty. They looked down on the experimental physicists and the observers whose only job was to confirm the high-brow conclusions they had reached, and those who were not able to confirm them were thought to be incompetent. Observing astronomers came under heavy pressure from theoreticians. The result was the development of a cosmological establishment, like that of the Ptolemaic orthodoxy, which did not tolerate objections or dissent.[1]

As an alternative to this orthodoxy, Alfvén advocates a return to a strictly empirical approach, one that doesn't sweep inconvenient observation under the rug when it conflicts with dogma. "The difference between science and myth," he wrote, "is the difference between critical thinking and the belief in prophets, between *'De omnibus est dubitandum'* (Everything should be questioned—Descartes) and *'Credo quia absurdum'* (I believe because it is absurd—Tertullian)."

But Alfvén's remained in the early eighties a voice crying in the wilderness. His antimatter cosmology still had some significant problems, though no more than the Big Bang. But more to the point, it represented a huge leap beyond what was known and an even larger leap beyond what was accepted. As of 1980 Alfvén's cosmic triple jump remained a research program, not a reality: up to the scale of the solar system its detailed theory had been confirmed by observations from Voyager and other space probes. But at the galactic scale there was only Alfvén's theory of radio galaxies' origins, which remained unelaborated and unconfirmed by observation. Because the second jump was not more than half complete, the third jump was more problematic. Alfvén and his colleagues had not developed antimatter cosmology to the point where detailed comparison with observation could be made, and they were the only ones working on it. Nor had antimatter theories been extrapolated from the laboratory—no one *had* produced enough antimatter.

Alfvén's handful of plasma cosmologists remained an isolated group in Sweden, but this was hardly a new situation for Alfvén: twenty years earlier his ideas on the solar system had been dismissed, and now they held the field, with researchers around the world elaborating them, and new observations supporting them every month. For cosmology, change was on the way.

GALAXIES IN A COMPUTER

The first critical steps in the second cosmic jump, from solar system scale to galactic scale, were taken by Anthony Peratt. Peratt had been one of Alfvén's graduate students in 1969 and 1970, when he assisted Alfvén and Arrhenius with their book on the origin of the solar system. Ten years later, in 1979, while working at Maxwell Labs, an aerospace-defense contractor, he experimented with a device called Blackjack V—the largest pulsed-power generator in the world at the time, capable of momentarily producing ten trillion watts of electrical output, five times more than the world's *entire* generating capacity. Blackjack pumped this enormous power through wires which instantly vaporized into filamentary plasmas, giving off an intense burst of X-rays. This was the machine's purpose: the X-rays simulated the effects of an exploding hydrogen bomb on electronics and other equipment.

Peratt was studying the plasma in Blackjack with extremely high-speed X-ray photography. What he saw surprised him: initially the plasma filaments moved toward one another, attracted by one another's magnetic fields. But then they merged into a tight helix—from this spiral form the most intense X-rays emanated.

"I learned from the literature that Bostick had observed the formation of objects like spiral galaxies back in the fifties," Peratt recalls. "And of course from my work with Alfvén I knew of his theories of the importance of filamentary currents and plasma in the cosmos. Here were the two of them together—filaments forming spirals and something that could be studied right in the laboratory" (Fig. 6.6).

Even the best measurements of a plasma are limited, however, as Peratt well knew. By a fortunate coincidence, less than a year earlier Oscar Buneman, a leading plasma physicist at Stanford, had developed SPLASH, a new program for simulating plasma. Instead of approximating plasmas as fluids—the MHD approach, whose limitations Alfvén had emphasized—the Buneman model is an accurate three-dimensional particle-in-cell approach. (This means that the computer follows each electron or ion step by step as it moves from one "cell" or grid point to the next, in accor-

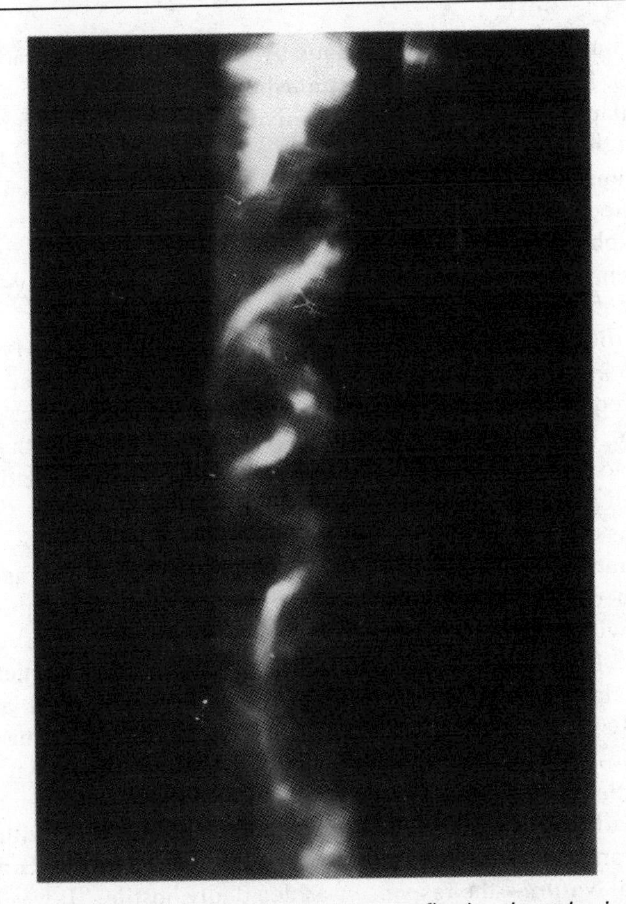

Fig. 6.6. In the Blackjack V experiments, current flowing through wire filaments creates a plasma, which twists itself up into a spiral form. In this photograph the plasma is seen by the X-rays it emits.

dance with the forces applied to it by all other particles and fields present.) Peratt described it as "the Rosetta stone for plasma." He designed a simulation of the Blackjack filaments, starting from the simplest case of two filaments side by side, each with an electrical current and magnetic fields running up its axis. (Peratt knew that the magnetic field must be axial if the huge currents, more than five million amps, were to flow at all.)

"I was due to give a talk in Boston at an American Physical Society meeting on both the experimental and simulation results.

The simulations were being done by one of Buneman's grad students, Jim Green. He had just finished a computer movie of the simulation results and this was delivered to me at the conference hotel the night before my talk. I sat on the bed in my hotel room and unrolled the 16mm computer images for a first look. I was stunned." They showed the filaments in cross section: two circular blobs moved slowly toward each other, then started to rotate, stretch, and merge into a perfect miniature spiral galaxy—the *exact* shape of a typical "grand design" spiral.

"I didn't know what to do with it," Peratt remembers. "My first reaction was, 'Wow, the code is wrong.' But I went ahead and showed it in my talk anyway." When he arrived back in San Diego, he called Alfvén, who quickly invited him over. Although he had never been a fan of either computers or simulations, Alfvén was impressed and encouraged Peratt to apply the simulations directly to cosmic problems.

Peratt ran more simulations, varying parameters such as the distance between the two filaments. He then compared the results with those of real galaxies. "Once I found Halton Arp's *Atlas of Peculiar Galaxies,* it was beautiful. I could link up each picture of a galaxy with some stage of one of my simulations and I knew exactly what forces—electromagnetic forces—were shaping the galaxies" (Fig. 6.7). Excited, he submitted the findings, with illustrations, to *Physics Review Letters,* the high-visibility journal that had published his initial results on the simulations (without mention of galaxies). The paper was rejected. He tried the British journal *Nature*—the response was decidedly hostile. "It was just miserable. They said, 'No, no, there can't be magnetic fields on these scales, it takes tons of copper to carry all this current, and so on.' But now I was really interested."

By now Peratt was specializing in three-dimensional plasma simulations, so he took a job at Los Alamos National Laboratory, where he had access to the world's largest concentration of supercomputers. With the help of James Green of Stanford and Charles Snell, he began his galactic simulations. Alfvén had already postulated that force-free filaments, pinched together from currents flowing toward the center of a galaxy, can initiate star formation, so Peratt reasoned that larger filaments—stretching for hundreds of millions of light-years—can similarly pinch together vast clouds of plasma to initiate galaxy formation. He therefore

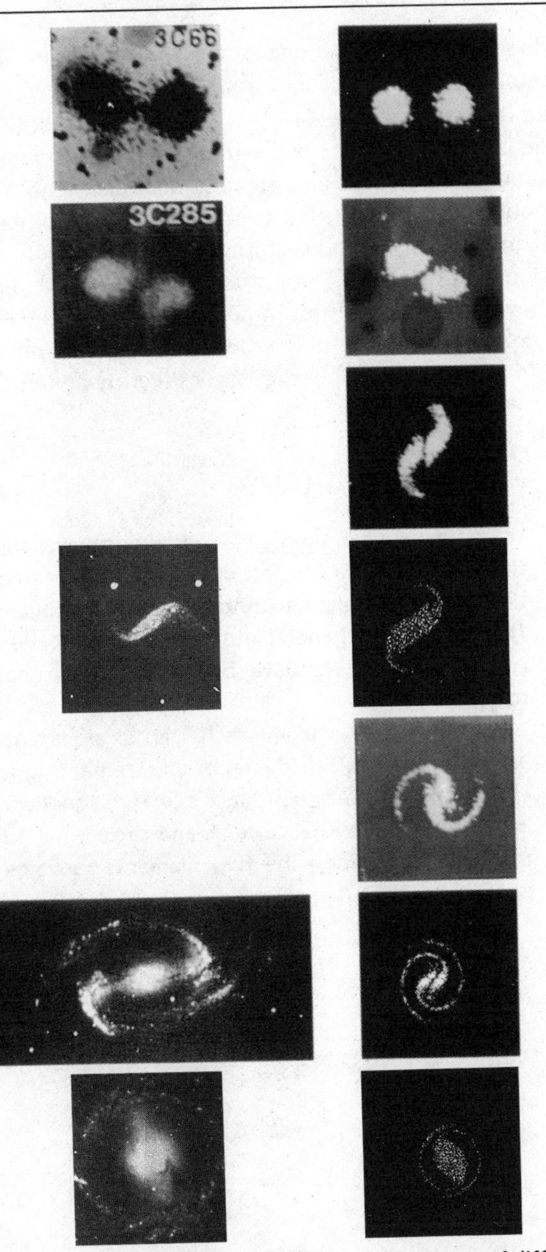

Fig. 6.7. Peratt's computer simulations (right) mimic the shape of different types of real galaxy (left). From top, galaxies are 3C66, 3C285, NGC3187, NGC1300, and M95.

modified Alfvén's model of a galaxy spinning in a magnetic field, producing inward flows of current.

Peratt knew from the Blackjack experiments that the phenomena he had observed derive from the interaction of more than one filament, so he created two gigantic filaments with a density and magnetic field typical of a galaxy. He also eliminated gravitational interaction entirely—the galaxy was entirely confined by electromagnetic forces. This was an unwarranted simplification, he knew, but relatively unimportant, because the key was the rotation of plasma in a magnetic field. The whirling clouds can be held together by either gravity or electromagnetic forces.

HOW A GALAXY FORMS

What happened in the simulations is worth explaining in detail. It's simplest to start off by looking at the earlier simulation, where the plasma stretches along the whole filament. This is the more general situation, since it shows how the vortex filaments we have been discussing actually come to be.

Each of the two filaments has an axial current running along the magnetic field lines of an external magnetic field. Their own currents also produce circular magnetic fields. The interaction of the axial currents and circular fields produces the pinching action—the two filaments converge toward each other (Fig. 6.8a).

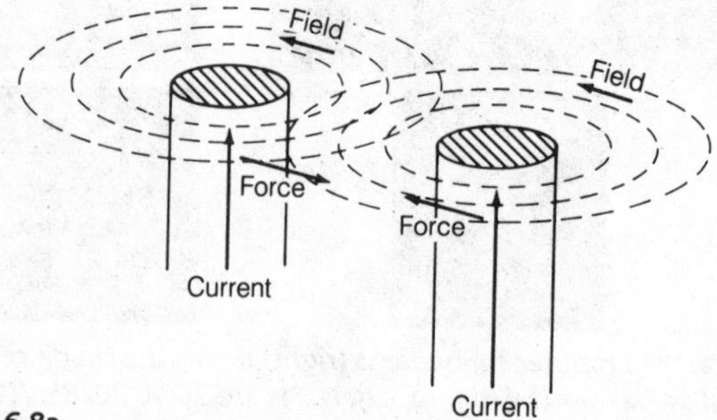

Fig. 6.8a.

As they move through the vertical magnetic field, electrons are pushed to the right and ions to the left (Fig. 6.8b),

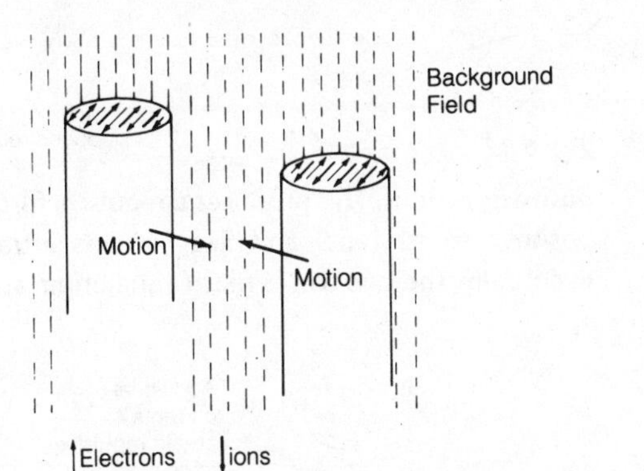

Fig. 6.8b.

producing additional currents. The electron excesses on the right are forced upward as they move in, through their interaction with the circular or azimuthal field of the other filament; the ions on the other side are forced downward. Since the electrons move far faster than the ions, the axial currents in both filaments shift off to the side (Fig. 6.8c). Now the force

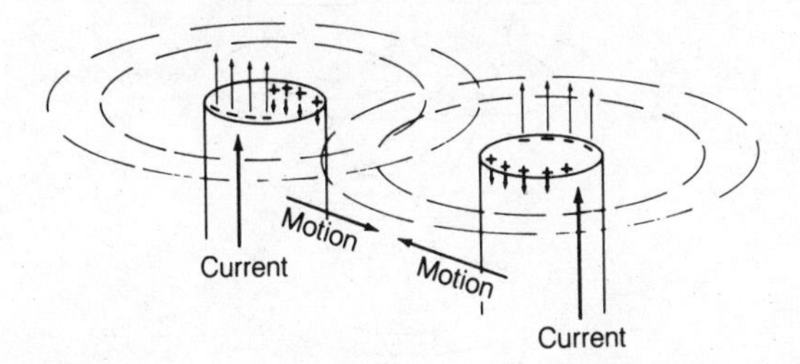

Fig. 6.8c.

between the two currents still attracts them along a straight line. But this is no longer a line between the centers of the filaments, so they start to move obliquely, rotating around each other as they move inward, producing a twisted pattern (Fig.6.8d). Finally, when the two filaments get closer and move faster around each other, the excess charges on the inner edges of the filaments begin to move past each other in op-

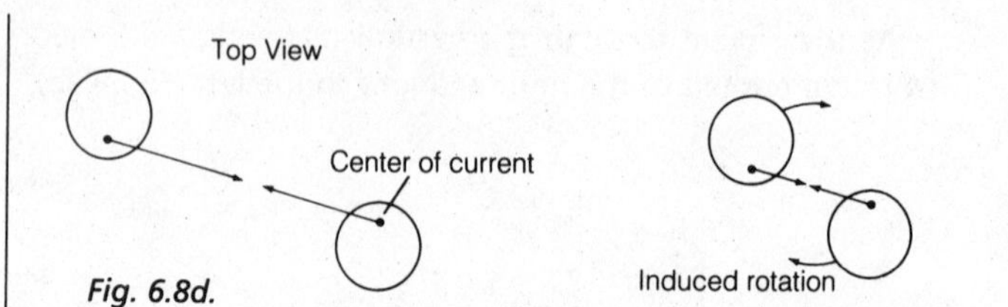

Fig. 6.8d.

posite directions. This produces a repulsive force, because opposite currents repel and like currents attract (Fig. 6.8e). Eventually, the two forces reach equilibrium and the contrac-

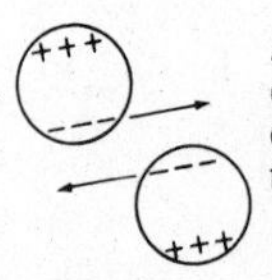

Fig. 6.8e.

tion stops. The two filaments have been twisted into a single large filament, which is now rotating, ready perhaps to merge with another (Fig. 6.8f).

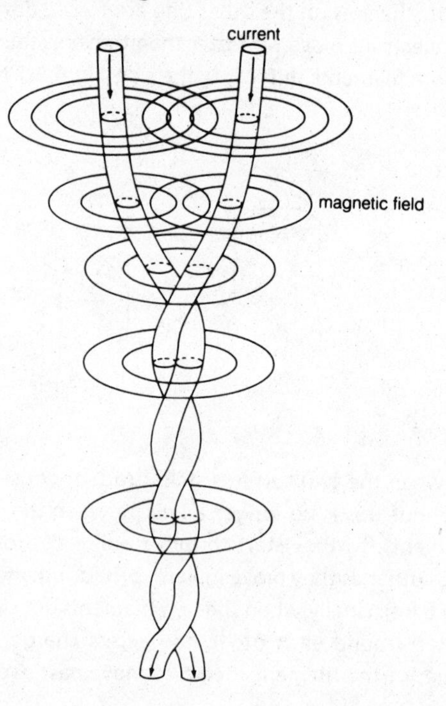

Fig. 6.8f.

The situation with the plasmas, the galactic-size plasma clouds of the simulation, is somewhat more complex. The two clouds, pushed by the pinch force, approach each other. In this case too the ions and electrons are forced in opposite directions, producing a downward force on one side of the blob and an upward force on the other side (Fig. 6.9a). These

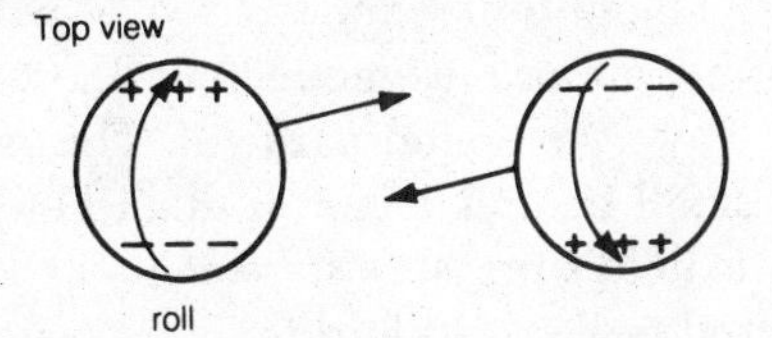

Fig. 6.9a.

unbalanced forces start the clouds rolling. Since the electrons are pushed far faster than the ions, a circular current begins to flow in the same direction as the rolling motion. As in the case of the filaments, the additional currents push the blobs obliquely, so they roll around one another. They distort and stretch in the process, because the forces on the inner parts are more powerful than on the other parts (Fig. 6.9b). The

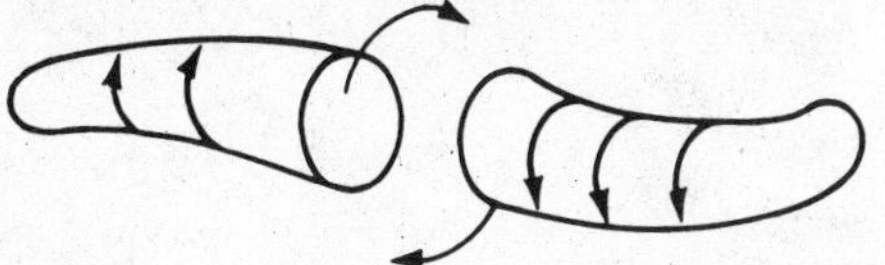

Fig. 6.9b.

two rolling currents create opposing magnetic fields, so the blobs repel each other, bouncing off a cushion of magnetic field (Fig. 6.9c). The pinch force prevents them from separating again, and they end up as a single rotating object—the galaxy.

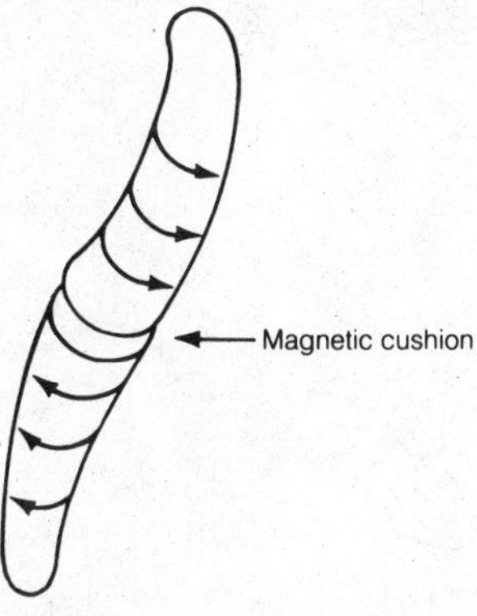

Fig. 6.9c.

Peratt and Green scaled up their model's results from laboratory scale to galactic scale. The currents were no longer millions of ampere but 10 million trillion amps, and the filaments were 300,000 light-years apart instead of a few millimeters (see Figures 6.8 and 6.9). The new simulations showed something very curious: electrons trapped in a central "cushion" in the magnetic field radiate great amounts of radio waves by synchrotron process (that is, by spinning around magnetic field lines, they are forced to radiate, as is any accelerated particle). The radiation is confined to a very small central region of high fields. At the same time the magnetic field's pressure starts to break small plasmoids off from the central region and hurl them outward. They rapidly form a powerful beam of energy, coming out from the center of the protogalaxy in both directions (Fig. 6.10).

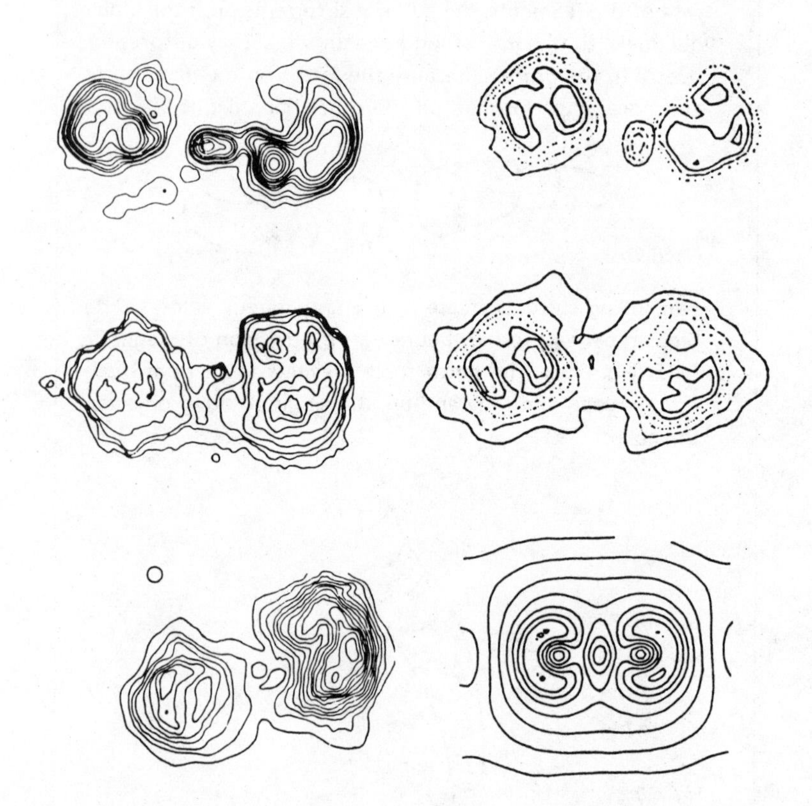

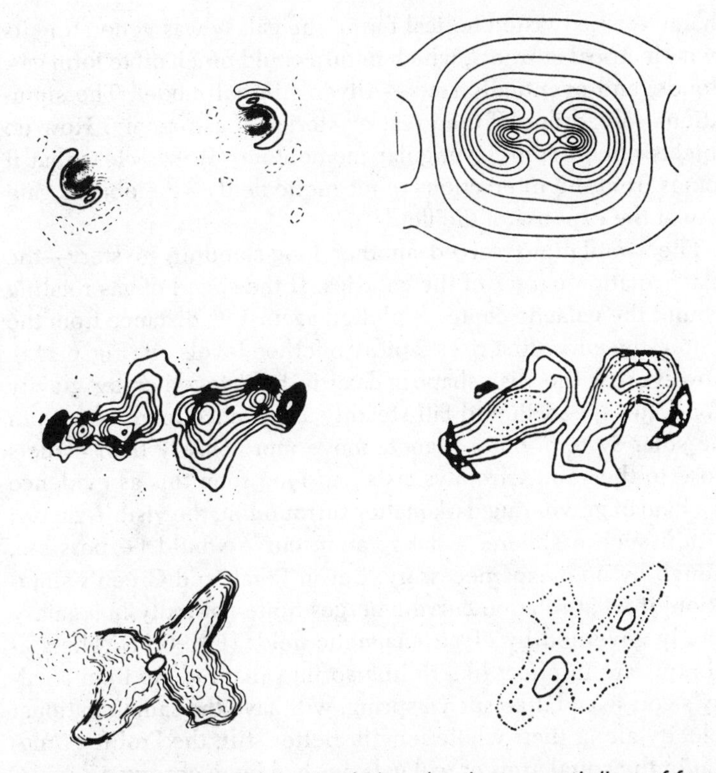

Fig. 6.10. In this version of the simulation, showing magnetic lines of force, the two blobs have merged, creating two high field areas around a lower field central sump. The central area then starts to emit blobs of magnetized plasma at high speed in a double jet. For comparison, at left are maps of radio galaxies, showing the similarity in form to the simulations.

Peratt was extremely excited by the new results. The central radio source and emerging jets looked exactly like quasars and active galactic nuclei that emit such jets—which had long been observed, and which Alfvén had theorized plasma processes can generate. Evidently there is no need for a black hole at the galactic center to generate such energy, because trapped magnetic energy, squeezed by the pinch effect, can do the trick even better.

As the simulation continued, the resulting galaxy's rotation started to induce currents flowing in opposite directions along the spiral arms, pinching them into filamentary shapes and the characteristic twisted helical form. The galaxy was generating its own electrical currents, which in turn could pinch off to form gas, clouds, and eventually stars—Alfvén's initial model. The simulations solved one of the great mysteries of astronomy: How do objects obtain spin, or angular momentum? It was clear that it comes from the interactions of magnetic fields—the objects gain spin at the expense of the fields.

The simulations solved another long-standing mystery—the "flat" rotation curves of the galaxies. If the speed of gas rotating around the galactic center is plotted against its distance from the center, the curve first rises rapidly but then levels off (Fig. 6.11a). However, if the disk-shape galaxy is held together by gravity alone, the speed should fall steadily as distance increases. As in the solar system, outer planets move more slowly than planets close to the sun. Astrophysicists had long seen this as evidence of a halo of gravitating dark matter surrounding the visible galaxy: within such a sphere, a flat rotation curve would be possible, though by no means necessary. But in Peratt and Green's simulation, the flat rotation curve emerges quite naturally in a galaxy wholly governed by electromagnetic fields (Fig. 6.11). The spiral arms are in effect like rolling springs that radiate from a galaxy's core—and, like such a spring, will have the same rotational velocity along their whole length. Better still, the "rolling" motion in the spiral arms of real galaxies had been observed.

Peratt's simulations accurately matched observations of ordinary galaxies *and* radio galaxies. Laboratory experiments with electron beams at Los Alamos confirmed that the same phenomena apply to currents from microamperes to mega-amperes, a range of a trillionfold. Another trillionfold jump from that brought Peratt to the galactic scale of his simulations. The entire package was finally published in 1983 in a small astronomy publication, *Astrophysics and Space Science.*

There was no response from conventional astronomers. Undaunted, Peratt decided to publicize this work on his own, and wrote a series of articles for the widely read amateur astronomy magazine *Sky and Telescope.* The first was provocatively titled "Are Black Holes Necessary?" and was published in July of 1983—to no avail. Peratt remained the only researcher actively working on galactic-scale problems. "It was a bit frustrating," he admits. Plasma cosmology seemed somewhat short-handed.

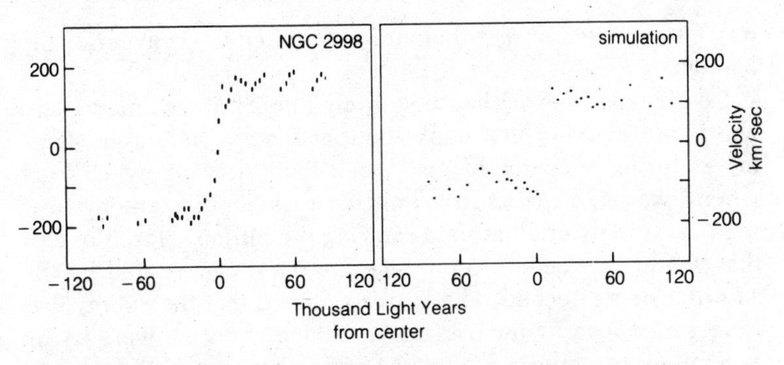

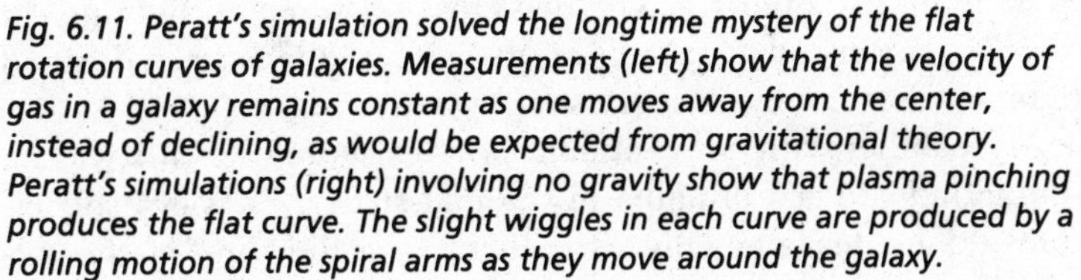

Fig. 6.11. Peratt's simulation solved the longtime mystery of the flat rotation curves of galaxies. Measurements (left) show that the velocity of gas in a galaxy remains constant as one moves away from the center, instead of declining, as would be expected from gravitational theory. Peratt's simulations (right) involving no gravity show that plasma pinching produces the flat curve. The slight wiggles in each curve are produced by a rolling motion of the spiral arms as they move around the galaxy.

In August of 1984, however, Peratt received a manuscript from a scientist unknown to him. The paper presented a detailed model of quasars and active galactic nuclear theory which was based on Alfvén's theory of the galactic generator. It was, however, an analytical treatment, using calculations rather than simulation, and it concentrated on the fine structure of the quasar—too fine for Peratt's simulations. (Quasars appear to be only a light-year across, compared with the one hundred thousand light-years of a galaxy and the ten thousand light-years cell-size of his simulation.)

Peratt was enthusiastic—finally someone outside of Alfvén's immediate circle was working along similar lines. Peratt sent a warm note to the scientist. "It would appear that 'ice-age' in science is beginning to thaw," he wrote.

I remember the wording of the letter quite well, since it was addressed to me.

▪ A PECULIAR CAREER

If Peratt had thought that my paper was a crack in the astronomical establishment, he soon learned he was wrong, for I was about as far from being established as one could get. I was not in any way a typical scientist, although I had started out conventionally enough.

Like many others, I had begun my interest in astronomy as a child, and my very first book about astronomy helped to shape my enduring interests. It was a book about the sun, which I got when I was eight. What struck me most was an illustration showing a seemingly endless train carrying the billions of tons of coal that would be needed to produce the energy generated by the sun in a single second. The book explained that the energy was in fact produced by nuclear fusion, which scientists were trying to tame for use on earth. The two areas—astronomy and nuclear fusion—fascinated me as I grew up, and by college I had decided to be a research physicist or astrophysicist.

My conflict with conventional physics started when I was an undergraduate at Columbia in the mid-sixties. Physics itself interested me, learning why things happen as they do—mathematics was merely a tool to understand and test the underlying physical concepts. That was not the way physics was taught; instead, mathematical techniques were emphasized. This is almost exclusively what students are still tested on, and obviously what they study the most.

I went on to graduate work in physics at the University of Maryland, intending to get a doctorate. But after a year, I left. I couldn't reconcile myself with the mathematical approach, which seemed sterile and abstract—especially in particle physics, in which I had considered specializing.

After leaving school in 1970 I began to work as a science writer —first for *Collier's Encyclopedia* and then free-lance, writing technical reports and magazine articles. This kept me in touch with the latest developments in astrophysics, controlled fusion, and particle physics, among other things; my work was an opportunity to complete my education in physics. I especially learned about plasma physics, which had not been touched on at Columbia or Maryland.

The seventies were the heyday of the Big Bang cosmology, but I was skeptical of it and the associated developments in high-energy physics. I knew from my Columbia days that there were fundamental contradictions in particle theory which had been swept under the rug (see Chapter Eight). The Big Bang's universe, wound up in the beginning and steadily running down, seemed wildly unscientific, and I knew that its theorists had never resolved the fundamental problem of the initial source of energy. It seemed far more likely to me that the universe had always existed, its evolution accelerating over the aeons.

I thought a great deal about problems that interested me in physics and cosmology, but I was busy earning a living. So it was not until 1981 that I actually began serious scientific research. The origin of that first project dated back to 1974, when I met Winston Bostick while we worked with a group advocating greater funds for controlled-fusion research.

Bostick's research centered on a fusion device called the plasma focus. It was the inspiration for my first astrophysical theories. The focus—invented independently in the early sixties by a Soviet, N. V. Filippov, and an American, Joseph Mather—is extremely simple, in contrast to the huge and complex tokamak, a large magnetic device that has long dominated fusion research. The focus consisted of two conducting copper cylinders, several centimeters across, nested inside each other (Fig. 6.12). When a large current is discharged across the cylinder, a remarkable sequence of events ensues.

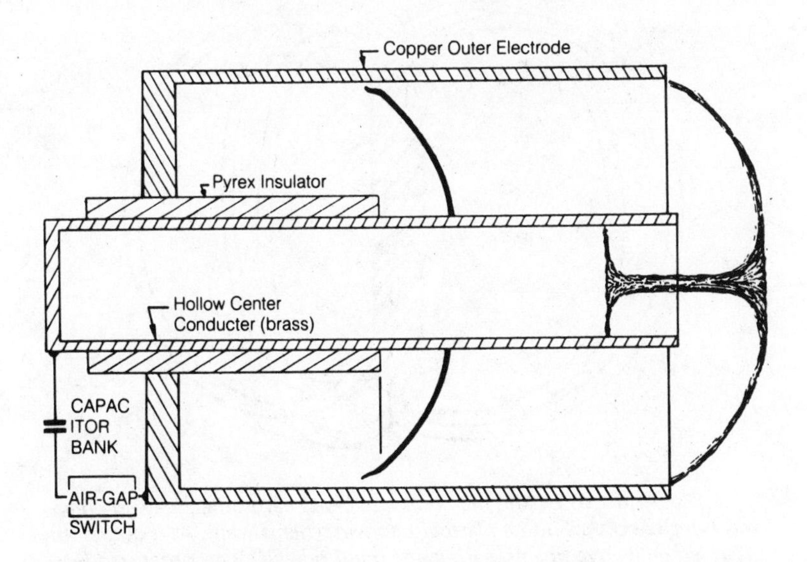

Fig. 6.12. A plasma focus device.

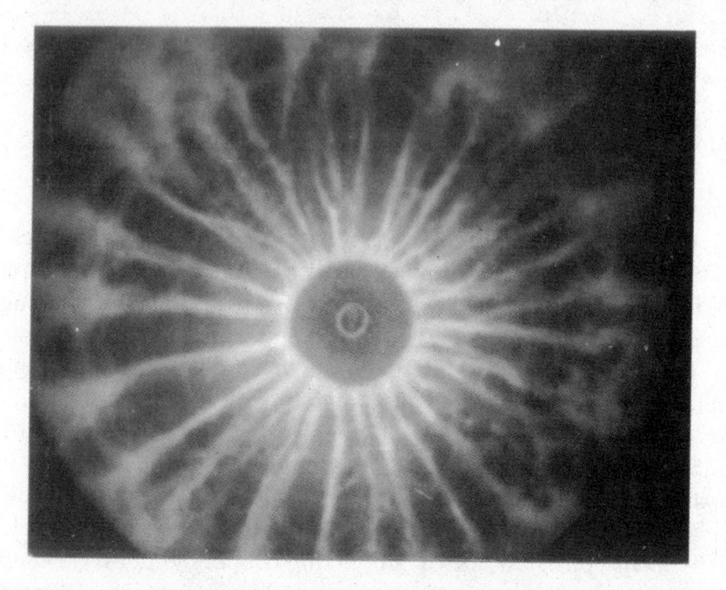

Fig. 6.13a. As the sheath carrying the inward-moving current forms, pairs of vortex filaments are generated.

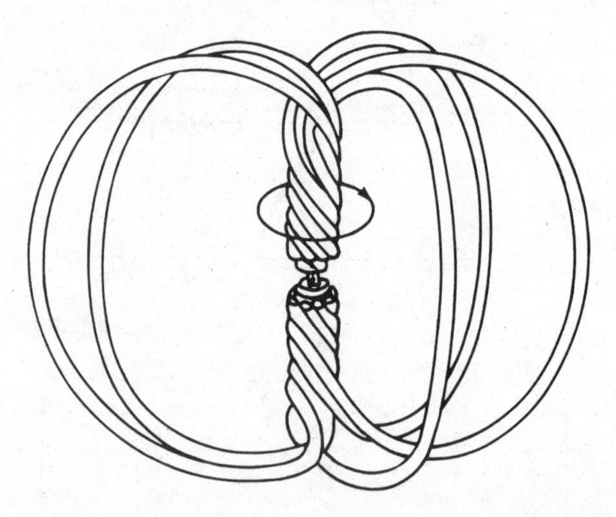

Fig. 6.13b. At the focus, the filaments annihilate each other, leaving only one, which necks off into a plasmoid, shown schematically. As it decays, the plasmoid emits two beams, each made up of tiny filaments organized into a helical pattern.

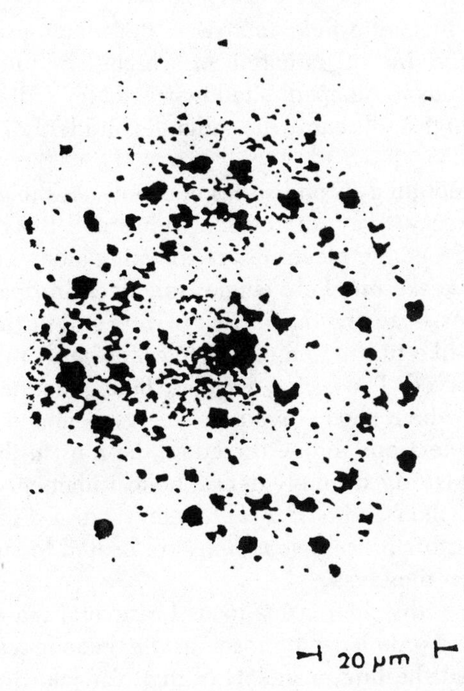

Fig. 6.13c shows the impact made by the beam of electrons on a plastic foil. The beam is about forty microns in radius, while the individual filaments are only one to two microns in radius.

The current rapidly ionizes the plasma and forms into eight or ten pairs of force-free filaments, each a millimeter in diameter, which roll down the cylinder, propelled by the interaction of their currents with the background magnetic field. When they reach the end of the cylinder, they fountain inward (Fig. 6.13a). Each pair, consisting of two vortices rotating in opposite directions, annihilate each other, leaving only one survivor to carry the entire current. This survivor pinches itself off into a doughnut-shaped filamentary knot—a plasmoid (Fig. 6.13b).

The plasmoid, only a half-millimeter across, now contains all the energy stored in the magnetic field of the entire device—a million or more times bigger in volume. For a fraction of a microsecond, as the plasmoid continues to pinch itself, it remains stable. But as its magnetic field increases, the electrons orbit in smaller circles, giving off radiation of a higher frequency. Because plasma tends to be opaque to low-frequency radiation and transparent to high-frequency, the radiation suddenly begins to escape.

This sets in motion a second series of events. As the electrons radiate their energy away, the current drops and the magnetic field weakens. Since the electrons are traveling along magnetic-field lines, the weakening field tangles the electrons' path up as its shape changes—causing the current to drop still further.

The result is like turning off a switch, as in the double layers Alfvén had observed. The falling magnetic field generates a huge electrical field, which shoots two high-energy beams out of the plasmoid—the electrons in one direction, the ions in the other. The beams consist of extremely dense, helical filaments, each a micron (one ten-thousandth of a centimeter) across (Fig. 6.13c). In the course of this process some ions are heated to such high temperatures that they fuse.

I was fascinated by the plasma focus for several reasons. For one thing, it was a promising approach to very economical fusion —it doesn't need the huge magnets of the tokamak. But it also dramatically demonstrated plasma filaments' capacity to compress matter and energy. While at the time I wasn't aware of Alfvén's extensive work, Bostick introduced me to his own ideas of how such filaments must have been relevant to galactic formation.

A few years later, I began to think that the plasma focus provides a model for another cosmic phenomenon—quasars. Over hundreds of thousands of years quasars radiate ten thousand times more energy than an average galaxy of a hundred billion stars, yet appear to be no more than a light-year or so across, compared with a galaxy's hundred thousand light-years. Their power density (power per cubic light-year) is a million trillion times larger than that of a galaxy.

How can such a small object generate so much energy? Conventional wisdom claims that a black hole is at work, but, among other objections, there are cogent reasons to think that any object massive enough to power a quasar will break apart before it collapses into a black hole. In any case, new observations had raised another mystery.

Beginning in 1978 high-resolution radio maps revealed that a radio galaxy's nucleus emits narrow beams of energy which connect them to outlying radio lobes. Then in 1980 a huge new radio telescope—the Very Large Array (VLA), consisting of twenty-seven dish antennae spread over miles of New Mexico desert—revealed to observers that the same jets emanate from the hearts of quasars.

It occurred to me that a plasma focus and a quasar are two processes, wildly different in scale, but identical in form and dynamics. Both consist of an extremely dense source of energy that emits diametrically opposed jets giving off high-frequency radiation. A plasma focus can increase the power density of its emissions by a factor of ten thousand trillion over that of the incoming energy—comparable to the ratio of a quasar to a galaxy.

But how can a galaxy generate an electrical current? I knew about disk generators and calculated that a galaxy rotating in a magnetic field will generate a current flowing toward its center sufficient to power vast plasmoids—a process Alfvén had proposed four years earlier, I later found out. Since the currents must flow out along the axis, they will arc around, as in the plasma focus—a similar geometry leads to plasmoid formation in both cases.

While these ideas were crystallizing I was also studying the evidence, accumulating since 1978, for the existence of filament-like superclusters of galaxies. I had seen Peebles's "Cosmic Tapestry" poster, showing the galaxies strung along lacy threads. Why couldn't these filaments of galaxies be larger versions of the filaments in the plasma focus and the filaments that I hypothesized to form in galaxies? They would produce magnetic fields in which galaxies, as they rotate, would produce the plasmoids that make up quasars or active galactic nuclei.

In early 1981 these were still theories without the quantitative detailed work needed to prove them. But I had no time—my wife, Carol, and I had just had our first child, Kristin, and my writing provided our only income. An opportunity soon arose, however, for paid scientific work, at least indirectly.

At the time I wrote primarily for *Spectrum*, the journal of the Institute of Electrical and Electronic Engineers (IEEE), the international organization of electrical engineers. *Spectrum* covered everything having to do with the field, including such topics as controlled thermonuclear fusion. It did not deal with pure science, only technology, but the editor I worked with, Ed Torrero, wanted articles on scientific developments that are related to technology. Previously he had studied plasma physics and knew of my interest in it and in astrophysics, so he suggested that I write an article (or find an expert to write one) about the connection between plasma science in astrophysics and in controlled fusion.

I immediately agreed. Bostick and his close co-worker, Vittorio Nardi, were involved in fusion research and had applied their work to astrophysical problems, such as solar flares, so they were logical coauthors for the article. Moreover, if I worked with them, as I had with many expert authors, the required research would overlap with what I needed for my own scientific ideas. Equally important, I hoped that Bostick or Nardi would work with me on my scientific hypothesis—it needed experienced help, and I doubted that any paper would be published under my name alone.

It turned out that they too had been thinking about the connection between the quasar and the plasma focus. Bostick, however, was too busy, but Nardi agreed—I was to do the main work, with his input being mostly advice and criticism. I began work in the fall of 1981.

Nardi was an ideal partner. Although he had little time for the project, his contribution was invaluable: he is an extremely rigorous scientist, with a high standard for theoretical work, so his careful criticism of my work forced me to dig deeper and build a firmer foundation for my theories.

Initially I tried to demonstrate the validity of the theory by finding scaling laws that can quantitatively relate what has been observed in the laboratory to what has been observed at the galactic scale. I needed what are called scale invariants—quantities that don't change as the scale of the phenomenon increases. At first I proceeded empirically, but Nardi showed me Alfvén's *Cosmic Electrodynamics*, where he had laid out theoretically derived scaling laws (see Chapter Five).

It was clear that the key invariant is the velocity of matter under the influence of a magnetic field. For the initial stage of the plasma focus device or of a galaxy's collapse, the typical velocity is about 160 km/sec; but for the final stage, plasmoid or quasar, the velocity is much higher, around 10,000 km/sec. As Alfvén points out, the effective resistance of plasma—the ratio of voltage to amperage—is also scale invariant. It also seemed true that, for a given scale, certain quantities must remain unchanged —the quantity of magnetic energy, and the ratio of density to magnetic field strength.

With these few relationships, I was able to make quantitative predictions about quasars and their jets based on laboratory data, and roughly to explain their magnetic field strength, vast energy, and relatively small size. I put together a scenario that describes the main events in the formation of the galaxies.

A MODEL OF A QUASAR

Both theoretical studies and computer simulations had shown that any plasma with sufficient energy will create vortex filaments, and that these filaments will grow without limit, as time and space allow. Force-free filaments, those with the most twist, will grow fastest and thus come to dominate. They will pinch plasma together, forming thick, dense ropes. These filaments will grow until they became self-gravitating: gravity will then break them up, producing blobs of plasma spinning across the field lines of the huge filaments. This, in turn, will generate inward-flowing currents that will produce new sets of filaments, thus repeating the cycle, spinning an ever-finer web of matter.

The first filaments will be the supercluster chains. These will give birth to protoclusters, which in turn will generate galaxies. Finally, the galaxies will produce stellar clouds, which will condense into stars. At each stage the inward-flowing currents and the background magnetic field will brake the spinning plasma, allowing further contraction of the protocluster, protogalaxy, or protostar.

The energy taken from the rotation and gravitational contraction of the object will go into the creation of the dense plasmoid and will be released in the beams the plasmoids create as they decay. A quasar is thus the birth cry of a galaxy, the means by which the excess energy of rotation, which must be removed if the galaxy is to collapse, is carried away in the form of the energetic jets.

Once the galaxy forms, the same process at a lower rate fuels the repetitive formation of small plasmoids at its nucleus. The process is today generating stars in the dense filaments of the spiral arms.

The theory can explain the source of a quasar's immense power. The ultimate source is the rotational energy of an entire galaxy, augmented by the gravitational energy released as the galaxy contracts. This energy is converted to electrical power by the disk-generator action and concentrated in the smaller filaments moving toward the galaxy core. The filament pinches into a plasmoid that, for the largest quasars, might be a hundred light-years across. The visible quasar, though, is far smaller (Fig. 6.14). This is the region, a light-year or so wide, where each individual subfilament that composes the plasmoid is bursting apart as it radiates its energy and powers the emitted jets. Just as a hydroelectric dam draws power from the water falling in a river valley, the quasar is drawing energy immediately from the plasmoid's magnetic field, a million times larger in volume, and ultimately from the entire galaxy. In this way the energy gained by the collapse of the galaxy is expelled as electrical energy in the quasar jets. Without the elimination of this energy, the galaxy would never form at all.

(The model can also account for the way that the quasars and their smaller brethren, active galactic nuclei, emit jets in only one direction at a time, switching after several thousand years. One beam is electrons, which radiate energy so quickly it will not extend beyond the limits of the quasar itself. The other is protons, which radiate far more slowly. The high-energy protons will, in turn, accelerate electrons along the way, and these will radiate radio waves, making the jets visible to radio telescopes. But only a fraction of the energy will be lost, so the protons will produce extensive jets.)

By this time my research had made me aware that most astrophysicists do not believe space can carry the currents I proposed, that the resistance of space is so low that the currents will dissipate instantly—objections Alfvén had encountered for decades. I wasn't aware of Alfvén's refutations, but I felt that I had a good solution: electrical currents must expend energy in forming vortex filaments, and this produces an effective resistance—just as in ordinary resistance, electrical energy is converted into the ran-

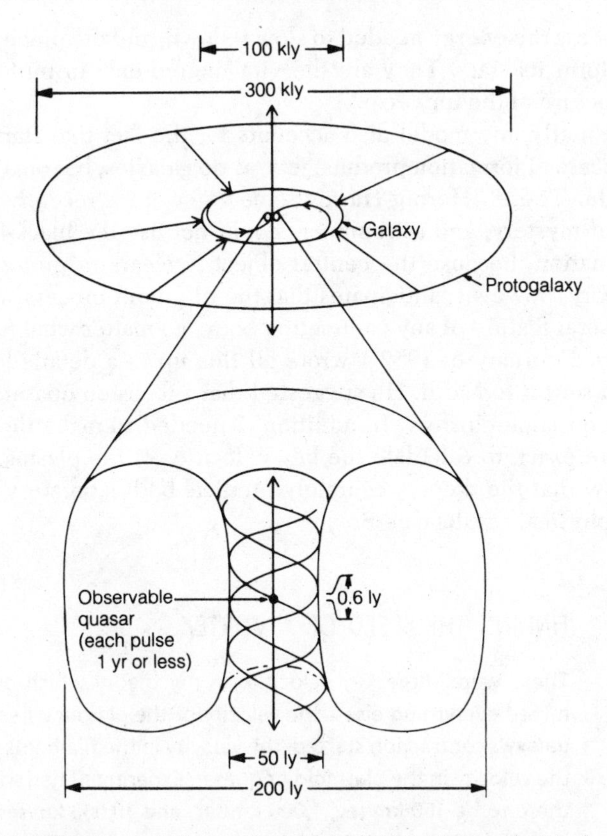

Fig. 6.14. A collapsing, rotating protogalaxy (a) generates an electrical current that spirals in toward the center and leaves along the axis. At the center the filament carrying the current pinches into a plasmoid (b) about 200 light-years across. The plasmoid, in turn, generates a powerful electrical field along its axis, accelerating beams of protons and electrons. The small region from which these beams emerge, the region of highest magnetic field, is only half a light-year across. This is the visible quasar.

dom motion of heat (this is what makes an electric light work). The vortices work as motors, turning electricity into motion, just as a galaxy as a whole works as a generator, turning its motion into electricity. Without the vortices, the magnetic field in which the galaxy rotates will be "frozen into" the galaxy's motion, its patterns disrupted; as a result, the galaxy won't lose enough angular momentum to contract and form stars. The vortices allow the galaxy to roll through the background field, generating in the process the energy needed to slow it down, and the pinch needed to form its stars. They are the vital ingredients in building the structure of the universe.

Finally, my model also accounts for the fact that stars in the process of formation produce jets, as do galaxies, but on a smaller scale. These "Herbig-Harrow objects" were a recently discovered mystery, and astronomers could not use the black-hole explanation, because the central object is clearly a protostar. My theory, however, anticipates that the plasmoid process will be a general feature of any contracting body, no matter what its size.

In February of 1982 I wrote all this up in a detailed outline and sent it to Nardi. He suggested that I focus on quasars alone, not on superclusters. In addition, I needed to make the theory more exact, to calculate the key velocities of the plasma, and to show that the theory accurately predicts both laboratory and astrophysical results (see Box).

FINDING THE SPEED OF A VORTEX

There were three key velocities in my theory which determined everything else—the velocity of the plasma when the galaxy's contraction started, the velocity in the filaments, and the velocity in the plasmoid or quasar. Experiment had shown these to be 160 km/sec, 1,000 km/sec, and 10,000 km/sec, respectively. Why these speeds rather than some others? The answers turned out to be critical not only in understanding quasars but in comprehending the large-scale structure of the cosmos as well.

In the case of the plasmoid, for the plasma to be transparent to synchrotron radiation, the characteristic velocity of the electrons must be near the speed of light. The ion velocity is related to that of the electrons by just the square root of the ratio of their masses. Now, the proton is 1,836 times as massive as the electrons, so the velocity should be about forty-odd times less. Relativity corrects this to a predicted velocity of 9,900 km/sec.

This was an important value, since the velocity of gas in quasars had been measured using the Doppler shift and it averages just 9,500 km/sec—just what my theory predicted.

But what about the initial conditions and the filaments? How exactly did the filament form? Filament formation appeared to be a resonance phenomenon. Resonance occurs when two processes operate with the same frequency—like the pushes of a child on a swing. At a certain velocity, the frequency of an electron spinning in a magnetic field would be equal to that of the ions bouncing back and forth in the plasma. This speed would be 1,836 times less than the speed of light—the ratio of the electron to the proton masses, or very close to the observed conditions for filament formation. But what mechanism allowed the electrons to push the ions? And what determined the velocity of the filaments themselves?

For several months in 1982, I struggled with this problem without progress. After a roundabout process, I finally figured out that the key to all of the critical velocities was plasma instabilities—oscillations in a plasma which would spontaneously grow, passing energy first into the filaments to make them form and then out of the filaments to stabilize them. Plasma oscillation was an area where much work had been done, especially by Soviet fusion researchers.

The Soviet work showed that indeed the instabilities would do just what I had guessed. In essence, when the plasma velocity reached the first critical value, 160 km/sec, the stream of electrons would set up shock waves that would push the plasma ions into helical motions, creating the vortices. The vortices would then grow by the pinch mechanisms (and by the mechanism I later learned Peratt described in detail) until they reached the second critical velocity, 1,000 km/sec. Then a similar shock mechanism would slow the electrons in the filament, transferring energy to the background electrons. The initial conditions had a velocity equal to the ratio of the electron and proton masses (m_e/m_p) times the velocity of light (c). The plasmoid had a velocity $(m_e/m_p)^{1/2}c$ and the filaments I calculated would then be neatly halfway in between $(m_e/m_p)^{3/4}c$.

Now that I had rigorously calculated what the key velocities were, I could show that all the key quantities, such as magnetic field strength and radio emission, observed in the quasars were accurately predicted by the theory.

When I completed this paper in early 1984 I was anxious to publish it, since I had learned that I wasn't alone in this research. Nardi had lent me a copy of Alfvén's newly published book *Cosmic Plasma*—in it I saw that Alfvén had, in 1978, developed the same notion of a galaxy as a disk generator and had also applied it to the problem of radio galaxies, attributing the newly discovered jets to the currents produced along the axis.

While Alfvén had anticipated my thesis, I was more gratified than worried by this—being scooped by a Nobel laureate, and by only a few years, is no disgrace for a novice. In addition, my work was still needed. Alfvén's hypothesis had been characteristically general. In particular, he had not used the disk generator to produce the intense quasar or galactic nucleus at the center of the jets. Instead, he theorized that this is caused by an entirely different process—the collision of an ordinary star and an antimatter star. This I did not consider at all plausible, since observation made it clear that the jets and the "central engine," either quasar or galactic nucleus, are connected. They must both have the same cause.

Generally, I was extremely encouraged that an authority such as Alfvén was thinking along the same lines, and also that he had swept away objections to powerful currents in intergalactic space. But I was much more ambivalent in July of 1983 as I read Peratt's *Sky and Telescope* article. Here was detailed work, using a model extremely close to mine. I anxiously read his technical paper, wondering if now I had really been scooped.

Again I was reassured. Peratt's work was overwhelmingly a simulation, while my work was analytical—detailed calculations using exact equations. Such approaches tend to be complementary, their conclusions supporting one another. Moreover, since his simulation had limited resolution, the Los Alamos work could not explain why quasars are so tiny—this was the heart of my model. In other ways, his work tremendously supported my own (and mine supported his) because, by independently using very different methods, we came to identical conclusions.

However, seeing that there was work along these lines made me eager to publish.* In July I submitted a new draft to Nardi.

* I later learned that another researcher, P. F. Browne at the University of Manchester in England, was also working along similar lines, using magnetic filaments to explain phenomena ranging from quasars to superclusters. Browne's theories differed significantly from mine, however, in a number of ways, such as how the filaments were generated.

He thought it vastly improved and had no major objections, but was still unwilling to publish under our joint names. When I wanted to go ahead, he agreed that the paper be published under my name alone, with just an acknowledgment to him (this was, in fact, an accurate reflection of his thesis-advisor role).

I had no idea who would accept such an unorthodox paper from an uncredentialed, unaffiliated author. I decided to send the paper to Peratt, see his reaction, and then ask him for advice.

THE THAW

I was ecstatic to receive his letter. By phone he told me he thought the theory to be well founded. And although my work hardly indicated a thaw of the orthodoxy, his prediction was nonetheless accurate: the day I received it both his work and mine received dramatic observational confirmation, displayed on the cover of *Nature*.

My paper had predicted the existence in every galaxy—including our own—of dense, powerful magnetic filaments, about a light-year across, looping toward the center and arcing out along the axis of the galaxy. Peratt had predicted the same currents, although in less detail, and the same high magnetic fields. On the cover of the August *Nature* were our filaments, a bundle of them arcing out of the center of our galaxy, the outer ones helically twisted around the straight inner ones—a textbook illustration of a force-free vortex. The filaments are a light-year across and nearly one hundred light-years long—exactly what both of our models predicted.

The filaments had been observed with the VLA radio telescope by a Columbia University graduate student, Farhad Yusef-Zadeh. As Yusef-Zadeh points out in the accompanying article, such filamentary forms cannot be confined by gravity or other forces—they must be force-free magnetic filaments.

It seemed that the second cosmic leap to the galactic scale had been achieved. Here was solid evidence that vortex filaments exist on the scale of light-years, and they dovetailed with Peratt's simulations and my own calculations.

From this time on, astronomers did indeed begin to pay attention to the significance of magnetic and plasma phenomena at a galactic scale—but enlightenment was to spread quite slowly, as I soon found out. Peratt had suggested that I submit to the journal that he had published in, *Astrophysics and Space Science.* Unfortunately, they rejected my paper summarily, the reviewer dismissing the analogy between galaxy and plasma focus as absurd. I then resubmitted, on Peratt's further advice, to a small plasma physics journal, *Laser and Particle Beams,* where my first paper was finally published in 1986.

Despite the revelation of Yusef-Zadeh's twisting filaments, the black-hole explanation of galactic nuclei and quasars has remained dominant. The filaments are simply dismissed as a fascinating mystery, or, in some cases, explained as a plasma beam generated by a black hole.

In 1989, however, new evidence developed which will probably doom the black-hole hypothesis. Gas and plasma near the center of galaxies has always been observed to move at a high velocity, up to 1,500 km/sec for our own galaxy, and similar or higher values for others. These velocities are generally treated as evidence for a black hole whose powerful gravitational field has trapped the swirling gases. But two scientists at the University of Arizona, G. H. and M. J. Rieke, carefully measured the velocities of *stars* within a few light-years of the center of our galaxy, and found the velocities are no higher than 70 km/sec, twenty times slower than the plasma velocities measured in the same area. Since the stars *must* respond to any gravitational force, their low velocities show that no black hole exists. The high-speed gases must therefore be trapped only by a magnetic field, which does not affect the stars. In addition, the currents Peratt and I hypothesized formed the galaxies have also been detected (see p. 49, Chapter One).

Certainly, as of this writing, the evidence for plasma processes shaping galaxies and the violent events at their core is becoming overwhelming, although it is still blithely ignored by most of the astronomy community. This attitude is changing, however: in August of 1989, Peratt was invited to present his work to an international conference of radio astronomers dealing with magnetic fields in galaxies.

SUPERCLUSTER FILAMENTS

With the galactic-scale work on its way to publication, the next step upward was the supercluster filaments, and in 1985 I returned to my earlier ideas. I knew that Peratt had imagined the galaxies to be strung along giant filaments of current, as I had, and I wanted to develop a quantitative theory about the large-scale structure of the universe. Why are galaxies, clusters, and superclusters the size they are?

Fortunately, the work I'd done in determining key velocities in filament formation would be central to such a theory of structure. Oddly enough, the origin of the crucial ideas lay in work I had done on an entirely different subject. Back in 1977 I had become interested in the role of vortices in the earth's weather system. Weather was in the news at the time because of the extraordinary droughts and extreme winters in much of the world. Of course, the main weather systems of the world—cyclones and anticyclones—are flattened vortices, and in my spare time I tried to learn their basic physics.

I found that vortices tend to share a common characteristic—the product of their density and radius is a constant; in other words, the ratio of their mass to surface area is constant. I was already interested in Bostick's ideas about galaxies being vortices. When I found that he had suggested that electrons are vortices too (discussed in Chapter Eight), I wondered if the mass-area ratio would be similar for these two extreme examples. To my surprise, the values are indeed similar—about five hundred grams per square meter for a galaxy and about thirty grams per square meter for electrons. (For comparison, a piece of paper has a mass-area ratio of about one hundred grams per square meter, so a layer of electrons weighs one-third as much as a sheet of paper, while a galaxy weighs only five times as much as a galaxy-size piece of paper!) Considering that the galaxy weighs 4×10^{71} as much as an electron, these numbers seemed pretty close. I wondered if there were some important relationship here that made the mass-area ratio a universal constant.

Now, numerical coincidences are somewhat dangerous things in science. Sometimes they are clues to fundamental underlying laws, at other times just coincidences. I didn't see how this could be a fundamental law—although I suspected it might be—so I just put the work aside.

But while I was working on my quasar model, I came across a 1981 study of the masses and radii of galaxies and galaxy clusters. In it two Dutch astronomers, J. Kaastra and H. G. Van Bueren, show that for each type of object, the ratio of mass to surface area is roughly constant, even though the smallest objects might be a thousand or ten thousand times less massive than the largest. The authors point out that the mass-area ratio of galaxy clusters is almost identical to the ratio for an electron, some twenty to thirty grams per square meter.

Obviously, there is indeed something to this ratio—a relationship based on *hundreds* of galaxies and clusters is no longer a coincidence, but a fact. Perhaps, I thought, there is some good plasma explanation. I had been studying Alfvén's 1963 book *Electrodynamics* and knew that he classifies the properties of plasma in part by the "collision length"—the average distance a particle travels before colliding with another one.

"Noncollisional" plasmas are those whose dimensions are smaller than their collision length—collisions are rare in such plasmas. I realized that such a plasma will not contract gravitationally—if the particles do not collide, they will stay in orbit just as the planets of the solar system stay in their orbits. For a mass of any sort to collapse gravitationally, the particles that make it up must collide with one another and radiate away their energy. Therefore, the radius of the mass has to be larger than the collision distance of its particles. The denser the plasma, the shorter the distance between collisions: so, for a given particle velocity, the collision length times the density is a constant—and thus so is the mass-area ratio of the collisional plasma. It seemed possible that this ratio's constancy relates to the fact that only a collisional plasma can contract.

Now, the faster a typical particle travels, the farther it goes before colliding with something and the larger the mass-area ratio of the plasma if collisions are to occur. Since the mass-area ratio of the electron seemed somehow to be critical, and is the same as that for galaxy clusters, I wondered what velocity a particle would need if the mass-area ratio of the plasma was to be the same as that of an electron. Roughly, the answer turned out to be 1,000 km/sec—just the velocity characteristic of vortex filaments, as measured in the laboratory.

This was the clue that led me to discover how filaments are formed by instabilities (once I had a mathematical expression for the key velocity I could guess a physical mechanism). A few years later, though, this seemed a possible hint as to how the structure of stars, galaxies, clusters, and superclusters is formed.

I knew that all filaments must have the same characteristic velocity—1,070 km/sec, to be exact. If they are to collapse gravitationally, then they must be collisional; but for the fixed velocity to apply, there must then be a fixed relation between their density and radius—a filament will break up into blobs about one collision distance across. Smaller blobs will not collapse at all, but large ones will break up while collapsing.

This all meant that the distance *between* objects that contracted out of the filaments, such as stars, galaxies, or clusters, should be inversely proportional to the density of the filament from which they contracted. Density times distance should be a constant.

On average, stars are separated in our galaxy by a few light-years. Before it condenses, a typical star's matter fills up a volume a few light-years on each side and has a density of a few atoms per cubic centimeter. Density times distance is about ten atoms per cubic centimeter times light-years. This product should be the same for all objects—stars, galaxies, and clusters.

A more convenient way of putting this is as a ratio of an object's mass to the square of the distance to its nearest neighbor. My prediction was that this ratio should be about one sixth of a gram per meter squared, or one solar mass for every fifty light-years squared. And voluminous published data on stars, galaxies, and clusters bore out my prediction: from stars to clusters, a thousand trillion times more massive, the ratio of mass to distance squared *never* differed by more than a factor of four or five. Knowing an object's mass, I could predict how far away it was from its nearest neighbor (Fig. 6.15).

This implied that all of these objects had indeed formed from vortex filaments whose characteristic velocities were very close to my calculated 1,000 km/sec. Any other value would yield a

THE COSMIC HIERARCHY

Object	Mass (Solar Masses)	Radius (Light-Years)	Distance to Nearest Neighbor (Light-Years)	Mass/Radius	Orbital Velocity (Km/Sec)	Mass/Dist.^{2}
STAR	.5	5×10^{-8}	7	10 million	330	.01
GALAXY	.3 trillion	30 thousand	4 million	10 million	330	.02
CLUSTER	600 trillion	30 million	175 million	20 million	460	.02
SUPERCLUSTER COMPLEX	20,000 trillion	450 million	1,000 million	40 million	660	.02

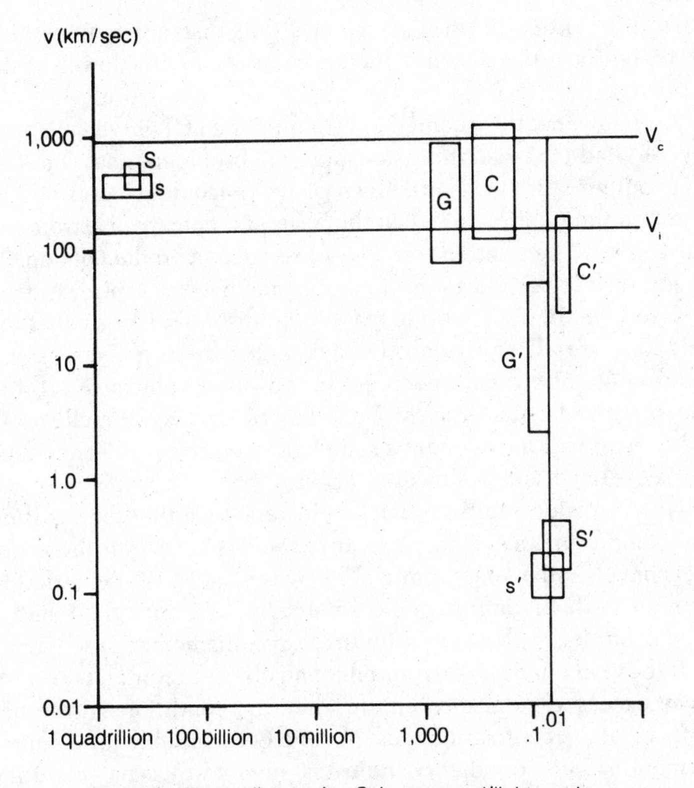

Fig. 6.15. The cosmic hierarchy. Objects in the universe are plotted in terms of two characteristics. One is their orbital velocity—the speed an object at their surface must have to stay in orbit. The other is the ratio of their mass to either radius squared or the square of the distance to a nearest neighbor. The boxes at right use the nearest-neighbor distance. The left-hand boxes use the radius of the objects: C for clusters of galaxies, G for galaxies, S for heavier stars (more than 1.8 times the mass of the sun), and s' for lighter stars.

DeValcouler's relation is indicated by the fact that none of the objects is located above 1,000 km/sec or so. Plasma theory shows that the two solid horizontal lines are the upper and lower limits for orbital velocities. The vertical line corresponds to plasma that are barely collisional at a velocity of 1,000 km/sec, the characteristic velocity of plasma vortices.

The diagram neatly encapsulates the structure of the observable universe. It implies that there is a limit to the size of the plasma filaments, indicated by the upper right-hand corner.

very different collision distance and thus distance between objects, because the distance increases as the fourth power of the velocity.

And now that this seemed to be a fundamental universal velocity, I had the solution to another problem—that of DeVaucouleur's limit. I had known of DeVaucouleur's and others' work on the relation between the mass and density of astronomical objects. The relations can be expressed as a product of density times surface area, a ratio of angular momentum (spin) to mass squared, or simply a ratio of mass to radius—but I knew that the relations are all equivalent to saying that these objects have a fixed range of orbital velocities. If you put an object in orbit at the surface of a star or at the boundary of a galaxy or a cluster of galaxies, it will have to move somewhere between 100 and 1,000 km/sec—never much slower or faster.

Now, considering that galactic clusters are a thousand trillion times more massive than stars, it seemed odd that all these objects have similar speed limits. This was what Alfvén had tried to explain with his antimatter cosmology. But I thought I had a much simpler explanation with the vortex filaments.

If these filaments are essential for an object to contract, to carry away excess angular momentum, then the conditions at the beginning of the contraction must obviously be suitable for filament formation. This, my theory indicates, involves plasma velocities around 160 km/sec—a lower limit on the orbital velocities of the final object, since as it contracts, its orbital speed will increase. But if the speed were to exceed the velocity of the plasma in the filaments themselves, then the filaments would be blown apart (as has been observed in the laboratory); thus when that speed is reached, an object stops contracting. Any speed between 160 and 1,070 km/sec will allow vortex filaments to develop.

Thus, if the universe was indeed sculpted by the counterpoint of gravitational contraction and the pinching of vortex filaments, its observed structure was inevitable. The size and mass of contracted objects and the spaces between them were determined in a simple way. Together, the mass-radius law for contracted objects and the mass-distance-squared law for their spacing also explains a major feature of the universe—the smaller the objects, the more isolated they are. Thus stars are separated from each other by distance typically ten million times their own diameters, galaxies by only thirty times their diameters, and clusters by about ten times their diameters. In short, my theory explained why space is so empty.

I was excited by these results—from the simple laws of plasmas and gravity the whole hierarchy of the cosmos could be understood. The model indicates that there are other patterns—for example, objects of a given type will share the same mass-area ratio when they have contracted. The distribution of stars in a galaxy was also explained. The denser the plasma, the smaller the objects formed: in the outer arms of a spiral, where the plasma is less dense, many stars will be large, bright ones, while in the inner, denser regions they will tend to be smaller and dimmer.

Perhaps most significant, the theory implies that there is, in fact, an upper limit to this hierarchy of vortex filaments. As filaments grow larger from an initial homogeneous plasma, they will eventually form a single vortex, with a radius of a collision length, that is *also* self-gravitational, having an orbital velocity equal to 1,000 km/sec. No larger filament should develop. If a filament were to form in less dense plasma, its gravity would distort particle paths and prevent further growth while the vortex is still smaller than a collision distance, and thus incapable of contracting. In such plasma, which would have a density less than a particle per thousand cubic meters, galaxies and stars will not form. Denser plasmas, on the other hand, have filaments that will become self-gravitating at a smaller size.

The diameter of the maximum primordial filament would be impressive—about ten billion light-years across. I roughly calculated that it would gravitationally compress itself to about one-fifth this size, keeping its filamentary shape. It would then break up into a couple of dozen smaller filaments spaced some two hundred million light-years apart—the supercluster filaments, which then gravitationally contract into clusters, and then into galaxies, and so on.

The theory indicates that supercluster filaments should be grouped into a larger structure about one billion light-years in radius and a few billion light-years long, an elongated filamentary shape. I submitted the completed paper on March 31, 1986, to a special issue of the *IEEE Journal of Plasma Science,* which Peratt edited.

I had hoped that the Hubble space telescope would confirm

this prediction when it was finally put into orbit, but fortunately I didn't have to wait; the next day, Tully's paper announcing the discovery of "supercluster complexes" appeared in *Astrophysical Journal*. (As in 1984, my sending a paper to Peratt seemed to magically call forth instant observational confirmation!) Tully's objects were virtually identical to what I predicted, although they had contracted a bit further. They are about six hundred million light-years in radius and nearly two billion light-years long, with an estimated density and orbital velocity that fit neatly into my relations.

The same week, moreover, *Nature* carried a paper by C. A. Collins, who had measured the large-scale motion of galaxies over a region nearly as large as that of Tully's supercluster complexes. He found streaming velocities of 970 km/sec, as close as I could want to my prediction of 1,070 km/sec. Such velocities over such a vast area by themselves imply the existence of an immense gravitating object—just the dimensions of the structure Tully observed.

THE BIG BANG AND THE *NEW YORK TIMES*

At this point there were definite grounds for challenging the Big Bang. As we've seen, Tully's discoveries were far too massive and ancient for the Big Bang, yet they meshed neatly with alternative models. It would take a trillion years or more to form the primordial filaments, but in a universe without a beginning, there is no rush.

Tully's results quickly became a hot topic in cosmological circles. However, any alternative to the Big Bang remained almost unknown, since plasma cosmology was routinely rejected by astrophysical journals, and our papers were published only in plasma physics journals, which astronomers never read.

I was, however, offered a golden opportunity to break the news. A friend of mine, Randy Rothenberg, had recently been hired as an editor specializing in science articles at the *New York Times Magazine*. In January of 1986 he called to ask if I had any good articles to write for the magazine. I immediately suggested an article on Alfvén and the plasma universe. Randy was enthusiastic and won over the magazine editors, who gave me a contract.

At last I was to meet Alfvén and Peratt, whom I had thus far known only by phone. In March I interviewed Alfvén in Washington, D.C., as he traveled from San Diego to Sweden. In his courtly way, Alfvén carefully went over his ideas on the problems of the Big Bang and the development of the plasma-universe concept. But he warned me that he thought it was premature for an article on the Big Bang, that instead the article should focus on his more established theories of the solar system. "Wait a year," he advised. "I think the time will be riper next year to talk about the Big Bang."

The advice turned out to be prescient, but I couldn't follow it. I had a contract with one of the most widely read publications in the world, and the *Times* editors were eager to see why the Big Bang might be wrong. The next month I was off to Los Alamos to meet with Peratt.

Peratt showed me his computer simulations of galactic formation, and we discussed our respective theories as he showed me and my family around the Indian ruins that surround Los Alamos.

On the last night of our visit I went to his house to try to see Halley's comet, then nearing the earth. It was just a fuzzy blur, but the night sky was spectacular through Los Alamos's clear, dry air. The Milky Way poured across the heavens in a brilliant arch, dotted here and there with globular clusters, which could be seen easily with the naked eye. This was the sky our ancestors saw before the coming of electricity—it takes no leap of the imagination to recognize why they believed that what happens on earth reflects that awesome spectacle in the heavens.

But I was soon to learn it was still difficult for people to change their views of the heavens. After much writing and rewriting, my article for the *Times* was ready in the fall. Randy conveyed the news that the article had not only been accepted, but was to be the cover story of the October 26 issue. I was elated—but not for long. Two weeks before publication the article was canceled. It had been routinely sent to the science section of the daily paper for review. Walter Sullivan, who twenty-five years before had proclaimed the confirmation of the Big Bang in a front page news article, had vetoed it. He had dismissed Alfvén as a scientist well known for his maverick ideas, ideas that, Sullivan asserted, had no support even within the plasma physics community.

For a week or so Randy and I scrambled to refute Sullivan's brief note. He dug through the *Times* morgue to show Alfvén's enormous prestige in the field, and I gathered up letters from various plasma physics notables. But the decision was made.

BIG BANG IN CRISIS

Alfvén's advice about restraint, based on his decades of scientific battles, proved to be a good one. In the course of 1987, for the first time, observations of large-scale structure began to call the Big Bang into serious question. At the same time, Peratt and I were formulating a detailed alternative to the conventional cosmology.

It was clear that any real challenge to the Big Bang would have to provide an alternative explanation of the two phenomena that Big Bang supporters rest their case on—the abundance of light elements, especially helium, and the microwave background.

Existing stars cannot have produced the 24 percent of the universe that is helium. At the rate they currently produce energy from fusion, only 1 or 2 percent of their hydrogen should have been burned to helium in the twenty billion years that our galaxy has existed. Therefore, say Big Bangers, the rest derives from a primordial explosion.

But there's a simpler answer, as I discussed in Chapter One. The larger a star, the hotter its interior and the faster it burns its nuclear fuel. If, in the early stage of galactic formation, a generation of stars considerably heavier than the sun formed, they all would have burned up in a few hundred million years, exploding as supernovas and scattering large quantities of helium.

There was now good reason to believe that the first generation of stars *was* more massive. In my models, as in Peratt's, stars would form inside, and in front of, the filamentary spiral arms as they rolled through the surrounding medium. The mass-area ratio shows that as a plasma's density increases, the size of the objects formed from it decreases. So as the galaxy contracted, the largest stars would have formed first and smaller stars with longer lives would have formed only when the density had risen.

Conventional theorists object that the most massive stars, giants that culminate in a supernova, *also* generate large amounts of oxygen and carbon. Yet the universe is only about .5 percent carbon and 1 percent oxygen, less than would be expected if such stars produced all the 24 percent helium.

My model provided a natural answer to this objection. As the filaments of the spiral arms slice through the plasma, they produce a shock wave, like that of a supersonic aircraft. Within the compressed material of this shock wave stars will form, as pinched currents flow through it. For stars more massive than ten or twelve times as massive as the sun, this process will continue outward from the plane of the galaxy—the plane of the filament's motion—until they blow up as supernovas, scattering oxygen and carbon. This will disrupt the shock wave that contributes to star formation, thus confining it to a rather narrow disk. Not many of those very massive stars will form, so oxygen and carbon production will be limited.

But stars with less than this mass will not explode. These more sedate stars will blow off only their outer layers—pure helium—not their inner cores, where the heavier elements are trapped. As these medium-sized stars, four to ten times bigger than the sun, form in the dense, inner regions of the galaxy, the shock wave will spread through the entire thickness of the galaxy. Consequently, helium production will far outweigh that of oxygen and carbon.

This model predicts the amounts of helium, carbon, and oxygen that a variety of galaxies will produce. The results are in close agreement with observation—almost any galaxy would produce about 22 percent helium, 1 percent oxygen, and .5 percent carbon. It is only after all these stars have burned that density will rise sufficiently for still lighter, longer-lived, and dimmer stars like our sun to form (Fig. 6.16).

Certain rare light isotopes—deuterium, lithium, and boron—cannot have been produced in this way, for they burn too easily in stars. But the cosmic rays generated by early stars, colliding with the background plasma, *will* generate these rare substances in the correct amounts as well. (This was an idea that scientists such as Jean Adouze in France had independently been arguing for.) There is simply no need for a Big Bang to produce any of these elements.

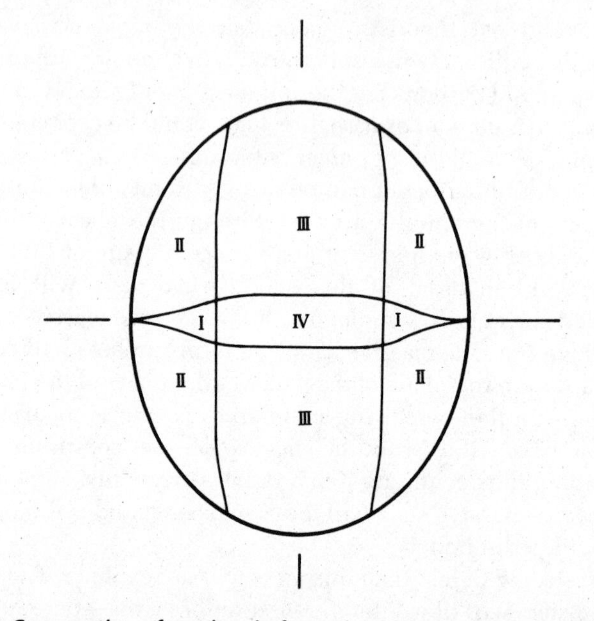

Fig. 6.16. Cross section of a galaxy in formation. Heavy stars that explode in supernovas and produce heavy elements, such as carbon and oxygen, are confined to the disk marked I. Lighter stars, releasing pure helium from the cylinder, are marked III and IV. The outer zones, marked II, are still pure hydrogen, but will be incorporated into stars as the cylinder of stars forming moves out.

The microwave background—the second phenomenon an alternative theory needed to explain—was a stickier problem. Its enormous energy wasn't hard to explain—Cambridge's Martin Rees had pointed out in 1978 that the energy released in producing the helium observed is just enough to produce the microwave background as well. As he calculated, the intense blue and ultraviolet light of the bright early stars would be absorbed by interstellar dust and reemitted as infrared radiation. My own work indicates that early galaxies would shine about five hundred times as brightly as our own galaxy in the infrared. (Rowan Robinson had, in fact, just observed such superbright galaxies with the Infrared Astronomical Satellite [IRAS]. Evidently they are forming galaxies, Johnny-come-latelies compared with most of the galaxies we observe.)

The real problem of the microwave background is not its energy but its smoothness. If it came straight from its source, various ancient galaxies, its intensity would fluctuate from one region to another, depending on the density of ancient protogalaxies in any given direction. But no such variations are observed. This, said conventional cosmologists, is clear proof that the microwave background must have come from some universal homogeneous event—the Big Bang.

The question in my mind in late 1986 was: How can the microwave radiation be scattered so as to become smooth and isotropic, as observed? The answer came to me as I drove through Princeton, after a visit to the physics library (perhaps Einstein's nearby house inspired me!). High-energy electrons spiraling around magnetic field lines within filaments, like any accelerated particles, generate synchrotron radiation—in this case, of radio frequencies. And Kirchhoff's law, a fundamental law of radiation, states that any object emitting radiation of a given frequency is able to absorb the same frequency. Thus, if the electrons in the filaments absorb photons from the background and then reradiate them in another, random direction, they will in effect scatter the radiation into a smooth isotropic bath, just as fog droplets scatter light into a featureless gray.

I found out shortly that Peratt had again beaten me to the punch. He calculated the radiation of his simulated scenario and found that the galactic filament produces about the same amount of microwave radiation as the microwave background. He too had reasoned that repeated emission and reemission might lead to isotropy and was about to begin detailed work with a colleague, Bill Peter. We agreed to keep each other informed so as to avoid needless duplication.

The only requirement for smoothing the background radiation was lots of small filaments, each with a strong magnetic field—a good description of the jets emerging from galactic nuclei. It seemed that every galaxy has these filaments, as my own theory indicates, and as was confirmed by observation (Fig. 6.17). On average, a photon of background radiation would encounter one of these filaments every few million years and be scattered. After several billion years isotropy would be complete. So the microwave background is not the echo of the Big Bang—it is the diffuse glow from a fog of plasma filaments, the hum from the cosmic power grid.

Moreover, this theory had testable implications. Specifically, radio waves will be absorbed by filaments traveling through intergalactic space. At long wavelengths, then—radio wavelengths—the filaments would appear to be black, except for a slight amount of energy they emit. So radio sources such as galaxies would be observable only through gaps in the filament thicket. Thus radio emissions from these sources would grow sparser with the distance of the source.

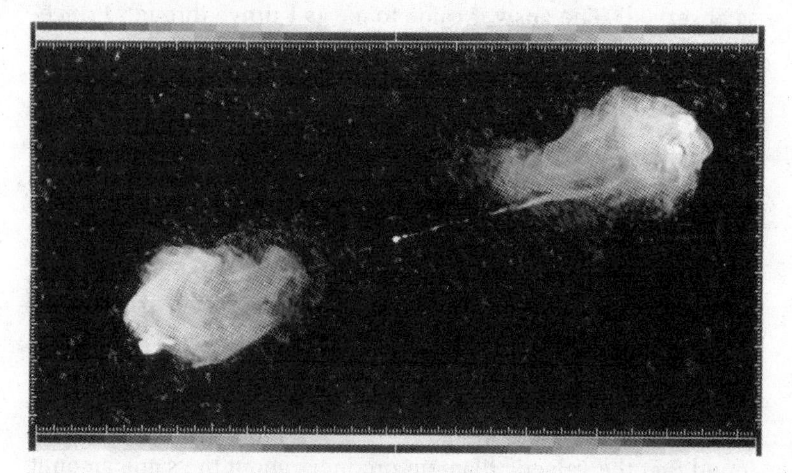

Fig. 6.17. In this radio telescope image of the powerful radio galaxy Cygnus A, the fine lines extending out from the central dot illustrate the galaxy emitting jets or beams that are force-free filaments. Such beams break up into amorphous clouds of filaments and eventually disperse through intergalactic space, cooling and forming the dense microfilaments whose electrons scatter the microwave background.

HOW THE FILAMENTS ABSORB ENERGY

While working on the absorption of radiation by the filaments, I calculated that the process would be very efficient. However, Peratt and Peter's conclusions were quite different—scattering would be far less effective. Their calculations involved the large supercluster filaments, which were less effective than the filaments radiating from galactic nuclei which I used—but not enough to account for the huge differences in results. After several months of worry I realized that they had assumed electrons move more or less randomly in a filament relative to the magnetic lines of force. In such a plasma, the effective temperature of the electrons is about equal to their energy—extremely high. When a photon hits such electrons they are as likely to be stimulated to emit another photon as they are to absorb the incoming photon. The two effects would almost entirely cancel each other out. I, however, had assumed a force-free filament, in which electrons move exactly along magnetic lines of force: these would generally radiate very little, since the lines of force curve only gently—the slower the curve, the less the radiation. When an electron does absorb a photon, however, it will jump off the line of force, circling around it in a tight spiral until it reemits a photon and jumps back down to the magnetic field line.

In this situation the effective temperature of the electrons will be their low-energy motion *around* the field lines, not their high-energy motion *along* them. Most electrons could not be stimulated to emit photons because they are too close to the field lines—it would take more than all their energy in moving around the magnetic field to emit. They will be able to absorb a photon only to reemit it later. Absorption will not be balanced by stimulated emission and will be far more effective than with random electron motion.

In more recent research, I realized that the jets from the galactic nuclei will spontaneously break up into finer and finer filaments, each pinching itself to stronger and stronger fields. (This phenomenon of finer and finer filamentation is observed in the laboratory and on the sun, where the intensities of the magnetic fields are close to those in the much larger galactic nuclei.) These finer but denser filaments further increase the efficiency of the scattering process—much the same way a given amount of water, condensed into fog droplets, scatters light far more effectively than the same amount evenly dispersed as humid air. It appears that most of the actual scattering occurs in tiny subfilaments no more than a few hundred meters across, having magnetic fields of twenty-five thousand gauss, comparable to the fields in sunspots or to strong artificial magnets on earth. (The earth's own field is about a gauss.)

This directly contradicts the conventional assumption that intergalactic space is completely transparent to radio waves. However, astronomers had observed that as one looks farther out into space, the number of radio sources increases much more slowly than the number of optical sources, and thus the ratio of radio-bright to optically bright sources decreases sharply. For example, a distant quasar is only one-tenth as likely to be radio-bright as a nearby one. Cosmologists have attributed this to some unknown, mysterious process that somehow caused the early, distant quasars to be less efficient at producing radio radiation, even though their optical and X-ray radiation is no different than that from present-day, nearby quasars. My model, however, provides the simpler explanation that the radio sources are there, but we can't observe them because their radiation is absorbed by intervening thickets of filaments.*

ORGANIZING THE OPPOSITION

I first presented my theory at the May 1987 IEEE International Plasma Physics Conference in Washington, D.C. Peratt had organized a session on space plasmas, including a half-dozen papers on galaxies and cosmology. My paper was well received. Fälthammar and Peratt both thought the model was valid, as did Alfvén when he later saw it. But I was most struck by the question asked by John Keirein of Ball Aerospace at the start of his presentation. Taking a straw poll of the hundred or so plasma physicists, he asked, "How many think the Big Bang is probably wrong?" About a third raised their hands. "How many think it is probably right?" Another third raised their hands. Evidently, the rest were undecided. Obviously, among plasma physicists, the Big Bang didn't hold undivided sway.

As I listened to the presentations by Keirein, Peratt, and Fälthammar, among others, I wondered why there were no science journalists here, attracted to what could be a hot story in cosmol-

* There may well be other mechanisms that help to smooth the microwave background. In 1975 N. Wickramasinghe, a longtime coworker of Steady-Stater Fred Hoyle, and others proposed that tiny iron whiskers a millimeter long but only a millionth of a centimeter across could strongly absorb and scatter radio waves and microwaves. Such whiskers may be produced in supernova explosions and widely dispersed in intergalactic space.

ogy. I realized that journalists are mainly attracted by new discoveries—observations are announced at *astronomy* conferences—and they asked theorists, conventional cosmologists, for their interpretations. And because observing astronomers never learn about the plasma work, which is not widely reported, they must be content with conventional responses.

By why shouldn't plasma theorists get together with observers, I wondered, especially with those whose observations contradict Big Bang assumptions? Theorists could introduce observers to a point of view far more consistent with their observations, while a small workshop bringing the two groups together would also attract journalists, thus a wider audience. Finally, plasma scientists would learn in depth about the critical observations. At the end of the session I raised the idea of an International Workshop on Plasma Cosmology with Peratt and Fälthammar, who agreed that it would be good to pursue. Alfvén, then in Sweden, later concurred, although he thought that it would be difficult to organize.

He was right. An organizing committee consisting of myself, Peratt, Peter, and Yusef-Zadeh made little headway because we couldn't find an institution willing to host such a controversial workshop.

However, our work received wider circulation when a revision of my article, originally written for the *Times,* was published in the popular science magazine *Discover* in June of 1988. A brief article, written by John Horgan, also appeared in *Scientific American.* But more significantly, there was a rash of new results. Byrd and Valtonen's work on dark matter was published. S. J. Lilly and then others reported the discovery of galaxies with extremely high redshifts which appeared to be older than the Big Bang! The discoveries led to widespread consternation among theorists —a full-blown crisis seemed to be brewing.

Interest in alternative theories started to grow, and leading observers like Tully, Shaver, and Valtonen expressed interest in the workshop. By this time, we had found a sponsor—the IEEE Plasma Sciences Society—and a place to hold it near the University of California at San Diego.

I sent out press releases to drum up the needed public attention, since without it I doubted that scientists would pay much attention. Both the *New York Times* and the *Boston Globe* sent reporters; and because Walter Sullivan was no longer the main

Times correspondent for cosmology, the assignment went to John Noble Wilford, who was new to this field, although a highly experienced science reporter. Wilford came with an open mind and was clearly interested in the possibility that conventional wisdom might be wrong.

The conference itself in February of 1989 was a success. Yusef-Zadeh's colleague Mark Morris reported on the latest evidence from the center of the galaxy that magnetic forces, not the gravity of black holes, must control the formation of filaments there. Rainer Beck, a leading West German radio astronomer, described his work on magnetic fields in other galaxies, showing how they contradict the conventional explanations. Valtonen summarized his and Byrd's work and the absence of "missing mass." Tully and Shaver presented overviews of the large-scale structure of the universe, from local filaments a few million light-years across to giant agglomerations a billion light-years across. And all this evidence was complemented by Jean-Paul Vigier, a leading French physicist, who described his and Halton Arp's work showing concentrations of quasars at specific redshifts.

The theoretical presentations from the plasma physicists were equally comprehensive. Alfvén gave an overview of the plasma universe, with Fälthammar and Timothy Eastman of NASA explaining the basic phenomena of plasmas and how they occur in the laboratory and in the solar system. Peratt followed with his description of galaxy formation. Between sessions, the two groups, observers and plasma physicists, engaged in lively discussions (Fig. 6.18). The observers were, in general, neither quickly won over (impossible in only a few days) nor thoroughly skeptical.

On the final day of the three-day conference I presented my own microwave theory, which again was well received. The final session centered on alternatives to the conventional explanation of the Hubble expansion, with presentations by Paul Marmet of the Canadian Research Council, Vigier, and Keirein.

At the end, I repeated Keirein's straw poll. I asked, "How many are sure, beyond reasonable doubt, that the Big Bang is wrong?" All the plasma physicists' hands went up, not unexpectedly. "How many," I asked, "think it is true, beyond reasonable doubt?" Only one observer's hand went up. Finally, "How many think that there remains reasonable doubt about the Big Bang?"

Fig. 6.18. After the February 1989 International Workshop on Plasma Cosmology, Hannes Alfvén (second from left) relaxes with Tony Peratt (center) and the author (right) at his home in San Diego.

All the remaining observers' hands went up. Evidently, the workshop had raised some doubts.

But it had done more than that. For the first time the data that conflicted with the Big Bang and the alternative plasma explanations were brought together in a single conference. And the press coverage was excellent: the *Times* ran a prominent article describing "The first serious challenge to the Big Bang in twenty-five years," and the *Globe* ran a detailed and accurate article. From now on, there would be no question that scientists and much of the general public would be aware that the Big Bang is not the only cosmological possibility.

One question raised at the conference worried me, however. After my presentation Tully had asked whether my theory that radio waves are absorbed as they travel between galaxies contradicts observations of nearby spirals. Because there is a correlation between a galaxy's infrared and radio brightness, but the infrared radiation is presumably *not* absorbed, the correlation should change with distance—*if* the radio waves are absorbed. In other words, infrared emissions should not drop with distance, but radio-frequency emissions should, just as a headlight is dimmed by fog. Yet, Tully continued, this isn't the case, so my theory is contradicted. I replied that, to the best of my memory, the correlation isn't so tight as to preclude the drop in radio intensity that I predicted.

Flying home from California I wondered if this might be a crucial test of my theory. In May I found a brand-new collection of such data on 237 galaxies, compiled by Nicholas Devereux of the University of Massachusetts: it showed some correlation between radio and infrared brightness, but with considerable scatter. The galactic distances, however, were not given in the paper, so I called Devereux and he kindly sent me the data.

There was indeed a clear correlation. When I took into account the fact that the density of filaments, and therefore the fading, is greater near to earth (because our galaxy is in a dense part of the supercluster complex), my model predicted that galaxies thirty to one hundred million light-years away should be about 3.4 times dimmer in radio frequencies than those nearby (allowing, of course, for the quantifiable loss of brightness due to distance). The data showed a fading of 3.9—a great agreement, considering the two-in-one-trillion chance that the correlation is accidental and that there is really no fading (Fig. 6.19).

Here was clear-cut evidence not only that my theory is accurate, but that the conventional explanation based on the Big Bang *must* be wrong. Such an absorption would greatly distort radiation originating ten billion light-years away, as the Big Bang theory claims the microwave background had. The cosmic background must, in fact, be generated locally, near to our galaxy, by an intergalactic medium that both absorbs and emits radiation.

I decided that there was finally evidence that conventional cosmologists could not ignore, and in July of 1989 I submitted a paper to the *Astrophysical Journal*, the leading astronomical publication. Previously this journal had not accepted plasma-cosmology papers since they were reviewed and vetoed by conventional cosmologists. But I figured that this observational evidence would be too strong to reject.

Not surprisingly, the reviewers did raise various objections to my claim that radio emissions were absorbed in intergalactic space and that therefore the conventional explanation of the microwave background is wrong. Did the dropoff in radio brightness continue to a larger distance? one reviewer asked. Perhaps

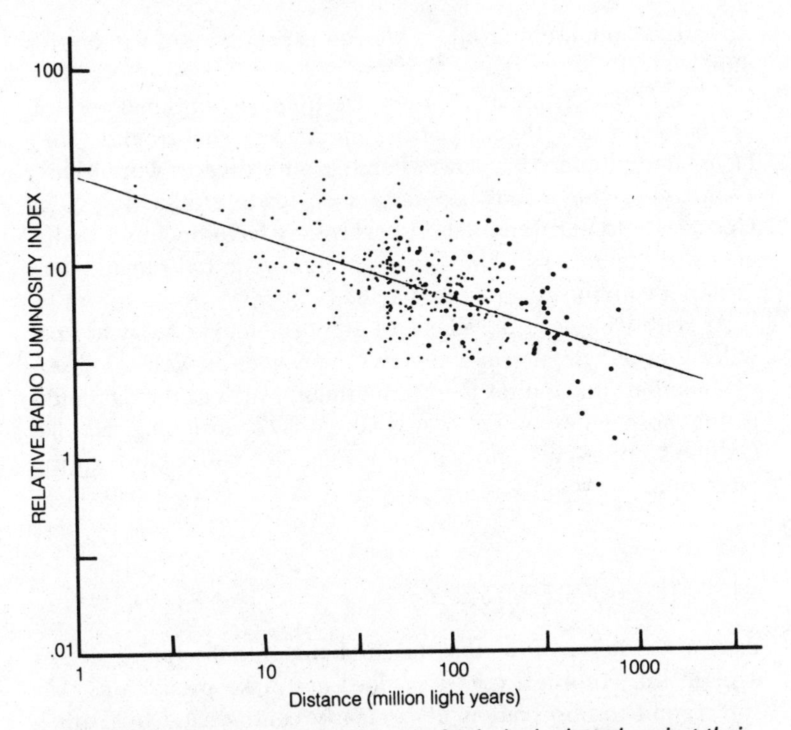

Fig. 6.19. The apparent radio brightness of galaxies is plotted against their distance. Radio brightness appears to fade rapidly with distance, implying that radio waves are being absorbed or scattered by the intergalactic medium.

brighter galaxies that can be seen farther off just intrinsically have less radio radiation, another suggested. "It is unthinkable that these conclusions are right," he mentioned, solicitously warning that the paper would hurt my reputation if it was published.

Over several months, I answered each objection in turn, using new data to show that indeed the decrease in radio brightness continued out to a billion light-years from earth and that even galaxies with identical infrared brightnesses were dimmer in the radio the farther away they were. Finally, in March of 1990, I called up the managing editor, Helmut Abt, and urged that the paper be published so that the issues could be debated more broadly than just among the reviewers and myself. He reluctantly agreed, commenting grimly, "We can expect a lot of flak on this one."

So far, there has been no response from conventional cosmologists to the new theories of the microwave background. They have simply ignored my own, Peratt's, and Alfvén's work. While I and others have sent papers to such leading cosmologists as George Field and Jeremiah Ostriker, they have declined to discuss the theories substantially, with noncommittal remarks like "I'm not convinced" or "You may be right."

As with Chapman decades earlier, cosmologists today are not willing to debate alternative views. This goes as well for observations that run counter to existing theory, such as the data indicating absorption of radio waves. Today's cosmologists, like Galileo's opponents four centuries ago, won't look through the telescope.

THE HUBBLE MYSTERY

Present evidence shows that the Big Bang, initially introduced to explain the Hubble expansion, does not make predictions that correspond to observation. It is clearly contradicted by Tully's observations of supercluster complexes and by the more recent confirmations of large-scale structures. Plasma cosmology *can*, however, explain this structure, as well as the abundance of the light elements and the microwave background. It can also explain phenomena that the Big Bang model precludes, such as the absorption of radio waves by the intergalactic medium and the formation of spiral galaxies and quasars. This returns us to the problem: What caused the Hubble expansion? The cosmological debate will not be resolved until this basic question is answered.

It turns out that there are actually a half-dozen or so competing explanations.

The first, of course, is Alfvén's antimatter theory. A second, which involves the least concessions from conventional cosmology, is based directly on general relativity. Einstein's equations allow for an infinite number of solutions, only some of which produce a singularity—a point of origin. Other solutions present a universe that contracts to a certain size, then smoothly turns around to expand again. It has no origin in time. What pushes the cosmos apart is the cosmological force discussed in Chapter Four —a universal repulsive force that resembles centrifugal force but operates in all directions equally.

This relativistic solution will work for a universe with as little matter as ours appears to have, yet the cosmos would be "closed" in space—finite because of the curvature generated by energy in the repulsive field. Since the cosmological constant—the strength of the cosmological force—can be adjusted in this model, the minimum state of contraction can be quite big—a billion light-years across, for example, so that large-scale structures which would have taken hundreds of billions of years to form might survive the minimum point of contraction.

This is what Alfvén has termed a "Tycho Brahe solution." Just as Brahe combined the heliocentric and geocentric theories, this one hangs on to a general relativist universe, reconciling it with a plasma universe of infinite duration. It isn't easily contradicted by observation, but neither can it be confirmed. The universe is expanding, in this view, because of a universal expansionary force—not much improvement over Aristotle's explanation for the moon's phases: Because the moon tends to have phases!

Another approach to the Hubble relationship denies that it represents a true expansion of the universe, a process in which the universe has become increasingly diffuse. The redshifts are real, but they do not signify what astronomers have thought. For example, light may lose energy as it travels through space, shifting it to the red. Or perhaps all objects, all space expands continuously, a certain tiny percentage a year. Distant objects are redshifted, yet no real expansion takes place, since the density of everything remains the same.

Many such theories have been elaborated over the past years, and all have problems. (A summary discussion of these theories can be found in the Appendix.) The question of the Hubble relation remains unanswered, and other fundamental questions about the cosmos must also remain unanswered until an adequate theory is found. Far more theoretical and observational work is needed.

But we are not stuck with the Big Bang by default. It is one explanation of the Hubble relation that is clearly contradicted by observation and can clearly be ruled out. There is no evidence that the universe had a beginning in time.

If the universe is truly infinite in time and space, then the implications go well beyond cosmology to the whole of our view of nature, to religion, philosophy, and society as a whole. It is to these implications that we now turn.

NOTE

1. Hannes Alfvén, "How Should We Approach Cosmology?" in *Problems of Physics and Evolution of the Universe*, Academy of Sciences of Armenian SSR, Yerevan, 1978.

PART TWO

IMPLICATIONS

7 THE ENDLESS FLOW OF TIME

Michele has left this strange world just before me. This is of no importance. For us convinced physicists the distinction between past, present and future is an illusion, although a persistent one.

—ALBERT EINSTEIN
on the death of his closest friend, 1954

The resulting dichotomy between time felt and time understood is a hallmark of scientific-industrial civilization, a sort of collective schizophrenia.

—J. T. FRASER, 1981

We were seeking general, all-embracing schemes that could be expressed in terms of eternal laws, but we have found time, events, evolving particles. . . . The irreversibility [of time] is the mechanism that brings order out of chaos.

—ILYA PRIGOGINE, 1984

Plasma cosmology is beginning to produce a fundamental revolution in our view of the universe—a return to the Ionian idea of a universe infinite in duration, evolving, not decaying, a universe characterized by progress. Such a revolution implies a radical change in our concept of time itself. Just such a radical change is already under way, inspired by developments not in cosmology but in thermodynamics. Indeed, it appears that these two revolutions in separate areas of science are about to merge in a broad transformation of the scientific view of nature.

Thermodynamics and cosmology are not, in fact, isolated disciplines. The Big Bang initially arose from ideas in thermodynamics—specifically, Eddington and Lemaître's interpretation of the second law of thermodynamics. If the second law says that the universe is continually running down, approaching the "heat death" of perfect equilibrium, they argued, then clearly the universe cannot have existed forever. It must have been "wound up" at a finite time in the past.

Conversely, if one asserts that the universe had no origin in time, then one must explain how it is that the universe has not completely decayed into uniform equilibrium in the infinite time that it has already existed. How is order maintained? How is progress possible?

So, to resolve the cosmological question of the evolution of the universe, I have to discuss the question of the second law of thermodynamics. Just as the second law of thermodynamics justifies the Big Bang, so the Big Bang, in today's cosmology, justifies the second law. It is the reason, in the view of many physicists, why time as we know it—with a past, present, and future—exists at all.

This is one of the deepest paradoxes of conventional physics today. According to all the laws of physics there should be no distinction between past and future, no direction to time. Since the second law says that entropy necessarily increases with time, and thus the past and future differ, the second law, too, is contradicted.

In relativity theory, for example, time is simply the fourth dimension—there is no more difference between past and future than between left and right, There is no flow of time: all the equations would look the same if time were reversed. Nor is this true of relativity alone. Newton's laws and the laws of quantum mechanics also are what physicists call "time reversible"; they define no unique direction for time. If one were to make a movie of two billiard balls colliding, for example, it would look just as credible if it were run in reverse.

But in the real world, there is a difference. If it is two raw eggs that collide and break in the movie, it would look absurd in reverse. The two eggs would assemble themselves out of a puddle and roll off. In the real world babies are born, never unborn, they grow up, never down, and eggs are scrambled, never unscrambled. These processes are all *irreversible*: time moves forward, toward growth or decay.

Hence the fundamental question: If "the laws of the universe" have no direction in time, why does the real world? Why do laws like the second law, which have a direction for time, work?

The conventional answer to this question is, strangely, the Big Bang. The Big Bang started the universe off in a highly orderly and regular state—a "perfect" state of very low entropy. Since the universe must run down through states of increasing disorder, closer to equilibrium (the state in which there is no flow of energy), the direction of time is defined. Time is just the direction "away" from the Big Bang. If there was no Big Bang, there would supposedly be no difference between past and future. The universe would be at equilibrium, and no event would distinguish past from future. But the unique event of the Big Bang, so symmetric in space, creates an asymmetry in time.

Thus, if there was no Big Bang—as seems to be the case—we have further questions: Why does time move forward? Is there a difference between past and future, or is it, as Einstein believed, merely a persistent illusion?

The importance of the answers extends far beyond their role at the center of a consistent cosmology. They strike at the heart of some of the greatest mysteries faced by science, philosophy, and religion—the questions of the nature of human consciousness, the relation of mind and body, and free will. The distinction between past, present, and future is basic to our experience of consciousness—we are conscious in the now, we remember the past, but we cannot know the future. It also is central to our idea of free will, for it implies that our actions in the present affect the future, that the past is fixed but the future can be changed. How can these ideas be reconciled with a concept of physical laws in which past, present, and future all exist equally and cannot be distinguished?

THE PARADOXES OF TIME

The basic answers to these questions have been formulated by Ilya Prigogine and his colleagues over the past twenty years. Prigogine, a Russian-born chemist raised in Belgium, received the Nobel Prize for his work in reconceptualizing thermodynamics. In his view, the paradox arises from a misunderstanding of time and of nature. He believes that there is no real tendency toward decay in the universe—on the contrary, order tends to arise out of chaos, the universe tends to move toward greater complexity and faster rates of evolution. The universe doesn't need to have been wound up because it isn't running down.

Nor is there a contradiction between the time-reversible laws that operate on the atomic and subatomic levels, and irreversible laws that operate on larger scales. Time, Prigogine argues, is irreversible at all levels—the reversible laws of physics are only approximations. In reality, temporal irreversibility is "built into" the universe from the tiniest particle to the mightiest galaxy. Time is not merely another dimension, it is the history of the universe.

What exactly is wrong with the conventional physics' understanding of time? Let me begin by describing these views a bit more exactly. There are, in fact, two views of time for two different worlds. First is reversible time for a world of changeless perfection. For this world real time—time with past, present, and future—does not exist. In this world, Einstein's world, the entirety of time is laid out like a map in four dimensions. The year one billion B.C. and one billion A.D., as well as 1991, all exist with equal reality. All is predestined.

In this world, an object's history is described by its "trajectory," a line drawn in four dimensions from the start to the finish of its existence. The line does not come into existence point by point. It exists in four dimensions, and describes where the particle is in the three spatial dimensions for any value of the fourth dimension, time.

Where does this timeless world of perfect predictability exist? Either in the world of the very small—atoms and elementary particles—or in the giant world of the heavens—stars and planets moving in accord with the laws of gravity. We have encountered this changeless, perfect world before—Plato's world of eternal ideas, the ancient and medieval world of the perfect, unchanging heavens.

The second world, and the second conventional definition of time, applies to everyday events on earth. Here time is measured by decay: energy is expended, work is done, but every effort to create order creates more disorder, and the world falls to absolute ruin. Here again we are on familiar ground—the medieval and ancient concept of the mutable world beyond Eden, after the Fall.

Yet these notions of time are generalizations derived from physical laws based on millions of observations, laws that form the basis of present-day technology. They work spectacularly well: electromagnetics, quantum mechanics, and both Newton's and Einstein's laws of gravity are clearly time-reversible—without true time. In the case of gravitation, the laws are actually used as time-reversible. The planets' trajectories really do look just as reasonable run backward as forward.

The laws of thermodynamics have perhaps even wider use. In nearly every technology, engineers take into account the dissipation of heat and the very real limits on the way energy, including heat, can be put to use, such as in generating electricity. Used together with the laws of electromagnetism and quantum mechanics, thermodynamics can accurately predict a huge range of phenomena.

These laws are unquestionably valid and the ideas of time derived from them are also useful in many cases. But they have been grossly overgeneralized and applied to the universe as a whole. The situation was much the same with the ancients: they saw heavens that do seem, for the most part, unchanging, and things on earth that are, virtually without exception, subject to decay. They generalized this scheme into a theory of the cosmos and transformed it into an ideology. They were wrong, and the conventional view of time is equally wrong today.

This view of time rules out the primary tendency of universal evolution—progress. The cosmos evolves from chaos to order, developing more and more complex entities, in an ever-accelerating movement *away* from a final, eventless equilibrium. Conventional physics views any change as a necessary regression, as devolution toward equilibrium. Yet if we look at the long-term tendency of evolution, reality is just the opposite—the universe winds up, not down.

Let's begin here on earth. Our planet derives nearly all its energy from the sun. After five billion years, it should be quite close to equilibrium, its temperature a constant, all chemical reactions halted—like the moon. But in reality, we know from the fossil record that earth, under the influence of life, has moved steadily away from equilibrium—energy flows on earth have *increased* over time. Living things depend on a constant flow of energy, and the total mass of living material on earth—the biomass—has clearly increased over time. Moreover, the rate at which each gram of biomass processes energy, its metabolism, has increased greatly. The most recently evolved types of organisms, such as mammals, have much higher metabolisms than earlier types, such as reptiles, fish, and invertebrates.

As a direct consequence of the increasing energy flows controlled by living things, the earth has moved away from chemical equilibrium as well. The existence in our atmosphere of 20 percent oxygen, an extremely reactive chemical, is possible *only* because plants use sunlight to break down carbon dioxide. The atmosphere's oxygen content has increased—the earth is moving away from chemical equilibrium.

To be sure, the tendency toward equilibrium is supposed to hold only in "closed systems," and because the earth is heated by the sun, it is not a closed system. But we can consider other large-scale evolutionary processes which can be treated as closed systems. For example, if we take a volume of air the size of a room and compress it into a small space, it will rapidly reach equilibrium—an even distribution through the space. However, if the room is astronomical in size, the evenly distributed gas will contract under its own gravitation. If it has little angular momentum or heat to start with, it will contract quite far, until it is compressed into a small volume. As it contracts it will heat itself up, so that at a later time, far more energy will flow from the contracted object to the rest of space than at first. This is what has happened in the gravitational contraction of objects like stars.

We've seen plasma behave similarly, even neglecting gravity. With sufficiently high initial energy, plasmas will naturally evolve from a homogeneous, evenly distributed state with small currents and energy flows, to a highly inhomogeneous, filamentary state with large currents and energy flows. Energy has not been created—it has simply been organized by the natural self-pinching of electrical currents.

These are the processes that have taken place as the universe evolved toward its present state—an accelerating trend toward an ever more differentiated, inhomogeneous condition, with more energy flows ever farther from equilibrium. Of course, proponents of the conventional view of thermodynamics, including such Big Bang theorists as Stephen Hawking, have reasoned that, because the overall disorder of the universe *must* continuously increase, any increasing order of matter must be compensated for by a greater increase in the disorder, or entropy, of the gravitational field. In this view, a smooth, homogeneous field is highly ordered by definition, while a gravitationally "dimpled" one is disordered.

But this is merely playing with definitions. Why is a homogeneous field "low entropy" and a homogeneous distribution of matter "high entropy"? The key point, which avoids questions of the definition of "order" or "disorder," is that the universe, just like our own planet, appears to be moving away from an equilibrial "heat death" toward higher energy flows, away from a homogeneous distribution of matter toward increasingly complex structures, and away from slow change toward faster rates of evolution. The universe we observe is simply *not* decaying; the generalization of "the law of increasing disorder" to the entire cosmos is unsupported by observation.

The notion of reversible laws as the underlying basis of the cosmos is equally problematic. If the true laws of the universe have no temporal direction, as Einstein believed—"past, present, and future are but an illusion"—then they are the products of human perception. There is no real "now" except insofar as our consciousness deceives us.

But, as Prigogine points out, almost everything we observe in the world either grows or decays. In particular living organisms, including *ourselves*, are clearly the products of an evolutionary process that is unidirectional, that somehow separates past from future. "Are we ourselves—living creatures capable of observing and manipulating—mere fictions created by our imperfect senses?" he asks in *Order Out of Chaos*. "Is the distinction between life and death an illusion?" Whatever physicists may argue in their journals or classrooms, it is a rare one that can honestly answer "yes."

The idea that all times have equal existence poses in the sharpest possible form the dualism of today's physics, for it rules out human consciousness as an object of scientific inquiry. It becomes, as it was for Descartes, the "ghost in the machine." For in our consciousness there is a now, a past, and a future. Since we are physical beings, that consciousness must either be incorporated in some way into our understanding of the universe, or it is forever relegated to a supernatural realm, a spiritual world beyond the ken of science.

A world governed by timeless, unchanging laws is reduced to an automaton, as predictable as a machine. Like any machine, "an automaton requires an external God," as Prigogine puts it—for without evolution, without creative time, nothing can explain the origin of the cosmos and its laws.

By denying progressive time, physics denies not only the consciousness of the physicists themselves but also the possibility of explaining the universe without recourse to the supernatural. And by denying human consciousness—the fundamental basis of all human experience as an object of science, even in principle—modern physics draws a gigantic chasm between the way physicists view the world and the way most people do. To banish consciousness is to banish all the qualitative richness of nature that consciousness perceives. The realm of the senses vanishes, leaving a disembodied and silent world of frequency, amplitude, and, above all, pure number—the dead and dull world of "unemotional" science which has turned so many away from the scientific enterprise as a whole.

As I discussed in Chapter Four, neither the conception of time as decay nor the notion of a timeless world based on eternal mathematical laws evolved in isolation from general, cultural, and political history. In the late nineteenth and especially in the twentieth century, these concepts arose from a society in the midst of titanic convulsions, one in which the progress of previous centuries seemed to have been superseded by a return to chaos. The world of decay seemed a pessimistic description of the real, historical earth, while the timeless world seemed a refuge. As Einstein put it, "one of the strongest motives that lead men to art and science is flight from everyday life with its painful harshness and wretched dreariness and from the fetters of one's own shifting desires. . . . Man seeks to form for himself a simplified and lucid image of the world and so to overcome the world of experience by striving to replace it to some extent by this image."[1] By trying to flee from the all too real world of the present century, of Auschwitz and Hiroshima, however, the conventional view of time has sharply restricted science's ability to describe the universe—the ultimate purpose of all science.

The result, as Prigogine emphasizes, is to alienate man from nature. If there is no tendency toward evolution or progress in nature, then human existence itself is nothing but a meaningless accident, and humans are isolated in an indifferent and incomprehensible universe. In either a timeless or a decaying cosmos there is no room for anything that has value for humanity, no room for consciousness, joy, sadness, or hope. The universe becomes, in the words of Alfred North Whitehead, "a dull affair, soundless, scentless, colorless, merely the hurrying of matter, endless, meaningless."

ORDER OUT OF CHAOS

What is the alternative to this conventional view of time? Why doesn't the world decay into chaos? Prigogine's explanation begins from the limitation of Boltzmann's original work on thermodynamics. As I mentioned before, Boltzmann proved his theorem —disorder always increases—only by presupposing a high degree of disorder, the random movement of atoms and molecules. Starting in 1967 Prigogine contended that only in such random conditions, already very close to equilibrium, does this law actually hold. Any slight deviation from equilibrium is immediately disrupted by random molecular motion.

But if the system were already far from equilibrium, if there were significant flows of energy through it, it would *not* tend to return toward equilibrium. In fact, it would move away from it, creating order and structure in the process.

Prigogine had been fascinated with this process since the beginning of his career in the forties, but it was not until the sixties that he solved the problem. The key he discovered was the growth of fluctuations through instability.

The simplest example of this idea is a pot of water being heated on a stove. If the heat is turned on very low, the energy flows are small and the movement of the water molecules remains random. Heat disperses through the water by conduction—through these random motions. But as the heat on the stove is turned up, at a certain critical point the water's motion suddenly becomes highly organized—convection occurs. If the water is initially motionless, the convection forms extremely regular hexagonal cells of water, called Bernard cells. The motion of the water becomes unstable, allowing tiny fluctuations to grow. A small amount of hot water rises, pushing aside cooler water above it, which moves downward, producing a circular motion. Energy from the flame then heats the cool water, causing it to rise in the wake of the first droplet, and so on. As more water joins the circulation pattern, it spreads, developing more cells around it. Rapidly more energy is entrained in the pattern, until the entire

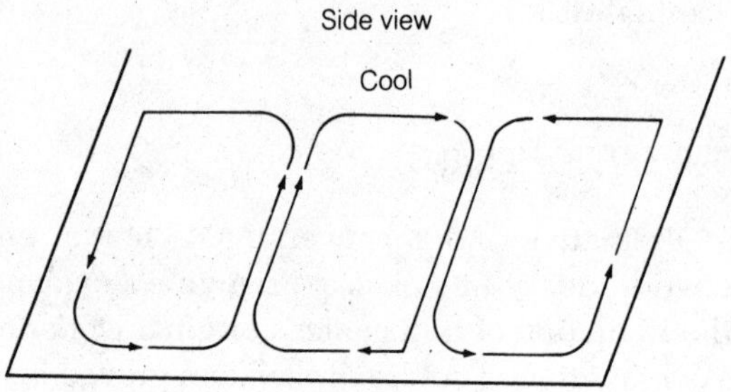

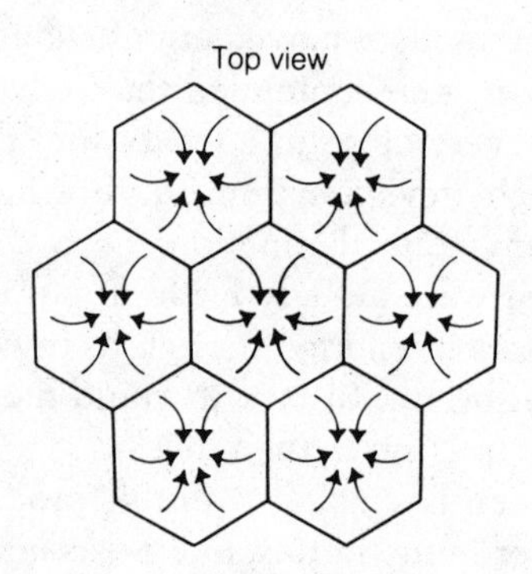

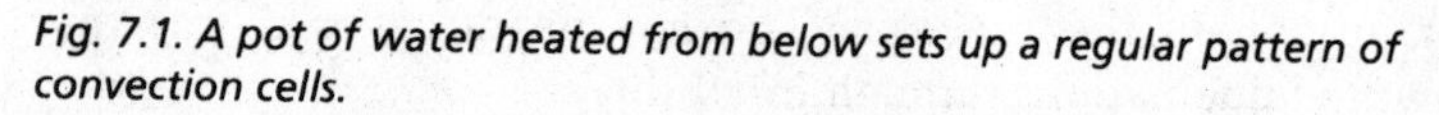
Fig. 7.1. A pot of water heated from below sets up a regular pattern of convection cells.

volume of water is filled with circulation cells—all of the water is moving in a coherent pattern, no longer random. Order has arisen from chaos (Fig. 7.1).

The instability that creates the order is generated by energy flows. If the initial energy flow is small compared to the water's random motion, and if the water is thus near equilibrum, a fluctuation is dispersed before it has time to grow—starved for lack of new energy, the system remains stable. But if a lot of energy flows—in our example, the heat is turned up—it is captured by the fluctuations so quickly that they grow exponentially, and the system becomes unstable.

Conventional thermodynamics, which ignores instabilities and their ability to create order, works so well in technology because engineers deliberately design machines that either avoid instability entirely or control it so that fluctuations cease growing. But where fluctuations and instabilities can't be ignored—for example in the turbulent flows of weather systems or in the unstable plasmas of fusion devices—conventional thermodynamic approaches break down.

Such unstable systems, which are the most common types in nature, are not easily predictable because the motion of individual particles is governed in part by the global behavior of the system—water molecules are forced to participate in the motion of the far larger circulation cells. The relatively small and subtle effects of distant particles' motion can be magnified by instability until these small effects override the immediate, random jostling of neighboring molecules.

Conversely, instability can magnify the motion of a small group of particles, or even that of a single one, developing new patterns of motion. Generally, when instability occurs, more than one possible pattern can result, and it is the action of relatively small "seed" portions of a system that determines what actually happens. As in social systems where the action of an individual can, at a critical moment, change the course of history, so for physical systems: individual subunits can affect the behavior of the whole. In this way a system develops a *history*.

The creation of order through instability is an open-ended process. The fluctuations that grow through instability create order by "capturing" energy flows. Thus the Bernard cells capture the flow of heat energy from the stove to the water, converting it into motion. But these structures in turn increase the overall energy flows. Once the water starts to convect, the flow of heat is far faster and more efficient: the system is pushed farther away from equilibrium, and new instabilities and new structures can arise.

For the simple example of a pot of water only one set of fluctuations and one structure can develop. But more complex systems can develop through a series of stages. Each stage begins with the growth of instabilities and the formation of new structures. These structures grow as large as they can until a new steady state develops. But since more energy flows are then available, new fluctuations can take advantage of them, setting into motion a new cycle (Fig. 7.2).

It is this generation of new sets of instabilities and fluctuations, new ways of capturing energy, new modes of evolution that allow the universe as a whole, and systems within it, to move further away from equilibrium. This is not an automatic process. At times, new instabilities will not emerge, and the existing system runs down as the source of energy flow is exhausted—returning toward equilibrium.

But as soon as some new instability develops—possibly else-

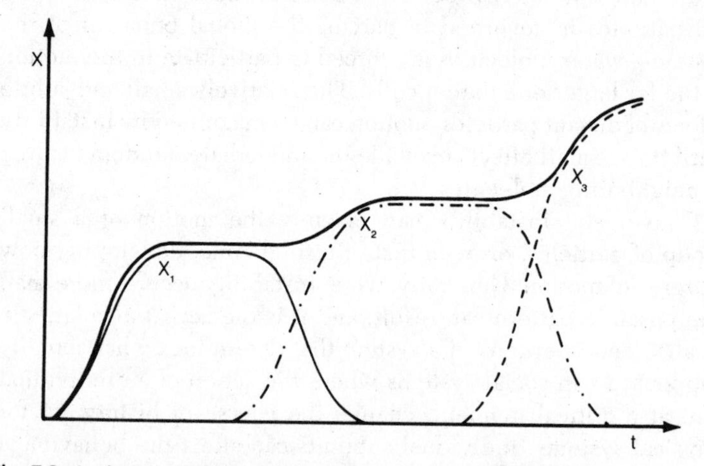

Fig. 7.2. A given set of instabilities can grow only so far (curve X_1) before saturating and decaying. But on the basis of the organization and structure formed by the first set of interactions, a new set comes into being (x_2) and rises to a higher level of energy flow. This sequence continues without limit in the course of universal evolution. (Illustration from Prigogine and Stenger, Order Out of Chaos.)

where—energy flows are again captured. Since those processes that capture energy the most efficiently grow the fastest and increase energy flows the fastest, there is a long-term trend *away* from equilibrium.

THE EVOLUTION OF THE UNIVERSE

The easiest way to understand what may seem to be rather abstract concepts is to apply them to the actual history of the universe, so far as it can be reconstructed at present. (While the basic concepts just described are Prigogine's, the examples I will discuss are others' work.) To simplify matters I will ignore the important question of the role antimatter may have played. While this could radically change the story's details, it won't alter the general form, and it will advantageously limit the description to processes directly studied in the laboratory.

The story's beginning depends not on some "initial conditions" ordained by reason, but merely on the present state of scientific knowledge. A convenient starting place is a universe filled with a more or less uniform hydrogen plasma, free electrons and protons. We'll also ignore whatever inhomogeneities exist as the result of earlier stages of evolution that brought this plasma into being. At present, we have no real knowledge of what such processes were. It is reasonably certain that the plasma had motion and energy, thus electrical currents and magnetic fields flowed through it.

This leads to the first stage of evolution (Fig. 7.3). As discussed in the previous chapter, plasmas are generally unstable to fluctuations, and the first fluctuations were formations of tiny plasma filaments or vortices. These vortices will attract others moving in the same direction and will tend to grow, capturing larger and larger flows of energy. Tiny trickles of electricity will flow together into mighty rivers. Indeed, theoretical studies have shown that force-free filaments capture energy more rapidly and release it as radiation more slowly than any other plasma structure.

Over an immense period of time the plasma will develop a web of ever-larger filamentary vortices, becoming less homogeneous, and will have increasing levels of energy flow. But, like any single mode of growth, it has inherent limits. Filamentary

THE EVOLUTION OF THE UNIVERSE

Stage of Development	Duration (Years)	Maximum Power Density (ERG/SECXCM^2)
Electromagnetic (growth of magnetic filaments)	2 trillion	10^{-34} (magnetic filament)
Gravitational-electromagnetic	1 trillion	one millionth (quasar)
formation of		
superclusters	100 billion	
clusters	10 billion	
galaxies, stars	1 billion	
Nuclear	20 billion	
formation of elements	(to present)	1,000
planets		(center of star)
Biological (on earth)	4 billion	
single-celled	2 billion	30,000 (mammal)
multicelled	600 million	
land dwellers	300 million	
mammals	60 million	
hominids	2 million	
Social (on earth)	2 million	
humans	50,000	
agriculture	10,000	
Bronze Age—irrigation agriculture	3,000	10^{31} (so far) (plasma focus device)
Iron Age—specialized agriculture	1,500	
feudal agriculture	1,000	
capitalist-industrial	400	

Fig. 7.3. While the Big Bang universe starts out fast and slows down, the likely real evolution of the universe steadily accelerates as its power density increases, going from evolutionary stages of trillions of years to hundreds or less for social evolution.

growth is limited by its characteristic velocity—around 1,000 km/sec. As the vortices get larger their rate of growth slows down. At a certain point, the growth of the vortices will be balanced by their energy loss through synchrotron radiation from the protons, which will come to carry most of the current: the vortices will cease to grow and begin to decay.

If we can imagine a deductively oriented physicist magically transported to this earlier cosmos, he would be able to describe the essential processes in terms of a simple set of laws—Maxwell's laws of electromagnetism. Knowing these laws, he would predict that within about ten quadrillion years or so, the universe would start to run down, irreversibly converting organized electrical currents into random synchrotron radiation.

But that didn't happen. After a mere one or two trillion years, something new happened. By this time (again measuring from our arbitrary start of the story, not from some first instant of time) the largest filaments had grown to about five billion light-years in radius. At this point a new instability suddenly appears—gravitational instability. Earlier, when the largest concentrations of matter were far smaller, gravitational forces were no more than subtle "corrections" of the dominant electromagnetic interactions. Our physicist would have been justified in ignoring such minor phenomena. But as larger energy flows move our system farther from equilibrium, *new* instabilities become possible. A slight gravitational contraction of the filaments increases their gravitational force, causing more contraction. As we've seen, such contraction induces new systems of filaments, drawing current toward the center of the contracting body. The gravitational energy generated by the contraction is partially converted to electrical energy—quasars and the galactic nuclei shoot forth glowing jets, feeding energy back into the vortices' magnetic fields. This magnetic field energy then interacts with the plasma of other contracting bodies to generate new filamentary systems, which distribute the developing angular momentum, pinch the plasma, and allow the release of more gravitational energy.

A new cycle of instability has started. The universe has become a vast power grid, converting gravitational energy into electricity, which powers the compression of more matter and the release of more energy.

This new cycle of fluctuations creates not a single set of structures, but a whole hierarchy—superclusters, clusters, galaxies, and stars. Each stage of contraction has its own natural limits, producing concentrated bodies with a certain range of orbital velocity, as we've seen. Each body in turn breaks up into subbodies, each with its own filamentary currents, each giving rise to its own progeny. In effect, the "waste matter" of one cycle's release of gravitational energy becomes the "raw materials" for the next cycle.

The result of this gravitational-electromagnetic stage of evolution is the production of a complex and ordered system of entities, ranging from stars to galaxies to superclusters, each pouring out concentrated electrical energy. The degree of concentration of energy flows from this process is enormous. In the primordial vortices, energy flow amounts to only about 10^{-34} ergs/sec per cubic centimeter of space (the power of a hair dryer in the volume of the solar system), but in the concentrated outbursts of a quasar, energy flow is about one-millionth of an erg/sec per cubic centimeter (a hair dryer per cubic kilometer)—a colossal increase by a factor of ten thousand trillion trillion.

This evolutionary stage lasts perhaps another trillion years, but moves through its cycles more and more rapidly, forming supercluster complexes in hundreds of billions of years, superclusters in tens of billions of years or so, and clusters and galaxies in mere billions of years. As in Prigogine's laboratory experiments showing that chemical reactions can occur nearly simultaneously in widely separate parts of a reaction chamber, so in our metagalaxy, clusters and galaxies and stars spring into being more or less simultaneously throughout space.

Like the first stage this epoch has inherent limits, particularly for the density of objects. At the density of stars the collisions of ions and electrons become so frequent that filaments no longer form (the plasma in stars is quite different from that found elsewhere in the cosmos). Again, our imaginary physicist would conclude that the end of the universe is at hand. The process of contraction would peter out in a few tens of billions of years, leaving cold cinders of stars, or even black holes, and in a few hundred trillion years the huge filaments of current would radiate their energy away. Indeed, he would gloomily contend that the new round of evolution had only brought the inevitable end closer.

And again his deduction would lead him astray. A second, nuclear, revolution begins in the hot center of stars, bringing into play a third set of "laws of the universe" and a third set of interactions, fluctuations, and instabilities. In low-density space nuclear reactions virtually never occur—for most purposes they can be neglected, and in no case are they self-sustaining. But in the hot, high-density stellar cores, the end products of the previous evolutionary stage, collisions between protons will lead to the slow production of deuterium, then helium, and the release of sufficient energy to maintain the reaction and to support the star's structure. The universe is suddenly filled with light.

Enormously higher energy flows are generated—1,000 ergs/sec per cubic centimeter (enough power for a light bulb every cubic meter), a billion times the power density of a quasar. This energy is not "used" just once, only to wander away aimlessly, but it is recaptured repeatedly. The multi-MeV photons released by a single fusion reaction heat ten thousand ions to the temperature required for fusion. These ions then emit 1 KeV X-rays, which work their way up through the star's structure, their pressure supporting the star's weight against collapse, even after it has lost nearly all its angular momentum to its planets.

From the surface of the star plasma glowing at .5 eV (five thousand degrees) sends out light into the universe. Part of this is absorbed by interstellar and especially intergalactic dust, which is heated to around 7° K. The infrared photons emitted by this dust, in turn, heat electrons trapped in the filaments snaking out of galactic cores, adding to the 3° K cosmic background radiation that bounces back and forth among the trillions of jets and filaments.

This background radiation presses the filaments and plasmas of a given agglomeration of matter—a supercluster complex or a collection of such complexes—outward at several thousand kilometers per second. (This expansion, by the way, can contribute to the observed Hubble expansion.) As the plasma crosses existing magnetic fields it generates tremendous new electrical currents. Possibly as much as a tenth of all energy now being liberated in the stars' nuclear fires is thus converted to electricity. This is a colossal thermonuclear generator.

The expanded electrical currents now complete the cycle by pinching new supplies of plasma together to create new galaxies and to generate new quantities of fusion power. The magnetic fields as well help to create the filaments in existing spiral galaxies which lead to the formation of new stars in old galaxies.

As with the gravity-driven stage of evolution, nuclear-powered evolution involves a series of substages. When hydrogen is exhausted within individual stars, its by-product helium then becomes, at higher temperatures and pressures, fuel for the production of carbon and oxygen. When all the fuels for a star are exhausted, its explosion in a supernova scatters the elements to the surrounding interstellar medium—fuel for new stars.

The initial generation of stars burns hydrogen to helium quite slowly. But in subsequent generations, carbon and nitrogen act as catalysts, enormously accelerating the reactions. Again the debris of one stage fuels a subsequent stage.

Thus a series of interactions, generating fluctuations which are magnified by instability, has driven the cosmos from a homogeneous and random hydrogen plasma to the differentiated and dynamic universe of stars, galaxies, and planets, each made up of a hundred different chemical elements. There is no tendency toward decay, nor is an external power needed to generate order. Order and complexity come into being through natural processes governed by a series of interactions—electromagnetic, gravitational, and nuclear. In short, chaos begets order—progress.

Just as our cosmos has evolved, so has our imaginary physicist —he is now a conventional cosmologist or thermodynamicist contemplating the existing universe. But he is still saying the same thing. "In less than a trillion years all the hydrogen and other light elements in the universe will have burned to iron, fusion will be impossible, and all the stars will gutter out into endless night. Each of these stages has just succeeded in bringing the end a hundred times closer."

By now it's clear that, at each stage, he has assumed that he knows all the possible interactions that can produce energy flows in the universe. It is the myth of final knowledge. But he was wrong at each stage and is wrong today.

First of all, we do know that there are energy sources other than fusion. Annihilation energy provides about one hundred times more energy per unit of mass. But more important, we do *not* know the limits, if there are any, of energy production. Yes, we know that under present circumstances the maximum energy derivable from a piece of matter is defined by $e = mc^2$—mass times the speed of light squared—which yields 10^{21} ergs/gram. But our universe appears to have a very significant amount of energy tied up in existing matter. Where did *that* energy come from? Is there more?

Cosmology has dodged this question by hypothesizing that this matter-energy comes from the gravitational energy of the Big Bang. However, as we've seen, this requires that omega equal 1—which it clearly does not. Gravitational energy amounts to between one-hundredth and one ten-thousandth of the energy tied up in observed matter (and there's no reason to assume that substantially more matter exists). This is wholly insufficient. We simply *do not know* where the energy in matter derives from, and we do not know whether and under what circumstances it can be captured or released. Until we do know, we cannot set "scientific" limits to the energy available in the cosmos.

In the evolution of a system, the flow of energy is just as important as its generation. Energy cannot be created or destroyed, it can only be captured and released. The rate of energy flow, thus the distance from equilibrium and the degree of organization in a system, depends on the rate at which energy is captured or released. There is no limit to the number of times a given amount of energy can be recycled within a system, and therefore no limit to the energy flow that can be generated.

In technology, we recycle energy to a limited extent, although as a rule energy is used only once, as in a car engine. However, in taking the salt from sea water (an extremely energy-intensive process), one method is to evaporate some water by heating it, and to channel the resulting steam around pipes containing incoming sea water, thereby preheating it. In this way a single unit of energy does work repeatedly until all the water has evaporated and recondensed as fresh water, leaving a salt precipitate.

Natural processes do the same thing—reusing and recycling energy, creating new flows of energy. Thus there is no inherent limit to evolution away from equilibrium, even with a fixed supply of energy, *so long as a process can continually increase the efficiency with which it recycles the energy.*

LIFE

Such a natural process developed on earth some three and a half billion years ago—life. Even prior to the development of life, inorganic processes had already developed a form of evolution. What distinguishes life, though, is reproduction: individual life forms can replicate their respective ways of processing energy, thus producing a continual increase in the efficiency of this process through biological evolution.

From the standpoint I have explained, the origin of this fourth epoch of natural evolution, the origin of life, is not so mysterious. Experiments have shown that amino acids, the building blocks of organic proteins, naturally form when certain chemical mixtures are exposed to concentrated energy, as in a lightning flash. These molecules form because they are most efficient at rapidly capturing any energy made briefly available. The same helical forms so efficient at capturing energy in a plasma are also most efficient for molecules, such as the biologically vital DNA and RNA. Obviously, more complex products of such molecular interactions, like protocells, can also form by efficiently capturing energy long before they can accurately reproduce themselves—that is, duplicate individual variations in their interactions. Without relying on either miracles or amazingly unlikely events, such chemical evolution can build increasingly complex molecules and systems, systems that gradually acquire more accuracy in reproducing themselves, and thus become living organisms.

As conditions change, these organisms develop a great variety of energy-capturing processes. Their ability to preserve many of these processes allows them to build up ecosystems of enormous complexity, and thus enormous efficiency. The continual multiplication of various types is a general tendency of universal evolution. The first epoch results in a purely hydrogen plasma with only protons and electrons, and the nuclear epoch gives rise to ninety-two different chemical elements. Chemical evolution generates hundreds of thousands of different compounds, and biological evolution produces tens of millions of species and innumerable individuals, all slightly different from one another. This variegation develops processes to capture almost all available energy many times over before it finally escapes back into space.

In the rain forest, for example, this tendency toward recycling reaches an extreme development: every bit of energy must be kept in constant use in a living organism or it will be washed away, in the form of nutrients which contain it, by the continuous rains. Every erg of sunlight captured by a plant and turned to organic food is reused countless times as it passes through herbivores, carnivores, scavengers, insects, insectivores, fungi, and bacteria.

Within living organisms, especially multicellular organisms, a similar complex division of labor allows the internal reuse of energy. In humans and other animals, energy from food is stored as ATP molecules, which fuel a multitude of chemical reactions such as the synthesis of protein for muscle tissue, or the digestion of other food. Energy passes in and out of a given metabolic cycle repeatedly until it is finally degraded into heat. This heat, however, contributes by maintaining body temperature at the ideal level for further chemical reactions and energy capture.

Thus life in no way contradicts a general tendency toward decrepitude, but, as Prigogine puts it, "appears as the supreme expression of the self-organizing process" of the universe.

Life continues and accelerates the progress of evolution as measured by increasing densities of energy flow. Amazingly, the human body generates energy thirty times more rapidly, per unit of volume, than matter at the sun's core. The sun's enormously higher temperatures are generated because its huge bulk does not allow the heat to rapidly escape. And if one divides the energy production rate per unit volume by the relevant body's temperature, one finds the rate at which individual particles give up or capture their energy: the rate of energy transformation in the human body is *three million times* higher than at the sun's core.

This greatly concentrated energy flow is the result of three billion years of advancing biological evolution, in which both individual organisms and the biosphere's multiplicity have steadily increased the rate of energy flow. As with the prebiological stages, biological evolution has progressed through a series of stages, each with its own mode of energy capture, its own interactions, and its own inherent limitations. Each stage has been superseded in a rapid revolution by a higher stage, which has provided broader access to energy, a higher internal metabolism, and a more efficient reuse of energy both in the ecosystem and in the individual organism.

The increasing complexity of the various interactions increases the rate at which new interactions develop. As a result, the rate of evolution—the rate of energy flow increase—steadily accelerates.

The earliest organisms were one-celled creatures, procaryotes, whose modern-day descendants are, for example, bacteria. Procaryotes, however, derive their energy from fermentation, a relatively inefficient process—its waste products, alcohol for example, are quite complex, thus contain a good deal of energy. These organisms lasted for over *two billion years* before they were superseded (since then evolution's rate has clearly accelerated).

The next evolutionary generation, primitive procaryotic photosynthetic organisms (cyanobacteria), used solar energy as fuel to convert carbon dioxide to food, producing oxygen as a by-product. Fermentation of the resulting food by other procaryotes returned a limited amount of carbon dioxide to the atmosphere, tying most up in waste products like alcohol. As a result, a gradual depletion of carbon dioxide and a buildup of oxygen—poison to the procaryotes—ensued. Without a new mode of existence, a crisis for the entire biosphere would have developed.

But the rise of atmospheric oxygen made possible a new, more energy-intensive mode of life based on respiration, the oxidation of food to obtain far greater energy. A billion years ago, more complex single-celled organisms called eucaryotes developed, which could utilize oxygen. The oxygen level stabilized and carbon dioxide was returned efficiently to the atmosphere—in oxidation, the ultimate by-products are carbon dioxide and water. A crisis was avoided, and a much higher level of energy flow results.

A second revolution occurred six hundred million years ago when multicelled organisms first developed (Fig. 7.4). With their internal division of labor they processed food more efficiently. One aspect of this cellular specification was the development of organs permitting mobility: such organisms no longer passively waited for their food, but were able to go after it.

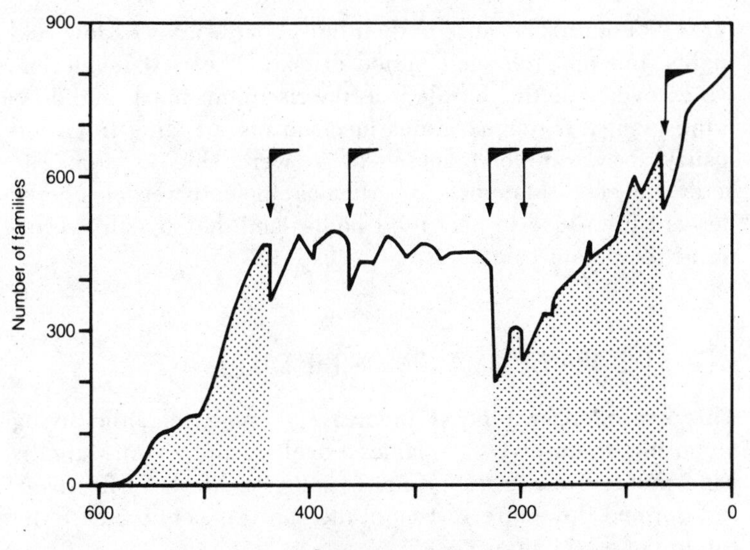

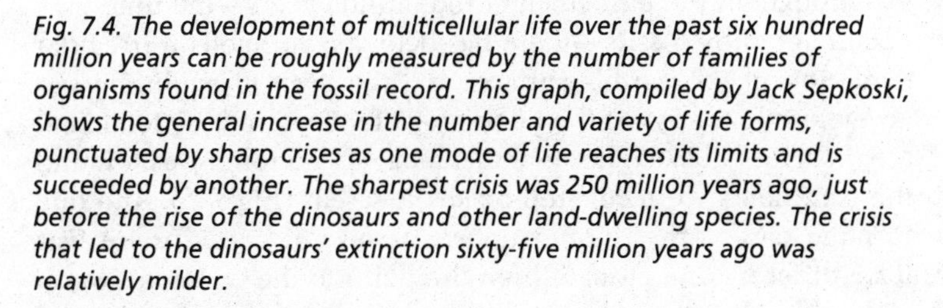
Fig. 7.4. The development of multicellular life over the past six hundred million years can be roughly measured by the number of families of organisms found in the fossil record. This graph, compiled by Jack Sepkoski, shows the general increase in the number and variety of life forms, punctuated by sharp crises as one mode of life reaches its limits and is succeeded by another. The sharpest crisis was 250 million years ago, just before the rise of the dinosaurs and other land-dwelling species. The crisis that led to the dinosaurs' extinction sixty-five million years ago was relatively milder.

From this, a third major change was the development of land-dwelling plants and animals—ferns and amphibians. This opened up huge new resources of energy. Sea-dwelling creatures are far more numerous in areas near the shore, where nutrients and sunlight are both available. Even today the deep ocean is a near desert compared with coastal waters. The land, in turn, provides both sunlight and nutrients over a much larger area. Amphibians, however, were limited to coastal areas, because their reproductive cycles still required the temperate environment of the oceans.

The next great revolution gave birth to reptiles, which no longer needed to live in proximity to water and so could live anywhere on land. They too—even the giant dinosaurs—were limited by their inability to survive rapidly changing temperatures. Dinosaurs, because of their bulk, could survive a few cold nights, but no prolonged period of cold. The next revolution, which overcame this hurdle, was the rise of mammals and flowering plants. Through various mechanisms—ranging from dormancy to warm-bloodedness—these were able to propagate across the earth. Mammals in particular efficiently recycle energy derived from food to heat their bodies and thereby survive extreme temperatures.

THE TRANSFORMATION OF THE EARTH

This evolutionary process involves far more than the living organisms themselves. As James Lovelock and Lynn Margulis have pointed out in their Gaia hypothesis, life has effectively transformed the entire surface of the earth so as to increase the capture of solar energy.*

Throughout the past six hundred million years—the time well documented by fossils—there has been a continuous if irregular evolution of the earth's surface, and the pace of evolution has accelerated rapidly. At the beginning of this period the total land area was less than a quarter of that today. Land relief was low and the continents were covered by shallow seas (Fig. 7.5). And due in large part to the uninterrupted expanse of ocean, there was little difference in climate from the poles to the equator—warm and dry, with little wind and little rain. Life was restricted to the ocean, concentrating mostly in the shallow seas where it was further limited by the small runoff of nutrients from the small, low-lying land areas.

In the course of this period, despite great oscillations (including the prior ice ages), the general trend has been a vast geographical differentiation. Continents no longer have shallow seas and there is a sharp shift from dry land to deep ocean at their margins. The climate is sharply differentiated, with large temperature gradients from poles to equator, intense winds, and far

* In the following, I go somewhat beyond Dr. Lovelock's conception. He describes Gaia as a homeostatic process in which life, through various feedbacks, maintains a constant, optimal environment. Critics have argued that "optimal" is too vague a concept. Instead I here use Prigogine's concept of increasing energy flows as the criteria for optimization. Life then tends to change its environment so as to increase the captured energy flows, thus continually modifying the environment, rather than maintaining it in one state.

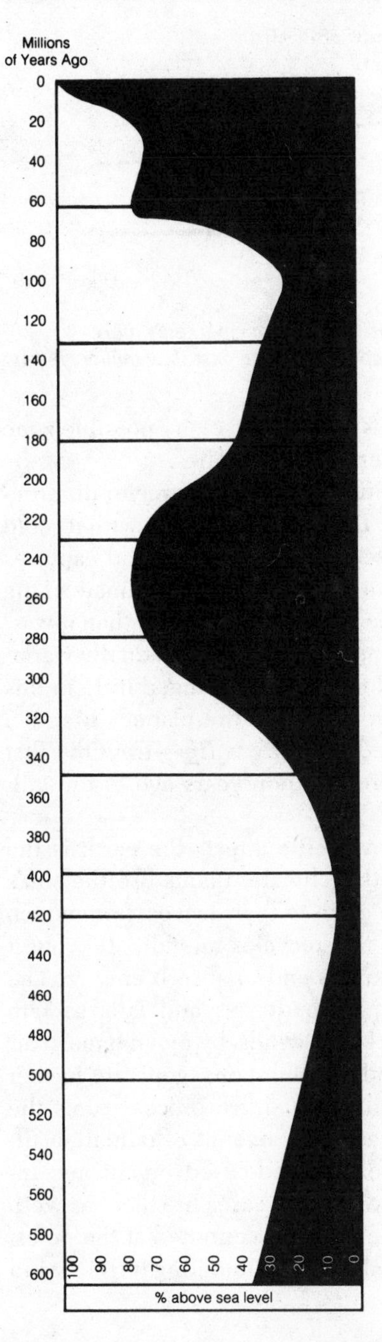

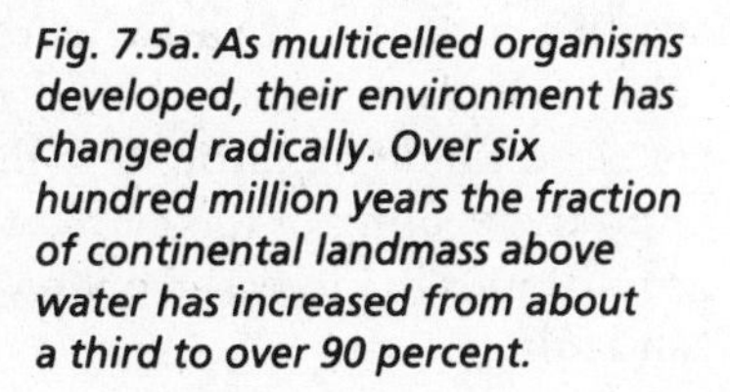

Fig. 7.5a. As multicelled organisms developed, their environment has changed radically. Over six hundred million years the fraction of continental landmass above water has increased from about a third to over 90 percent.

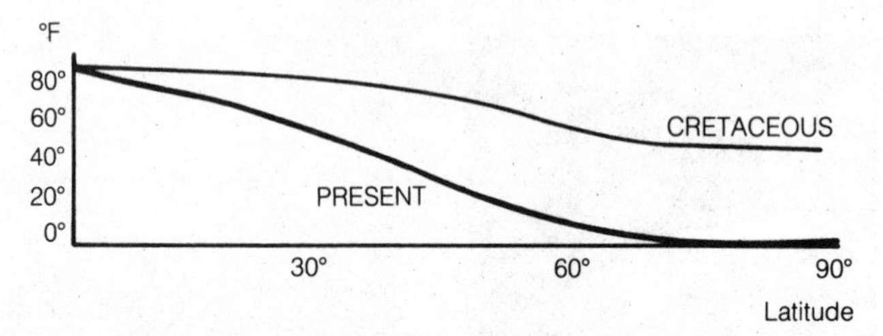

Fig. 7.5b. At the same time the climate has become markedly more differentiated, changing especially rapidly over the past sixty million years.

greater amounts of rain. Life has spread into every possible zone of habitation, sustained by larger energy supplies.

As with early stages of evolution, a greater degree of differentiation accelerates energy flow. The present climate, with its cold poles and hot equator, generates far more rainfall and captures far more of the sun's energy than the mild, uniform climate of the distant past—although the planet as a whole is cooler than it was.

Why? The sun can't be responsible—it is very gradually warming up. And the earth's internal forces have changed little in this period as well, little more than a tenth of the planet's lifetime. The only thing that has changed radically is *life*—from the first multicelled creatures six hundred million years ago to the millions of species today.

One of the most important ways life affects the earth is the action of vegetation on the earth's climate. Plants use the sun's energy to pump water from the earth to their leaves, from which the water evaporates. In effect this recycles rainfall, allowing it to be repeatedly exposed to the sun and to absorb energy. The energy released when the vapor condenses and falls as rain drives the earth's wind system. These winds, by moving moist air from ocean areas to the land and its mountains, generate further rain, releasing more energy still. The entire process cools the earth's surface and vastly increases the amount of rainfall available for plant life, and thus for all land-based organisms. Increased runoff of nutrients to the seas increases life there as well. And the cooling of the earth, which is concentrated at the poles, contributes to increased energy flows by stimulating further wind and rain.

For example, when grasses developed for the first time about twenty million years ago, they covered what had been huge desert regions in the interior of the continents; by trapping water in the soil and transporting it to the air they cooled the hot interior land. A rapid cooling trend started then, initiating the formation of an Antarctic ice cap—the first glaciation on earth in two hundred million years. The cooling produced wetter weather, increased vegetation, and still more cooling, culminating in the general ice age and interglacial climate of the past two million years.*

It is probable that biological processes contributed to higher energy flows through geological processes as well. For example, the development of flowering plants with their much deeper and extensive root structure one hundred million years ago may have contributed to the massive change in continental geography at the time of the dinosaurs' extinction thirty-five million years later. For the whole of the Mesozoic period, the dinosaurs—whose epoch this was—inhabited marshlands surrounding the shallow seas at the continental centers. A slight shift in the balance between erosion of the shorelines by wave action and the building up of deltas by vegetation-trapping sediments could have led to the rapid elimination of these seas, which did, in fact, occur sixty-five million years ago. This shift could have been caused by the development of more effective roots in shore vegetation, furthering sedimentation. The elimination of the shallow seas that covered most of the continental interiors produced a far more differentiated climate. The interiors became much cooler in winter and much hotter in summer. These temperature differences also increased winds and rainfall that contributed to the general increase in energy available for life—as did the great expansion of land area.

The changes in the environment caused by the continuing

* We are probably observing the reverse of this effect today in the deforestation of tropical rain forests. These forests recycle water two or three times, generating much of the energy that drives the winds of the rest of the world's weather. Eliminating recycling immediately removes the cooling effect, and by reducing energy flow into the atmosphere, reduces winds globally—and thus the rains created by the movement of moist air onto the continental interiors. It is possible that man's agricultural and deforestation activities contribute to global warming, rather more than do greenhouse gases like carbon dioxide, because this century's main warming trends have been in the periods from 1920 to 1940 and in the eighties. Neither period was one of industrial expansion, but both saw vast reductions of forest areas and, in the thirties, areas under cultivation.

evolution of life, in turn, accelerates that evolution. The more rigorous continental climates demanded organisms, such as mammals, able to adapt to rapidly changing weather, and totally eliminated the habitat that the dinosaurs had dominated for almost two hundred million years—contributing to, if not causing, their extinction.*

THE PROCESS OF EVOLUTION

Indeed, life on earth is a process that increases without limit the efficiency with which energy is used and reused. Over six hundred million years, evolution has not been a random, aimless process—the biosphere has progressed away from equilibrium, by the objective measure of energy flow, at an accelerating pace, in the process transforming the environment on earth. Confirming Prigogine's general model, life has evolved through a succession of different modes, each reaching its own inherent limits only to be superseded by another.

Although the sun's output, the energy source for this whole process, has changed little, the intensity of energy flow within the biosphere has continued to grow through increasing energy reuse without any obstacle to the process becoming evident.

But how can the evolution of millions of species collectively have enhanced the energy available to the biosphere as a whole? The process requires no conscious foresight on the part of the biosphere or an external creator. It does, however, require an evolutionary mechanism different from the one Darwin initially derived from Thomas Malthus's overpopulation theories. Many contemporary biologists have contended that evolution is less a matter of species competing for inadequate food supplies than a question of species evolving so as to increase the *total food supply* and thereby their own share of that supply. In effect, entire ecosystems evolve, not just individual species—it is the total energy flow through an environment in all its aspects that shapes

* In recent years some have attributed the dinosaurs' extinction to arbitrary, extrinsic events, like a comet colliding with the earth. I would only note that proponents of such a theory must explain how the comet could have caused the inland seas to drain from the continents, an event simultaneous with the end of the Mesozoic and the extinction of the dinosaurs.

evolution. For individual species cannot survive by themselves, but rely on a network of other species.

For example, as trees with more effective transevaporation evolve in a given locale, rainfall will increase more in that locale than in the biosphere as a whole. These trees and their enlarged rainfall constitute a small-scale fluctuation which will be magnified with time. The enhanced local rainfall will lead to the spread of the ecosystem, still more rainfall, and eventually a general shift in climate not only locally but globally.

Specific shifts can be sustained only if a large number of species evolve more or less simultaneously. For example, the development of grass greatly depleted atmospheric carbon dioxide; the effects would have been disastrous had it not been for the evolution of large grass-eaters that metabolize the grass, producing carbon dioxide as a by-product and returning it to the atmosphere. Similarly, flowering plants could not evolve without the simultaneous development of bees and butterflies to pollinate them. Such simultaneous evolution is not mysterious. Those local ecosystems whose species increase the overall energy grow, while those that don't perish or stagnate. Thus a competition does indeed exist, but it is a competition among local ecosystems to enlarge energy flows, rather than a fight over existing flows: it naturally leads to environmental changes that favor the further development of life.

It must be emphasized that this is not an automatic process; it is no longer the nineteenth-century concept of a linear progress ordained from on high. It is a long-term tendency and a trend that in no way precludes crises and lengthy setbacks. In fact, such crises are an unavoidable part of evolution. As shown in Figure 7.4, the biosphere, although it continues to expand, has suffered repeated crises and mass extinctions which occur when one global ecosystem has reached its limit and collapses.

A new system does not necessarily evolve swiftly. At times, one mode of life cannot expand, but no other mode can develop to take its place. In this case there is a regression, a movement back toward equilibrium, a decrease in energy flows until some new ecosystem can develop on the basis of new conditions. Such a broad setback occurred, for example, around 250 million years ago at the start of the Mesozoic period, soon after the development of reptiles.

As in the present period, the climate had become increasingly differentiated, with widespread glaciation, and the continental area had greatly increased—probably due to the formation through continental drift of a single giant continent, Pangaea. Its size reduced the ocean's moderating effect on the climate. However, no new plants or animals evolved to exploit the increased available energy. The existing ecosystem relied heavily on the mild coastal environments, which were reduced. As a result of the contraction of *usable* energy flows, a prolonged period of mass extinctions ensued.

The growth of the biosphere—as measured by the number of families or organisms—did not resume until the climate became mild and uniform as the continents weathered back into shallow seas. A new ecology dominated by the dinosaur developed, and with it the evolution of a more differentiated ecosphere.

SOCIAL EVOLUTION

During the last two million years the emergence of a rapidly altering climate, oscillating between ice ages and warmer interglacial periods, has provided the environment for species able to change their own behavior in a radical fashion.

The potential for socially mediated changes in behavior—learned, not inherited—already existed in the primates of the time. Today, monkeys living in the wild have been observed to alter both their behavior and their environment—to use tools, for example, as chimpanzees and gorillas sometimes do—and to learn from others who do so.

Two major changes gradually transformed limited capabilities into more complex social behavior—first, the ability to make tools, not just use them. According to paleontologists, the early production of tools, such as stone axes and choppers, is the key sign of the emergence of early human beings. By creating tools, our ancestors began the limitless transformation of their environment—for tools can be used to create other tools.

The second change was the development of true language—a symbolic language in which sounds describe events and concepts abstracted from concrete, immediate conditions. This allowed our ancestors to teach each other by means other than example and imitation. It eliminated the restrictions that still limit animal social learning.

The combination of language and tool making—mind and hand—characterized human beings, and started the third mode of evolution, social evolution. New modes of existence arose—new ways of human cooperation that allow new methods of modifying nature.

Just as biological evolution proved faster than physical evolution, so too social evolution is another great acceleration. But in several main aspects, social evolution continues the same general processes and shares many of the characteristics of the earlier modes.

Social evolution has proceeded through a number of successive stages, each involving an expansion of energy use. The first stage was that of the hunter-gatherers of the Paleolithic period, who relied on their environment as they found it for energy, in the form of food. The second stage, beginning around 10,000 B.C., was based on small-scale agriculture: the farmers harnessed the greater energy involved in concentrated agriculture, supporting populations ten or a hundred times greater than before. The limited use of animal energy to expand agricultural efforts generated further efficiency, thus further population growth.

Just as the first stage was limited by the availability of game, so the second was limited by the availability of the rainfall on which agriculture depended. This limit was overcome with the urban revolution in 4000 B.C., which mobilized the mass labor capable of developing irrigation-based agriculture—again leading to a rapid increase in population in the Bronze Age. This society too was limited by the availability of water suitable for intensive subsistence agriculture, but was in turn superseded by the specialized agriculture of the Iron Age. The civilization of Greece and Rome further exploited solar energy through specialized agriculture on an enormously extended scale—areas unsuited to grains were used for other crops, which were traded throughout the Mediterranean area and Europe.

All of these societies were based overwhelmingly on agriculture. They harnessed the vast majority of their energy in the form of food. Even the use of animal labor was limited; the dominance of human labor set strict limits on the available energy—there is only so much work a human being, even a slave, can do.

Medieval society was the first to overcome these limitations, by introducing widespread animal labor and inanimate energy sources—wind and water power. Mills and ships used these natural sources, not slaves, for power. Energy now was used not only for agriculture but increasingly for industry—for textile and clothing manufacture, above all.

While medieval society could again support a much higher level of energy use—thus a higher living standard for a larger population than previous societies—it too was limited in the power it could derive from wind and water, which remained secondary to human and animal energy. In their turn, these limits were overcome by the emergence of modern capitalist society: the main energy source shifted first to wood and then to fossil fuels, coal, oil, and gas, gigantically increasing energy consumption and the population that could be supported. Agriculture became secondary as the bulk of energy was poured into a range of new industries and new technologies.

Just as in biological evolution, where each different mode of existence was dominated by a different set of species, a different ecological system, so with social evolution each method of energy capture corresponds to a different form of social organization. Paleolithic hunter-gatherers were grouped into small clans, neolithic farmers into much larger villages, even towns like ancient Jericho, in which a division of labor and a specialization of tasks began to develop.

In the subsequent civilized societies, the division of labor was subordinated to the social hierarchy of those who worked and those who decided what work should be done. Thus the Bronze Age priests and priest-kings directed the work of the peasants, who collectively belonged to the pharaoh as semiserfs. The Iron Age commercial slaveholders commanded slave labor, side by side with a subordinated economy of free peasants and craftsmen. Medieval kings and lords ordered the work of serfs, while in the towns a new society of burghers and craftsmen developed. Such a division still persists between those—the bulk of society—who do the world's work and those—predominantly large-scale capitalists in the west and bureaucrats in the east—who decide what work is to be done, which factories are to be built and which are to be shut down. Obviously, with this economic differentiation of society comes an enormous evolution in the political structure of society—from the primitive democracy of ancient villagers to the divine kingships of the pharaohs, to the empires of the ancient world, to the feudal monarchs of the Middle Ages, to the democracies and dictatorships of today.

As with biological evolution these stages do not supersede each other gradually, but in sharp revolutions: societies that had been stable for centuries were overthrown in a matter of years or even days. Such revolutions are far from inevitably successful, though. At times, a society reaches its limits and falls back, as did Roman society, without, for a long time, being replaced by a succeeding form.

During such crises the outcome, whether progress or decay, often depends on individuals or small groups. As in Prigogine's model, seemingly trivial fluctuations—demonstrations, strikes, throwing tea into a harbor—in an unstable society can grow into the complete overturn of that society and the setting up of a new one, in the overcoming of the old limits. Equally, such fluctuations can die down—an uprising suppressed, a leader assassinated—and the old society can stumble on for decades or centuries, retreating into increasing chaos.

In social evolution, as in biological evolution, progress is not a smooth and automatic process but a long-term trend. Human progress is a fact as much as biological evolution is—not only by the objective measures of energy flow, population, or longevity, but by any rational yardstick human society is far better off today than in Roman times, when half the population was enslaved, or in ancient Egypt, when practically every member of the population was a serf.

At the same time, there are no guarantees if and when any given crisis in human history will be resolved. In sixteenth- and seventeenth-century Europe, the battle between lords and merchants led to the bourgeois revolutions and a new expanding capitalistic society. In sixteenth-century Japan, a somewhat similar clash ended in the triumph of feudalism and a long period of stagnation.

Yet the long-term trend still exists. When new, faster-evolving societies emerge, they spread everywhere—as did agricultural societies thousands of years ago, and as did capitalistic societies during the seventeenth, eighteenth, and nineteenth centuries. The upward curve is real despite the catastrophes and Dark Ages that litter human history.

This rate of evolution, of the creation of new interactions, depends directly on the number of different interactions already existing. Over the past twenty-five thousand years the increase in energy use has accelerated at a rate more or less proportional to the size of the population. Over that entire time it has taken on average about a billion human lifetimes to double the energy use. When the population was small this meant a slow rate of technological and social development. Now that it is far larger, the pace of change has vastly accelerated (Fig. 7.6).

This is an entirely reasonable relation. It means in essence that it takes just so much human labor to come up with the innovations needed for a given amount of technological change. Naturally, societies that make poor use of human intelligence, such as slaveholding societies, have relatively slower rates of social development, while those in which the opportunities for individual innovation are great, such as Elizabethan England, advance more swiftly in relation to their population.

Thus tested against what we know of three modes of evolution —physical, biological, and social—Prigogine's model stands up well. Progress is real. The movement away from equilibrium, the growth of energy flow as a process without limits, emerges as a natural, comprehensible phenomenon. Interactions grow by capturing energy, eventually reaching the natural limits for any given mode. They become unstable, and new interactions arise on the basis of the old, manifesting themselves initially as small fluctuations. Some of these fluctuations grow, capturing energy faster than the old interactions, and in sudden revolutions replace them with those that are more complex, capture more energy, and recycle it more efficiently.

Only when such new modes do not come into existence, when the old modes begin to exhaust the available energy, does a movement toward equilibrium occur. Instead of a cooperative competition between various processes in expanding available energy, competition becomes destructive over the shrinking energy flows. In biological evolution, this leads to epidemics and mass extinctions; in social evolution, to tyrannical self-devouring reigns like those of the late Roman Empire.

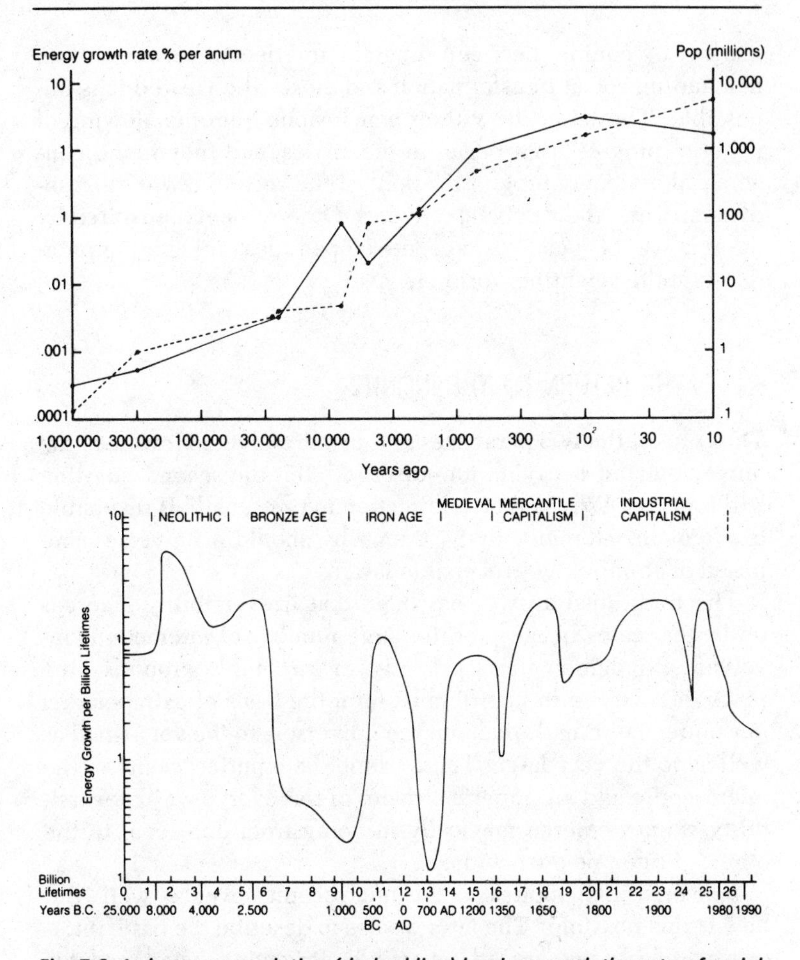

Fig. 7.6. As human population (dashed line) has increased, the rate of social evolution, measured here by the annual rate of energy growth (solid line) (estimated from archaeological and historical data), grows in step. (Top) If the rate of energy growth is measured per billion human lifetimes, it remains roughly constant over long periods, but varies in response to the rise and fall of succeeding forms of society. Societies that make poor use of human intelligence, such as the ancient slave societies, progress less rapidly for a given population level (bottom).

As Prigogine himself notes, it is no coincidence that his physical theory, with its emphasis on progress and revolutionary change, arose in the late sixties—a time of rapid and turbulent social transformation, within a troubled century. The sixties were marked by conflict between progress and decay, between those demanding social transformation and those who viewed it as impossible. The latter drew their conclusions from the slowing of material progress that began in the sixties, and they found congenial the gloomy prognostications of the second law of thermodynamics and the Big Bang universe. Others, though, inspired by the renewal of social advance, developed ideas integrating progress into the scientific world view.

THE RETURN OF THE INFINITE

Thus one of the two paradoxes of time is resolved: How does the universe avoid decaying into disorder? But the second question still remains: Why is there a direction to time at all? If reversible laws govern atomic activity, then why should a universe composed of atoms obey irreversible laws?

The traditional answer has been that irreversibility emerges on larger scales because of the large number of interactions involved. Prigogine rejects this answer on simple grounds—the processes we see in nature must form the basis of extrapolating our understanding throughout the universe—to the very small as well as to the very large. There cannot be a perfect realm of the microscopic and an imperfect realm of the everyday. "Irreversibility cannot emerge magically in going from one level to the other," Prigogine once wrote.

Irreversibility applies to the microscopic level as well. But how is this possible? The laws we use to describe the basic interactions of physics are all reversible. Prigogine developed his answer only in the early eighties. His breakthrough was to reintroduce the infinite into the natural universe.

What does it mean for interactions to be time-reversible? As in the example of billiard balls colliding, if we exactly reverse the velocities of all particles at a given instant, the system will revert to its starting position—the same events will occur if the movie is run forward or backward. Prigogine points out, though, that this reversal of velocities is impossible to accomplish, even in theory—it requires an *infinitely* precise knowledge of the velocities of each particle involved. *If the system is unstable,* any error, *no matter how tiny,* will prevent a system from returning to anywhere near its starting point. Time's irreversibility derives from two basic facts: all systems except very simple ones *are* unstable, and we live in a universe in which space is continuous, thus *infinitely divisible*. That infinite divisibility means that we can neither determine the future of any but the simplest systems nor reverse time.

To explain this concept, Prigogine first distinguishes between stable and unstable systems. As an example of a stable system, take two bodies interacting through their mutual gravitation—a planet orbiting the sun, for example. In this simple system a slight change in the velocity of the planet at one moment will lead to only a slight change in its orbit at any time in the future, even millions of years later. Such stability is a characteristic of a system that is truly reversible. We can, in theory, reverse a planet's velocity. If our initial errors in doing so are not great, the planet will retrace its orbital path almost exactly. For the exact same reason—the stability of the system when slightly changed—we can accurately predict the position of a planet for millions of years in the future—it is predetermined.

But such simple systems are rare. In general they are only approximations of nature or abstractions from it. To make the model a bit more realistic let us add one more body—a small comet, orbiting the sun, but influenced by the planet's gravity (as real comets are influenced by the planets, especially Jupiter). Now we have an unstable system: the planet will orbit the sun "forever," but after a finite number of orbits the comet will inevitably approach the planet so closely that it will be flung out of the solar system.

This system is unstable because any tiny change in the comet's velocity results in changes in position that grow exponentially with time. Thus, if there is a minuscule mistake in measuring the velocity today, in a relatively short time, a few orbits, it will be impossible to predict even roughly the comet's position or whether it will have been expelled from the solar system entirely. Similarly, if we reverse the velocity of the comet, any small error will grow so rapidly with time that there is no possibility of returning it along its original path.

In order to illustrate this, one of Prigogine's colleagues, T. Petrosky of the University of Texas, used a computer simulation to predict the number of orbits such a comet would make before being expelled from the solar system, consisting, in the model, only of the sun and Jupiter. He varied only the accuracy with which the orbit was calculated. If the velocities were calculated to a precision of one part in a million, the model showed that the comet would stick around for 757 orbits. When the accuracy was improved to one part in ten million the prediction was 38 orbits; one part in a hundred million, 235 orbits, and so on, down to one part in 10^{16}, 17 orbits. There was no tendency whatsoever for the predictions to approach a single solution with increasing accuracy—increases in accuracy had no predictable effect. Without absolute, infinite knowledge of the comet's velocity, and infinite precision in calculation, its orbit is simply unpredictable. Yet this is *not* an effect of "chance." At all points the orbit was under precise control of the laws of gravitation as programmed into the simulation. The unpredictability came from the instability of the three-body interaction.

Because all but the simplest real systems are unstable, the same reasoning applies to them. If we try to reverse the velocities of a particle, say, in a growing instability, to make it shrink again, our errors, no matter how tiny, will inevitably cause the instability to continue to grow. If a system near equilibrium is becoming more random and we reverse the velocities, errors will prevent the system from becoming less random—prevent the broken eggs of my earlier example from coming together and rolling off.

This doesn't mean we can't make useful predictions about the future. We can *if* the amount of time we try to predict is short enough. For unstable systems this time limit is the amount of time that passes between collisions of the particles that make up the system. For the comet this is a single orbit, but for a gas a tiny fraction of a second. We can, however, make useful *statistical* predictions; on average the comet will probably last about 150 orbits. And of course many systems are sufficiently close to stable that we can ignore their instability because the rate at which their instabilities grow is far longer than we need to worry about. For example, the orbits of the planets in our actual solar system appear to be unpredictable in excess of twenty million years from now. For all intents and purposes, in plotting a space mission the system is absolutely stable, predictable, and reversible.

Similarly, nearly all electrical and mechanical devices are designed to be stable, so time-reversible laws work well with them.

So we find that all our laws that assume time to be reversible are approximations of reality. Time would be truly reversible only if, at least in principle, we could reverse the motion of particles in any system and have them retrace their paths. But this, as we've seen, requires infinite precision and can't be done. This is *not* a limit to what we can know, it has nothing to do with subjective consciousness, rather it is a limit to what can be *done*. No theoretically possible physical process can reverse these velocities with infinite precision.

So temporal irreversibility derives from system instability. But all real systems are somewhat unstable, even microscopic systems. Some systems evolve so slowly that we can treat them as stable, but *only* abstract systems, isolated in our imagination from all other influences, can *be* absolutely stable. The problem of "reversible time," then, arises because scientists improperly abstract reality and believe their highly accurate equations to be absolutely, infinitely precise. *It is reversible time that is subjective,* an illusion, *not* irreversible time. The real world is continually coming into existence, created by an infinitely complex web of instabilities and interactions. As Prigogine puts it, "Time is creation. The future is just not *there*."

Time's irreversibility is based on the continuity of space, on its infinite divisibility. This can be understood by imagining what would happen if the universe were not like this. If space could be divided only into a finite number of steps, then by knowing, at a given time, exactly which step a particle is at, and how many steps it is advancing in which direction, it would be possible to predict *exactly* its future evolution, even in an unstable system. It would be equally possible to reverse its motion exactly and have time run backward. A universe-as-computer, a finitely divisible universe, would indeed have no future and no past. It would evolve only *if* some external or extrinsic force started it out with the right "initial conditions"—a Big Bang, for example. A universe that is finitely divisible on a small scale must then be finite on a large scale—finite in time.

(Scientists can use computer simulations for the very reason that they act as the external agent, inserting the proper initial conditions into the program. These are not generally very specific conditions, however, like the Big Bang, but simply suitable random conditions. Only if a scientist delicately "reversed time" by reversing the directions of particles in a previous simulation would the answers be grossly wrong.)

INFINITY AND FREE WILL

The idea of continuous space leading to irreversible time also potentially resolves the paradoxes plaguing the concepts of consciousness and free will. In our universe the future does not exist —there is a real now, the now of consciousness. The map of the future and the past laid out in the fourth dimension is merely an abstraction. Since one cannot, even with the greatest possible degree of knowledge, totally predict the course of the future, free will is indeed real. Even if a scientist were to know the exact location, to ten decimal places, of every atom in a person's brain, he or she could not determine what that person would do next.

This uncertainty is *not* the result of chance or of the quantum uncertainties we'll discuss in the next chapter. Rather, it is because the brain too is a complex and unstable system. Its behavior is determined by interactions whose outcome can be radically changed by infinitely small shifts. Again, this would not be the case if the brain functioned as a digital computer, as many theories of intelligence assume. Similarly, a digital computer, no matter how complex, cannot be intelligent, because its every action is precisely predictable—one can perform the same operation over and over without variation. However, experimental studies of the brain indicate overwhelmingly that it does *not* function in this manner. While neurons do send signals to each other along definite pathways, they also contribute to the formation of a general electromagnetic field within the brain—what is detected as the brainwave or EEG (electroencephalogram). This ever-changing field, in turn, affects what each individual cell does. Like any such unstable system, the brain functions nonlocally—that is, the behavior of every part is affected by every other part, the system acts as a *whole*. It is that total entity—the evolving cerebral flux and its electromagnetic fields—that must be connected with the phenomenon of consciousness, with self-perception (Fig. 7.7). And it is that overall unified system that determines a person's specific actions and his or her general makeup.

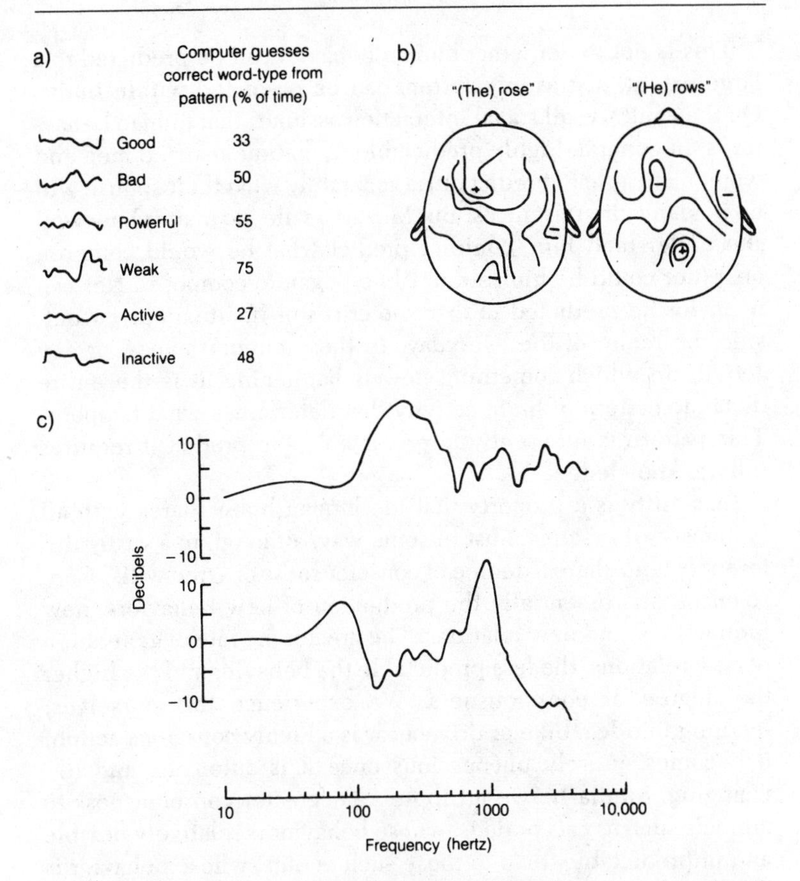

Fig. 7.7. Brain researchers have found that the overall pattern of the brain's electromagnetic field correlates with what the subject is thinking. In experiments performed by Robert Chapman and coworkers at the University of Rochester, subjects were shown words grouped into six categories—such as good words like "beautiful" and bad words like "crime." Each connotation gave rise to a distinctive pattern in time (a) that a computer could use to guess what category of words the subject had just seen. Plotting patterns in space over the surface of the brain, Warren Brown and Dietrich Lehmann working at the University Hospital, Zurich, Switzerland, found that the EEG, or brain-wave, patterns of many different subjects were broadly similar. Here, two different patterns emerge when an identical-sounding word is used with two different meanings (b). Different types of sensations produce markedly different brain-wave patterns, Erol Basar of the Lubeck Institute of Physiology found (c). Here an auditory stimulus (top) has an entirely different pattern than a visual stimulus (bottom). In all these cases a coherent electromagnetic pattern is produced by the brain as a whole, with billions of neurons acting in concert.

This is not to deny that human behavior can be predicted to a large extent, just as any system can be predicted within limits. On the contrary, all social interaction assumes that human behavior is, in general, highly predictable. Longtime married men and women are able to predict quite accurately what their spouse will do in many circumstances. But Mozart's wife, no matter how well she understood him, couldn't predict what he would compose next (nor could he himself, until he began to compose). Nor can behavior be predicted in extreme crises or in situations far outside the realm of the everyday. In these circumstances, or any activity in which something new is happening, it is the entire dynamic pattern of brain activity that determines what happens. This pattern is inherently unpredictable—to predict it requires infinite knowledge.

Instability is a property that the human brain shares with all systems—all systems must in some way, although to a vastly different extent, share a degree of consciousness or "free will." Consciousness is essentially the production of new behaviors, new interactions, and new relations. The greater the rate of generation of new relations, the less predictable the behavior and the higher the degree of consciousness. We experience this ourselves: learning to ride a bike or drive a car is a highly conscious action; it becomes entirely unconscious once it is automatic and unchanging. Similarly, we attribute a far greater consciousness to animals such as cats or dogs, whose behavior is relatively flexible and unpredictable, than to those such as ants, whose behavior is rigidly predictable.

Consciousness can be viewed as part of a continuum of natural phenomena—not something outside the range of scientific inquiry. Over the long run, scientists can pose such questions as "What is the sensation 'red'? What patterns of change correspond to this sensation?" just as today they can ask "By what mechanism does the retina detect light from a rose?" It may well take decades to come up with useful answers, given the complexity of the brain, but at least we can begin to consider consciousness as part of the natural world, not a parallel realm of the supernatural.

We find that the same instability that drives evolution and causes systems to move further from equilibrium is at the core of the phenomena of consciousness and free will. What allows evolution to go forward is the generation of new instabilities, new modes of capturing energy—modes that cannot be predicted beforehand. This same unpredictability and same production of new relations underlies consciousness.

We can understand a given set of relations and their instabilities and make predictions based on that understanding. This gives us the power to gain scientific knowledge and to harness it in technology. But we *cannot* predict what new relations or instabilities will come into existence, nor can we predict that nothing new will occur. It is for this reason that we cannot predict, as conventional cosmologists do, that the universe will simply run down. Ironically, the very unpredictability that makes it impossible to know in advance the outcome of any particular crisis in evolution also precludes the prediction that evolution itself will come to an end.

It is striking that the power of instability, the potential for the creation of an unlimited number of new modes of existence, is firmly linked to the infinite continuity of space. This is a profound vindication of Nicholas of Cusa's ideas of the infinite in the natural world, and his rescue of that notion from the realm of the mystical. Man's capabilities are infinite, human creative potential is infinite, because in a real way we contain infinity within us.

This is, it cannot be too much emphasized, a purely earthly idea of infinite, not the otherworldly concept generally associated in popular discourse with "the Infinite." In fact, there is nothing otherworldly about infinity, infinite numbers, or infinite space. The notion of the infinite continuity of space is fundamental to a large part of modern mathematics. Without it much of the mathematics used in everyday technology would become illogical and self-contradictory.

It has been nearly a hundred years since the great mathematician Georg Cantor put the study of infinite numbers, or "transfinite numbers" as he retermed them, on a rigorous footing. He showed that such transfinites can be manipulated and that they obey certain mathematical laws. Cantor proved that there are greater and lesser transfinites: the number of points on a line or in a space—the infinity of continuity that we have been discussing—is bigger than the number of all numbers that can be counted, for example.

It is curious, but not surprising, that a hundred years after

Cantor many people, including scientists, still trot out Aristotle's ancient arguments about the impossibility of the existence of infinities in the real, material world. Infinity, as we shall see, is still in many theologians' view a property only of the divinity. Mystical conceptions of numbers—and infinity is a form of number—are not new: their traces are in the words we use to describe classes of numbers—irrationals, transcendentals, and imaginary numbers. Irrational numbers (like $\sqrt{2}$) are not, after all, irrational, beyond reason: they can't be expressed as fractions, but they are useful if one needs to find the length of a side of a triangle. Transcendental numbers, like pi, which can't be expressed as a solution of an algebraic equation, are not transcendental with respect to experience: pi is essential if one wants the circumference of a circle. And imaginary numbers, like the square root of negative 1, are not figments of the imagination. They are used in the construction of electrical circuits for, say, a radio. So too transfinite numbers are needed if we are to understand the nature of time and space, and the very real unstable processes that govern the evolution of our universe.

Mathematicians have in the past ascribed such mystical names to ordinary numbers because they ascribed power to these symbols, just as our ancestors (and many people today) ascribe magical powers to the number 13. We saw earlier the origins of number magic in the theories of Pythagoras, and its influence persists to the present. But there is no more reason for excluding infinity from a scientific description of the universe than there is for skipping 13 when numbering the floors of an office building, although both practices remain common.

A REVOLUTION IN SCIENCE

Prigogine rightly considers his ideas to be part of a revolution in science, a replacement of the mechanistic universe with an evolving, progressive universe that includes humanity. The old view describes "nature as a mindless, passive mechanism that was basically alien to the freedom and purposes of the human mind. . . ." In contrast, the new view includes human consciousness and the progress and freedom characteristic of humanity within its purview and sees it as the most advanced expression of the infinite history of the material universe. The new view rids itself "of a conception of objective reality that implied that novelty and diversity had to be denied in the face of immutable universal laws . . . of a fascination with a rationality taken as closed, of knowledge seen as nearly achieved." It sees science as a form of history, as a description of the evolutionary stages of the universe and of the "laws" appropriate to each stage.

Such a view profoundly affects cosmology, entirely eliminating the need for any thermodynamic origin or for a Big Bang. Thus it isn't surprising that Prigogine's work remains controversial within thermodynamics and is largely ignored by cosmologists. Compared with Alfvén's cosmological heresies, however, it is far more widely known—especially after the publication in 1984 of Prigogine and Isabelle Stengers's popularization, *Order Out of Chaos*.

Prigogine himself has tried to elaborate a cosmos coherent with his view of time, a universe with no beginning, evolving from a disordered state. However, his cosmology involves, at least in its general outline, aspects of Big Bang theory—a period of extremely rapid evolution of the universe at high temperatures and densities. Until quite recently Prigogine was unaware of plasma cosmology, as Alfvén has been generally unaware of Prigogine's work.

Yet, despite the barrier of scientific specialization which hinders communications, these two trends—one from plasma physics, the other from thermodynamics—have converged in a fundamentally new view of the cosmos, a view that finds the universe to be progressively evolutionary, infinite in its capacity, and comprehensible both qualitatively and quantitatively.

NOTES

1. Albert Einstein, *Ideas and Opinions,* Crown, New York, 1954, pp. 224–27.

8 MATTER

Let us worry about beauty first and truth will take care of itself!

—A. Zee, particle physicist, 1987

Contemplation of superstrings may evolve into an activity to be conducted at schools of divinity by future equivalents of medieval theologians. For the first time since the Dark Ages we can see how our noble search may end with faith replacing science again.

—Sheldon Glashow, particle physicist,
Nobel laureate, 1987

Scientists live in the world just like other people. . . . They cannot escape the influence of the milieu in which they live.

—James Garner Murphy, Irish poet,
letter to Albert Einstein, 1932

There are no general guidelines to which we can cling. We have to decide for ourselves and cannot tell in advance whether we are doing right or wrong. Probably a bit of both.

—Werner Heisenberg, on his decision to
lead Nazi Germany's effort to develop an
atom bomb, 1939

If the Big Bang never happened, then the conventional view of matter must fundamentally change—just like the conventional view of time. Modern theories of the structure of matter, fundamental physics, have become so intertwined with the theory of the Big Bang that not one of the many shelves of popular books on modern physics separates the two subjects.

The Big Bang serves as the principal justification for constructing giant particle accelerators, such as the $10-billion supercollider, as the main tools used to probe the structure of matter. The vast energy of particle collisions in these accelerators, physicists argue, can reproduce the extreme conditions of the Big Bang, and thus penetrate the very origins of the cosmos.

The interdependence between fundamental physics and cosmology extends to method as well. Like cosmologists, theoretical particle physicists rely heavily on the deductive method, deriving their theories from the perfect symmetries of mathematics—which, because they aren't based on observation to begin with, can't be effectively challenged by experiment. If the Big Bang is wrong, then many of the basic ideas of fundamental physics are wrong as well. The same methods that have led cosmology into a blind alley have also simultaneously stalled the advance of knowledge of the structure of matter and energy.

THE SEARCH FOR BEAUTY

Fundamental or particle physics, the study of the underlying structure of matter and energy, focuses on the effort to unify the basic forces of nature. As far as is known, the interactions of matter can be described in terms of four forces: gravitation, electromagnetism, and two nuclear forces—the strong force responsible for keeping the nucleus together (the source of nuclear energy), and the weak force responsible for radioactivity and the decay of the nucleus.

As we've seen, over a century ago Maxwell unified two previously separated but related forces—electricity and magnetism—into a single force, electromagnetism, and elaborated its laws and many of its properties. Similarly, today's fundamental physicists hope to develop a theory that will unify all four forces, and thereby to explain the nature of the particles that make up matter—electrons, protons, neutrons, and a host of others.

In itself, this is a fine idea: science has frequently advanced by unifying hitherto distinct phenomena under a single theoretical concept. But it has also advanced by discovering new phenomena not covered by any previous theory. The problem in present-day particle physicists' search for such unified theories is that it is based overwhelmingly on certain mathematical concepts derived by pure reason, rather than on observation. Moreover, this theory is viewed not as the next step in an unlimited search for knowledge but as the Holy Grail of science, the final absolute knowledge that will explain the universe and everything in it, a Theory of Everything.

To most of today's particle theorists, their job is the search for Beauty. " 'Let us worry about beauty first and truth will take care of itself!' Such is the rallying cry of fundamental physicists," writes the fundamental physicist A. Zee, in his recent book *Fearful Symmetry*. "Some physics equations are so ugly we cannot bear to look at them, let alone write them down. Certainly, the Ultimate Designer would use only beautiful equations in designing the universe, we proclaim."

Like Plato twenty-five hundred years ago, fundamental physicists seek, through pure logic, the beautiful plan by which the creator designed the universe. And what is the criterion for beauty? "The system of aesthetics used by physicists in judging Nature draws its inspiration from the austere finality of geometry," Zee explains. "Following the ancient Greeks, who waxed eloquent on the perfect beauty of spheres and the celestial music they make, I will continue to equate symmetry with beauty." In this view he is echoed by dozens of other leading physicists who, as Zee puts it, attempt "to read [God's] mind by searching their own minds for what constitutes symmetry and beauty."

Zee explicitly returns to the methods of Platonic dualism, the methods that gave rise to the Ptolemaic system—the very methods discredited by the scientific revolution. To Zee, the basic mistake of the ancient astronomers, whose intellectual descendants battled Galileo and Kepler, was merely that they misunderstood the concept of symmetry: "The correct definition of rotational symmetry does not require circular orbits at all," he says of Ptolemy.

He does not criticize the deductive method of the ancients and the medieval scholars, however, since this method is the one used by Zee and most of his colleagues. He rejects the experimental method—what he calls the "nineteenth-century method of science of fooling around with frogs' legs and wires." Fundamental physicists are far above fooling around in laboratories! Zee writes, "In the silence of the night they listen for voices telling them about yet-undreamed-of symmetries," from which they deduce new theories, eventually checking to see if observation bears the theory out. Sometimes this last step is omitted as unnecessary, as we shall see.

The goal of this work is nothing less than a complete explanation of the universe, to be achieved "within the lifetime of many of those working today," as Stephen Hawking puts it. Such a Theory of Everything will explain not only the four forces, all the particles, the universe itself, galaxies, stars, planets, and people, but it will also be so simple a set of equations that it can be "written on a T-shirt." Or, as John Wheeler of the University of Texas puts it, "To my mind there must be at the bottom of it all, not an equation, but an utterly simple idea. And to me that idea, when we finally discover it, will be so compelling, so inevitable, that we will say to one another, 'Oh, how beautiful. How could it have been otherwise?' "

Such a theory will complete the main task of science, leaving only a mopping up of details, except for one major question, in Hawking's view: Why does the universe exist? Once we know the answer to that final question we will then achieve final knowledge; we will, in his words, "know the mind of God."

WHAT'S WRONG WITH A THEORY OF EVERYTHING?

But many fundamental physicists might ask, What's wrong with pursuing beauty? What's wrong with seeking a Theory of Everything? In other words: What's wrong with the deductive method?

Obviously there's nothing wrong with the study of beauty in nature. The love of nature's beauty has always been a powerful motivating force for scientific progress—to find beauty in nature, and to understand it. But the question is *whose* beauty—nature's or man's idea of what that beauty *must* be? Leonardo's concept of beauty or that of his "adversary" Plato? For Leonardo, both art and science were based on the close observation of nature's own beauty, as emerges so strikingly in his careful sketches of anatomy, of turbulent water, or in his paintings, where every plant is portrayed with botanical accuracy. He ridiculed the Platonists' idea of perfection based on the realm of ideas, a perfection that nature must conform to.

The difference between the two ideas of beauty is the difference between the marble sphere that perches in perfect symmetry atop a pedestal at the entrance to a garden, and a tree within the garden. There is no doubt which is more symmetrical, but is the sphere the more beautiful? On a more abstract level, look at the illustrations of Ptolemy's theoretical view of the universe as a set of perfect spheres and Kepler's solar system of elliptical orbits. Again, Ptolemy's is clearly the more symmetrical. But is this illusory solar system, which helped to stall scientific progress for nearly two millennia, more beautiful than the real solar system viewed by Galileo and described by Kepler—the real system with Saturn and its rings, Jupiter and its swirling spot, the solar system we eventually came to know through space exploration? It's hard to find symmetry in this solar system but not hard to find beauty in it.

Of course, these physicists may have a sense of beauty superior to Leonardo da Vinci's—after all, Mona Lisa's smile is definitely lopsided!

In fact, fundamental physicists do lay claim to a superior sense of beauty. They claim that the supreme beauty is in mathematical equations, which can be understood and judged only by an elite priesthood of reason. Anyone can see beauty in a tree, but it takes years of study to see beauty in an equation. All they need do to dismiss an idea is to say that its equations are ugly—no facts are needed.

In this we take a gigantic step back to the age of mythology and wisdom learned from authority. The idea that pure reason can divine the beauty that nature should have, and can derive scientific knowledge from that beauty, is an idea that *doesn't work.* This is the real lesson to be learned from Galileo, Leonardo, and Kepler. However beautiful and symmetrical, Ptolemy's system was sterile. It blocked the advance of knowledge of the heavens and prevented any application of that knowledge in navigation, or elsewhere. Like Aristotle's mechanics, with which it was allied, Ptolemy's cosmology shackled the human mind, and forced nature into the Procrustean bed of perfect symmetry. The deductive method failed and was replaced in a scientific revolution by a method that works, the empirical scientific method. As we've seen in the previous chapters, modern cosmology has repeated the sins of Ptolemy. Its insistence on concepts derived from pure reason and the rejection of the observational facts have led to a dead end.

This is not to say, though, that there is no role for deduction in science. There is a vital role, but only as one step in a cycle of endeavor that begins and ends with the observation of nature. From observation of new phenomena, a scientist can arrive at new hypotheses, new concepts that tentatively describe the phenomena—this is the phase of induction. Then the concept can be put into mathematical form, and consequences deduced from the theory—this is the deductive phase. The results are then compared with new observations.

But the cycle cannot end here, even if the theory is found to be valid—technology enters. A new theory is used in the real world—to develop new technologies, either directly for scientific work, or, more commonly, for economic purposes. These new technologies then lead to the observation of new and totally unexpected phenomena—thereby continuing the cycle.

As long as this cycle exists, as long as theory remains tied to observation and is applied in technology, roles exist for both scientists who prefer inductive reasoning and those who prefer deductive work, *and* for experimenters. Some are adept at formulating new theories, some at deriving startling predictions from existing theory.

The danger arises when deductive methods go off on their own, soaring away from the reality of observation and the observation of reality. This tendency toward such a split inevitably exists in science—very frequently within a single scientist. Kepler, before he arrived at the correct elliptical orbits, spent long years trying to prove that the orbits are perfect circles inscribed within the six perfect regular solids of the Pythagoreans. Only with great reluctance did he accept that this beautiful deductive structure is unreal, that it does not match the observed planetary orbits—especially that of Mars. It was only because he overcame his devotion to the beauty of his initial scheme that, in the end, he recognized the "ugly" ellipse as the true form.

By disengaging itself from observation and technology, the deductive method has repeatedly proven itself utterly sterile. No new discovery, no new observation can be made from deduction. In fact, since the "perfection" of existing theory is proven by logic, observations that contradict it are automatically rejected. We have seen this occur in cosmology and will see it occur again in fundamental physics. When the deductive method triumphs, science stagnates, threatening to stall technology and the advance of society generally, as occurred in the Middle Ages.

The pursuit of a Theory of Everything is but the most extreme expression of deductive method. It is the idea that all knowledge, *final* knowledge, can be derived from the human mind alone. The cartoon on the facing page ridicules this idea, but why is this strip funny? The reaction of the average person to the claims of the Theory of Everything correctly pinpoints its underlying absurdities.

First we laugh because the idea that a theory could explain everything, including penguins, is funny. Penguins are the products of evolution, the products of history. The idea that a theory could predict that evolution would take such a course as to arrive at penguins is absurd. Such an absurdity contradicts a common-sense understanding of a basic scientific truth: historical processes are unpredictable in any detail, and the universe is always evolving new processes in the course of its own history. One can no more predict penguins from the properties of elementary particles than one can predict the laws of biological evolution—or the Declaration of Independence, or the development of human society. They are all products of history and can be explained *only* by understanding how they evolved.

This is just as true for atoms as it is for penguins—atoms too are historical products. The nuclei of the atoms that compose us were forged in stars. Why, indeed, should elementary particles themselves, protons and electrons, be anything other than historical products, even though at this time we don't know their origins? They can be no more derived from pure reason than can a flightless water fowl. The Theory of Everything is a hopeless project because it implies, contrary to all experience, a universe without a history, one formed for all time at the beginning.

The second reason we laugh, of course, is because of the power of Oliver's equations to make Opus, the penguin, disappear and then reappear. Such a power of numbers is ludicrous—it is magic. Numbers, after all, are like words: they are symbols we

BLOOM COUNTY BY BERKE BREATHED

use to describe nature. To take mathematics as reality, as Hawking and other Theorists of Everything do, to believe the universe is formed by "breathing fire" into mathematical equations, is to believe in magic. A search for the ultimate mathematical reality inevitably creates fairy tales, not science.

It's worth emphasizing again that the deductive method, while it is dominant in cosmology and particle theory, is *not* the method of science today, or of physics in particular. There are, throughout the world, perhaps one or two *thousand* cosmologists and particle physicists out of five or six *million* scientists. The method of those millions of scientists—biologists studying the human body, physicists probing superconductivity, chemists creating new substances—is the same inductive, observational method used by Leonardo, Galileo, and Kepler. To many of these other scientists, cosmologists' and particle theorists' claims to create scientific truth by pure reason seem absurd. Unfortunately, cosmology and particle physics have a very high profile in the public's eyes, and it is their method that is routinely represented as typical of science on the shelves of bookstores and public libraries, on television, and in newspapers.

■ COSMOLOGY TAKES GUTS AND VICE VERSA

Clearly, there is a single method that connects cosmology and fundamental physics. But they are connected by content as well. For the Big Bang is, for particle theorists, the Golden Age of Perfect Symmetry, their Garden of Eden.

The real world we see today is not very symmetrical. The four fundamental forces work in very different ways with very different strengths. Gravity is by far the weakest force: the gravitational attraction between two electrons is 10^{42} times weaker than their electrical repulsion. However, gravity has an infinite range, and is always attractive, so it does become significant at large scales. Electromagnetism is much stronger—after all, a tiny magnet can overcome the gravity of the *entire* earth simply by making a piece of metal leap off a tabletop. Like gravity, electromagnetic forces are of infinite range, but they are both attractive *and* repulsive. The weak force and the strong force have extremely short ranges —about ten-trillionths of a centimeter, or the diameter of an atomic nucleus. And while the weak force is a hundred million times weaker than electromagnetism, the strong force is a thousand times *stronger* than electromagnetism. The strong force can also be attractive or repulsive, but the weak force is *neither*, mainly causing the decay of nuclear particles (Table 8.1).

The forces also act in asymmetrical ways. For example, most elementary particles have "spin," angular momentum, as if they were spinning around an axis. Theorists, certain of their beloved symmetry, thought all forces act on all spins alike. But they don't —the weak force is left-handed: when a nucleus decays and emits an electron, the electron's spin is preferentially pointed in the direction of motion, like a left-handed screw thread. In almost all cases, particles and antiparticles (the antimatter equivalents of particles) act as mirror images of each other—for example, a positron emitted in a decay is right-handed. Yet there are even

TABLE 8.1
THE FOUR FORCES

Force	Relative Strength (Electro-magnetism = 1)	Range
Gravity	10^{-43}	Infinite—attractive
Electromagnetism	1	Infinite—attractive, repulsive
Weak nuclear (radioactivity)	1/100,000,000	one ten-trillionth cm
Strong nuclear	2,000	one ten-trillionth cm—attractive, repulsive

exceptions to *this* rule. In one particular decay of a short-lived particle, the antiparticle decays faster than the regular particles. Even time-reversal symmetry (described in the last chapter) doesn't always hold.

Particles have properties that are, themselves, highly asymmetrical and complex. To begin with, there are a handful of stable particles, those that never decay. Two have mass—the electron and the proton, which is 1,836 times more massive. Photons and neutrinos have no mass. Neutrinos, moreover, hardly interact with matter—a neutrino could penetrate light-years of rock (were such a thing to exist).

But that's not all. In addition to these stable particles there are twenty-two or so relatively long-lived particles. They do decay, but their average lifetime is long compared with the time it takes for particles to collide—10^{-23} second. These particles are quite a zoo (Fig. 8.1). They range in lifetime from 10^{-20} second for the sigma-0 up to fifteen minutes for the neutron. (In a stable nucleus, the neutron remains absolutely stable.) They have masses from 207 times as massive as the electron up to 5,274 times as massive as the electron. Some are neutral, others charged. Particles less massive than protons (called leptons and mesons) decay into electrons, neutrinos, or pure energy; but heavier particles (called baryons) decay into protons. Mesons and baryons interact with the strong force, but the leptons are immune to this force.

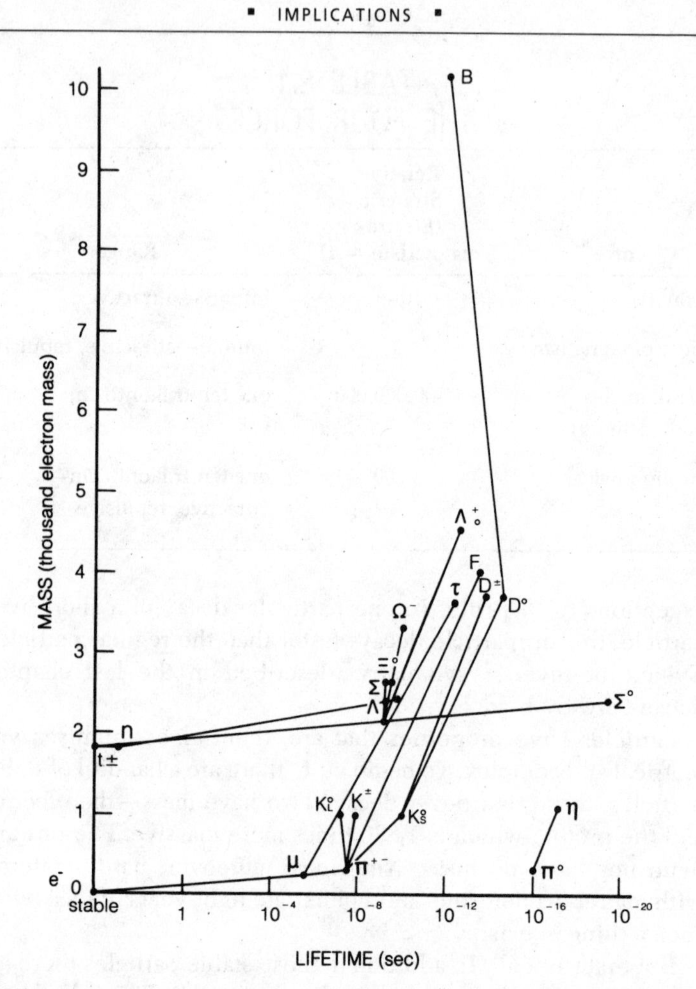

Fig. 8.1. Stable and relatively long lived particles. Lines indicate typical decay paths.

To make matters worse, when particles collide in accelerators, their interactions yield indirect evidence of short-lived states that exist only during the collision itself. There are hundreds of these states, called resonances.

In an effort to introduce *some* order and symmetry into this mess, particle theorists developed the "standard model" (Table 8.2). The model assumes that all the forces of nature are quantized—that each force is carried by particles. These force-carrying particles are exchanged by other particles, thereby

TABLE 8.2
PARTICLES OF THE STANDARD MODEL

QUARKS		LEPTONS	
(masses in electron mass units)			
Up mass 2–10	Down 6–10	Electron 1	Electron neutrino less than .000022
Charm mass 1,400	Strange 180	Muon 207	Muon neutrino small or zero
Top mass*	Bottom 4,800	Tau 3,491	Tau neutrino *

* not yet observed

FORCE CARRIERS

Photon, W, Z, 8 gluons (not yet observed)

generating a force. The model hypothesizes that mesons and baryons are made up of six different types of particles called "quarks." And then there are the six leptons: electrons, muons, tauons, and three neutrinos. A photon carries electromagnetic force, the W and Z particles carry the weak force, and no less than eight gluons carry the strong force—hence their name, because they glue one particle on another. Each of these twenty-four particles (not a noticeable improvement over the twenty-six "stable" particles) has its own mass and other characteristics. The theory has the disadvantage that no quark or gluon has been directly observed.

Only a little symmetry has been gained by this—the particles are now grouped into multiples of six, more or less. But to get the perfect symmetry they believe underlies messy nature, theorists have developed the idea of broken symmetry—the higher the energy, the greater the symmetry, they say. As matter loses energy, symmetry is spontaneously broken, producing the asymmetrical reality of an experiment. A favorite analogy is the freezing of water: at room temperature, water is symmetrical, it has no special "direction." But once it freezes, the facets of the ice give it asymmetrical "directions"—symmetry is spontaneously broken.

Here lies fundamental physics' crucial connection to the Big

Bang. The Big Bang is the golden age of perfect symmetry and ultrahigh energy. In the beginning was symmetry, fundamental physicists assume: all the particles and the forces were one, but as the Big Bang cooled in the first instants of time, asymmetry spontaneously occurred. First gravity separated out as a distinct force, then the strong nuclear force, and finally the weak nuclear force and electromagnetism. All this occurred in a tiny fraction of a second (Table 8.3). Similarly, in the beginning all particles were identical in mass and other qualities—now they're all different.

This concept of asymmetry's origin is obviously very different from Prigogine's theory. He hypothesizes that asymmetries arise as energy flows *increase*—and, moreover, that they are progres-

TABLE 8.3
HISTORY OF THE UNIVERSE ACCORDING TO THE BIG BANG

EVENT	TIME (sec)	TEMPERATURE (K)
Epoch of perfect symmetry (Supersymmetry)	before 10^{-43}	more than 10^{32}
Separation of gravity (GUT epoch)	10^{-43}	10^{32}
Separation of strong force Separation of quarks and leptons (inflation)	10^{-35}	10^{28}
Separation of weak force	10^{-12}	10^{16}
Formation of protons	10^{-5}	10^{13}
Formation of helium	250	10^{9}
Emission of cosmic background radiation, formation of atoms	2×10^{13} (700,000 years)	4,000
Formation of superclusters, clusters, galaxies, initial stars	3×10^{16} (1 billion years)	40° K
Present	5×10^{17} (17 billion years)	2.7° K

sive, historical developments. For particle physicists, though, all the asymmetries are built into the equations from creation onward, and merely become manifest as the temperature decreases and the golden age (which lasted 10^{-43} second!) recedes. God designs the asymmetries just to make the world interesting.

Thus the unification of the various forces occurs *only* in the remote past, at temperatures that can never be re-created. This premise is in sharp contrast to the most successful (in fact, the only entirely successful) theory of unified forces—electromagnetism. Maxwell showed that light, electricity, and magnetism are *in all cicumstances* aspects of a single underlying process. The same is true for the model on which all recent field theories are based—quantum electrodynamics. This theory unites electromagnetism and quantum mechanics. Although it has some difficulties, as we shall see, it still applies to the here and now, not the once-upon-a-time.

But if there was no Big Bang? Then there was never any period of high temperature and perfect symmetry, and there is no way of explaining in conventional terms how all these asymmetrical particles and fields came to be. In short, the foundations of fundamental physics and its whole tower of symmetry are tottering.

There is a more concrete connection between the Big Bang and fundamental physics, and more direct contradictions with observation. Since particle theories deal with higher and higher energies to achieve their higher symmetries, the predictions involved become more and more difficult to test. This has not seriously limited the first step of the unification program—the electroweak theory, which unites the weak force and electromagnetism. This theory successfully predicted new particles, the W and Z, at energies of 80 to 90 GeV (billion electron volts). It remains to be seen whether another vital particle, the so-called Higgs boson, exists at an energy of 1 TeV (trillion electron volts)—to be achieved by the superconducting supercollider now being built in Texas.

But to go beyond the standard model's particle zoo also means going beyond what can *ever* be tested with accelerators. The Grand Unified Theories (GUTs), which attempt to merge the strong, weak, and electromagnetic forces, make predictions about energies on the order of ten million trillion GeV, far beyond the reach of any conceivable accelerator. And it is at this point that fundamental physicists defer to the cosmologists, because such energies were supposedly achieved only in the Big Bang. So GUTs can be confirmed only indirectly, by observing the cosmic residue of the Big Bang.

However, the one testable prediction of the GUTs has already been disproved. As I mentioned in Chapter Four, all GUTs predict the decay of the proton. Because protons and electrons supposedly become equivalent at high energy, they should be able to change into one another—specifically, positrons should merge with pions, a type of meson, to become protons (the positron is supposed to turn into a quark). But according to the idea of time reversibility, any such process *should* reverse at low temperatures. Thus protons should slowly decay into pions and positrons, releasing large amounts of energy (Fig. 8.2a).

a)

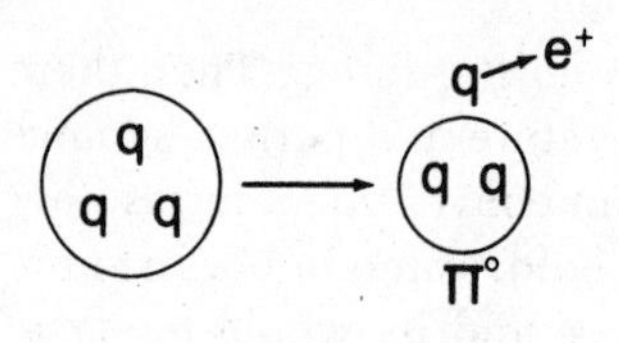

b)

c)

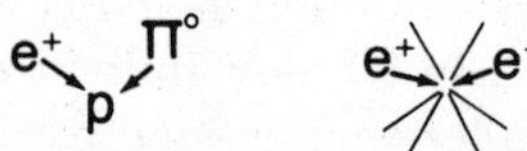

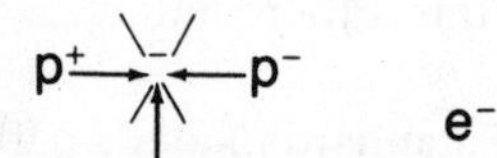

Fig. 8.2. According to GUTs (Fig. 8.2a), a proton (p^+) made of three quarks (q) will decay into a pion (π°) with two quarks and a positron (e^+). In the Big Bang (Fig. 8.2b), positrons and pions create protons, while antiprotons (p^-) and positrons are annihilated (Fig. 8.2c).

The Big Bang also requires interchangeable protons and positrons. Because matter and antimatter are created symmetrically, at the extremely high densities postulated by the Big Bang, all matter and antimatter would annihilate each other, leaving only energy—no universe. If, however, some positrons were to turn into protons, there would be an excess of protons and electrons left over after all the antimatter had been annihilated (Figs. 8.2b, c). This is another slender thread on which the Big Bang cosmos hangs.

But *do* protons decay? The GUTs predicted that they should, after an average life of 10^{30} years. So experimenters watched tons of water buried deep in mines for any sign of proton decay. They found none. The experiments showed that protons don't decay in even a hundred times the lifetime predicted by the GUTs. Protons are forever.

GUT theorists shrugged off these results. Obviously, they reason, the first theories were too simple. We now must come up with new theories that predict lifetimes longer than the lifetimes ruled out by experiment. Why won't they just admit that the proton is absolutely stable? Because the Big Bang tells us that there *must* have been conversion between protons and positrons. The proton *must* decay.

But if there was no Big Bang, then the only possible experimental test of the GUTs, proton decay, clearly disproved them. And if there is no proton decay, this is another strike against the Big Bang: the universe, then, is made up equally of antimatter and matter. And we know that some matter survived the alleged annihilation, so this implies that the universe never went through a state of such high density—and thus no Big Bang. GUTs and the Big Bang stand or fall together.

The subject of proton decay and antimatter is a good example of the extreme subjectivity of the notion of "symmetry" in nature. It could just as easily be argued that a universe with equal amounts of matter and antimatter is far more symmetrical than one with more matter, and that a stable, unchanging proton is far more "perfect" than one that decays. Yet GUT theorists have rejected these ideas because they don't fit with either the Big Bang or the theorists' a priori principles.

Again, it is futile to try to derive natural laws from aesthetic ideas. We can argue till the cows come home which universe is more beautiful, but only experiment and observation can determine which is *real*. If we ignore observation, as particle theorists do when they ignore the negative results of proton decay, they retreat toward medieval methods.

But if observation is our key criterion, then neither the Big Bang nor the GUTs are valid. GUTs supply the cosmologist with imaginary particles like axions, which allegedly fill the universe with dark matter. No GUTs, no dark matter, no Big Bang. Conversely, the Big Bang supplies GUTs with the extreme energies required for their theoretical symmetry, and many of their hypothetical particles. The two sets of theories rely on each other for confirmation, a form of cosmic circular reasoning. But the experimental tests that judge either theory invalidate both of them.

What is true for GUTs is even more true for the latest in theoretical fashions—superstrings and supersymmetry. Not content with unconfirmed GUTs, some theorists attempt to unify all four forces, including gravity, by postulating the existence of tiny strings, with length but no thickness. These are the underlying structures in the theory of all particles and fields—the Theory of Everything. However, the strings exist at such enormous energy levels that not a single verifiable prediction emerges from the theory. Again, theorists hope that some hitherto unknown effect of the Big Bang will provide evidence for the existence of superstrings. Without a Big Bang, the Theory of Everything is also left without visible means of support.

TROUBLES WITH QUARKS

If the latest theories—GUTs and superstrings—are stripped away from particle physics, the standard model with its quarks is left. Unfortunately, the problems don't end here: the standard model is not at all a satisfactory theory of nuclear forces, or of other structures of matter generally.

The theory arose as an attempt to simplify the zoo of particles discovered in the forties and fifties. Back in 1911 physicists believed that only two particles exist—protons and electrons. The neutron was discovered in 1930—it was a little heavier than the proton, electrically neutral, and a key constituent of the nucleus of the atom. Things seemed fine. The bulk of the mass of matter is contained in the nucleus, made up of protons and neutrons, while electrons swirl around the periphery of the atom. But this simple picture was spoiled as the cyclotron and other particle accelerators started hurling nuclei at each other with increasing energy, and scientists started to analyze the constituents of cosmic rays. New particles, all unstable, were discovered in the tracks they left on photographic plates and other instruments.

First came the muon, 207 times as massive as the electron. "Who ordered that?" nuclear physicist Isidor I. Rabi responded. Then came the pion, somewhat heavier, theorized as the carrier of the nuclear force. Then came an ever-increasing flood of particles.

By 1960 particle scientists were struggling to simplify this bestiary. Murray Gell-Mann noticed that the particles can be grouped together according to their properties in symmetrical arrays—the idea of perfect symmetry started to raise its head.

By 1963 Gell-Mann developed the idea that the symmetry of the groups can be accounted for if it is assumed that mesons and baryons are made up of smaller particles, which he called quarks, from a passage in James Joyce's *Ulysses*. Gell-Mann proposed the existence of three quarks, dubbed "up," "down" and "strange," which carried fractional charge—either one-third or two-thirds of an electron's charge. Two quarks together form a meson, three a baryon. Leptons—an electron, muon, and neutrino—and photons are left out of this scheme, but all the particles will be reduced to leptons, the photon, and the three quarks, a total of seven.

Complications in this neat picture developed immediately. For one thing, no matter how hard accelerators smashed protons against each other, no quarks came out—they were never observed. *Obviously*, theorists reasoned, there is a force between quarks that increases with distance—a confining force that never lets quarks go free. A second complication occurred when they realized that in some particles all quarks will spin the same way and thus are indistinguishable—which violates a fundamental postulate of field theory, that identical particles cannot exist in the same energy state. So the quarks were assigned a new property, arbitrarily termed "color." A quark can come in three "colors"—red, blue, and green. Three quarks had become nine.

Worse still, newer particles kept turning up uninvited, so new quarks were needed—a "charm" quark and a bottom or "beauty" quark. More neutrinos showed up among the leptons—a muon, neutrino, and a new massive lepton called the tauon.

To explain the nature of the strong and weak forces, still *more* particles were needed. A theory called quantum chromodynamics (QCD) was developed postulating gluons—also never observed—to carry the strong force. Another theory, the electroweak theory, described the weak field as merging with electromagnetism at high energy; it requires two more particles.

The synthesis of QCD and electroweak is the standard model, which had its successes. The masses of the W and Z particles needed to carry the weak force were actually predicted before the discovery of these particles in the eighties. The theories can make rough predictions of the mass of most particles and the lifetimes of some. Perhaps most significant, in particle collisions experimenters observed concentrated jets of particles coming out in certain directions. These, it was argued, show that unobserved quarks are hit in collision and then emit observable particles in the direction of the quarks' motion.

But the standard model has important limitations. For one thing, what it can predict pales before what it can't. The masses of all the quarks and the strengths of the interactions—a total of twenty constants—all have to be plugged into the theory, based on observation. Why these masses? Why is the proton, for example, 1,836 times as massive as the electron? Why are there so many particles? Why three generations of quarks and leptons? Who needs neutrinos anyway? The strengths of the field are even more puzzling. Why such different strengths? And where does gravity, 10^{42} times weaker than electromagnetism, fit in?

Unfortunately, like Ptolemy's solar system, the standard model requires many special assumptions to match observation even remotely. To be sure, it makes valid predictions—so did Ptolemy's system—within broad limits of accuracy. But it has no practical application beyond justifying the construction of ever-larger particle accelerators. Just as electromagnetism and quantum theory successfully predict the properties of atoms, one might expect a useful theory of the nuclear force to predict at least *some* properties of nuclei. But it can't. Nuclear physics has split with particle physics; nuclear properties are interpreted strictly in terms of empirical regularities found by studying the nuclei themselves, not by extrapolating from QCD.

What is more serious in judging the standard model is its own contradiction with observation. The model requires six particles to complete its symmetrical form, yet the "top" quark, the last of the six, has not been discovered in the range of energies predicted by theory. Another particle, the Higgs boson needed by the electroweak theory, is also AWOL. A major motivation of building the superconducting supercollider is to find the Higgs, but searches at lower energies have been unsuccessful.

The most serious contradiction with theory comes in a series of experiments done with spin-aligned protons. In a decade-long series of experiments, Alan Krisch and his colleagues at the University of Michigan have demonstrated that protons have a far greater chance of being deflected in a collision when their spins are parallel, instead of spinning against each other. What's more, they also deflect nearly three times more frequently to the left than to the right. In effect, the protons act like little vortices, pushing each other around (Fig. 8.3).

This seriously contradicts a basic assumption of QCD, that quarks act independently within a proton. This implies that a proton's spin should have little effect on a proton's motion. Each

Fig. 8.3.

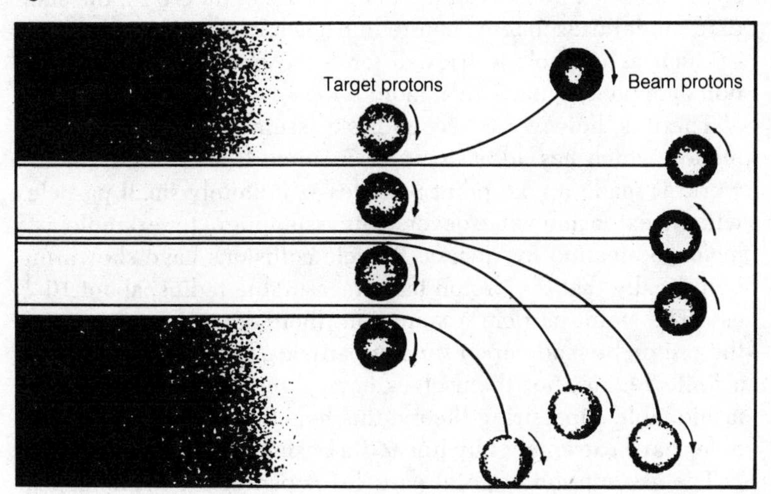

of the three quarks has a spin of one-half unit of angular momentum, so a proton's spin of one-half arises from two quarks spinning in one direction, one in the other. If two protons collide, it is the spin of the *colliding quarks* that should determine the outcome of the collision—in which case collisions of opposite-spinning quarks should be only 25 percent more common for opposite-spinning protons than for parallel-spinning protons. But the effects Krisch observed are far bigger—two or three to one. This strongly implies the spin is carried by the proton, not by the quarks—if they exist at all. In the view of many theorists and of Krisch himself, this clearly contradicts QCD.

Probably more important, QCD also predicts that spin effects, like all other asymmetries, should decrease at higher energies in accordance with the broken-symmetry approach of all particle theories. Yet Krisch's experimental results show that spin effects steadily increase with the energy of the collision. Evidently, spin effects are fundamental to the structure of matter—matter is, therefore, *inherently asymmetrical*. But as with proton decay, such contradictions have been ignored for the most part.

The standard model's problems aren't as profound as those of the wholly fanciful GUTs or the Big Bang. It rests on a mixture of observations and a priori assumptions. But it lacks predictive power and has been unable to advance beyond a rough correspondence with observed phenomena. Like the GUTs, the standard model tries to force nature into a neatly symmetrical pattern—much as cosmology tries to force the universe into a finite, homogeneous, symmetrical mold.

There is, however, a second basic assumption of the standard model, which has no basis in observation. It is claimed that the world is made up of "point particles"—infinitely small particles with no extension whatsoever. This assumption, for example, is a major motivation for quarks. Particle collisions have shown unequivocally that the proton has a measurable radius, about 10^{-13} cm. The point-particle assumption, therefore, necessitates that the proton be made up of smaller particles, swarming together in a finite region, but themselves having zero dimension. (In the fashionable superstring theory, this hypothesis is modified: particles have extremely tiny linear dimensions, but zero thickness.)

The assumption of point particles is part of the mathematical structure that underlies quantum mechanics and quantum electrodynamics, the most fundamental theories of the modern view of matter. So to understand the problem involved with the idea of point particles, we must strip away yet another layer of theory to look at these fundamental concepts—in particular, quantum electrodynamics, the theory on which all modern particle theories are modeled.

QUANTUM PARADOX

Quantum electrodynamics (QED) is the theory that melds quantum mechanics, electromagnetism, and special relativity together. It was developed in 1928 by Paul Dirac, and, unlike standard models, GUTs, and other recent efforts, it works spectacularly well. Some QED predictions are accurate to seven or eight decimal places. With only one arbitrary constant, Planck's constant, it can predict myriad physical situations. By incorporating special relativity corrections, it makes quantum mechanics exceedingly accurate and is used in a wide range of technologies whenever such accuracy is required (in most cases, ordinary quantum mechanics will do).

Since the theory is so successful, why have current physicists gone wrong by imitating it? The difficulty arises in the old problems of confusing the mathematical form of a theory with the physical processes it describes. Particle physicists took the mathematical form of the theory—its point particles and fields carried by particle exchanges—but left behind the physical reality of electromagnetism.

Unfortunately, this meant adopting exactly the aspects of QED where it breaks down and ignoring its contradictions and limitations. Although the theory is excellent in its predictions generally, the concept of point particles leads to contradictions—for example, it fails to predict the mass of an electron or any other particle. Since QED assumes that all particles are mathematical points, infinitely small, the electron is viewed as an infinitely small particle. Electrical force increases as distance decreases, so an electron theoretically has an infinitely high energy. Now this causes two related problems. First, there is nothing holding an electron together: like charges repel and it is *all* one like charge —it should explode. Second, and more serious, since it has infinite energy from its electrical field, QED predicts that it must also have infinite mass—which is nonsense. An electron has a mass of 10^{-27} gram.

Physicists were aware of this contradiction as soon as the theory was formulated, but initially ignored it. If the true mass of the electron is substituted in the equations, everything works out just fine. A mathematical trick was developed, "renormalization," which in effect subtracts infinity from the theoretical mass and adds the observed mass in. There's no justification for this, except that it works and allows the equation to be used accurately.

It would be silly to reject such an accurate theory just because *one* of its predictions is wrong. But it's equally silly to ignore this failure because one can "renormalize" it—however useful, *no one knows why this trick works.* There was no effort to ask, How can we eliminate the cause of these energies? Do they arise from our assumption of point particles? On the contrary, the mathematical trick of renormalization and the point particles themselves were embraced by particle physicists as the fundamental mathematical hypothesis of any theory that described any forces, including the nuclear forces and gravity.

When gravity is taken into account a new paradox arises. QED hypothesizes that the vacuum is filled with virtual particles, continually coming into, and out of, existence so fast that they are unobservable. Vacuum, therefore, has a vast energy density. We can't tap it because no lower energy level is available—just as we can't use water power at sea level, it has no lower place to go.

But general relativity says that energy, like mass, curves space. The gigantic energy density of the quantum vacuum should curve space to create a cosmological constant, an enormous repulsive field that would curve space into a sphere a few kilometers across. This obviously doesn't happen, so something's wrong with the theory. This problem is widely known in physics, but no attempt has been made to either perfect or to supersede QED.

The paradoxes of quantum theory extend as well to quantum mechanics itself, which was formulated in the twenties, only a few years earlier than QED. The paradoxes, again widely known, involve the relation between particles and waves that is basic to the whole theory. The equations of quantum mechanics describe waves, but the objects controlled by these waves are particles—point particles. A wave doesn't directly determine the exact position of a particle, only the probability that a particle will be in a given place. Over hundreds of identical experiments the particles will be found, on average, in just the distribution of places the theory predicts. But in a given experiment a particle's exact position is impossible to know in advance.

Thus in one famous experiment electrons pass through two tiny slits and are recorded on a fluorescent screen (Fig. 8.4). Each electron collides with the screen at a single spot, yet their overall distribution is a wavelike pattern across the entire screen. The same is true of photons in experiments with light: each photon lands in a particular spot, but the overall distribution acts like a wave.

How can this be? How can an electron or photon "decide" where exactly to go? The standard answer is that such a question can't be asked. When the electron is not being observed (when nobody's looking) it has no position—it is spread out over a volume of space. Only when a measurement is taken does the "wave

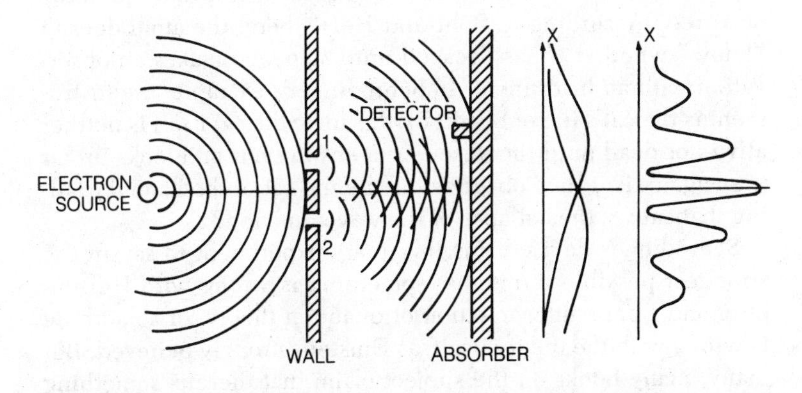

Fig. 8.4. In the double-slit experiment, a beam of electrons passes through two closely placed slits. Each electron makes a flash when it hits a fluorescent screen. The pattern of flashes is exactly predictable—a circular pattern that is just the same as the interference pattern created when waves pass through two narrow openings. Where the wave crests coincide, in this case of the Schrödinger waves, the electrons are more likely to land; where they don't coincide, electrons are less likely. However, there is no way of either predicting where an individual electron will go or even knowing which slit it passed through. In the Copenhagen interpretation of quantum mechanics, the electron has no position until it is "observed"—arrives at the screen.

function collapse" and the particle materialize at a single point—*as a result of the observation.*

This idea, developed by Niels Bohr and Werner Heisenberg, is just as bizarre as it sounds. It means that human consciousness has a direct impact on the electron. Until some conscious being observes an electron, it has *no* position. Observation magically—through no known law—makes the electron choose a spot to land.

By no means did all the founders of quantum mechanics buy into this astounding idea. Erwin Schrödinger, who developed the basic equation used today in quantum mechanics—the Schrödinger equation—ridiculed the idea with the famous cat experiment. Take a cat, Schrödinger argued, and put it into a sealed chamber with a vial of cyanide. A device breaks the vial with a hammer when a Geiger counter detects the decay of a given atom. By quantum mechanics it's impossible to predict the exact moment the atom will decay. The observer leaves and comes back the next day, opening the box to see whether the cat has survived this experiment. The question is: Which observation determines that the nuclear wave function has collapsed, that is, if the decay occurred? According to Bohr and Heisenberg, the atom doesn't "know" whether it has decayed until someone makes an observation with an instrument. In Schrödinger's example, the instrument is the cat's life or death. By this logic the cat itself is neither alive nor dead until the observer opens the box and looks in! Or then again, is it the "observation" by the cat of the hammer falling that causes the collapse of the wave function?

Schrödinger devised this imaginary experiment to say in the strongest possible terms that something is wrong with Heisenberg and Bohr's interpretation of quantum theory, or something is wrong with the theory itself, as Einstein strongly believed. But many, many books on the subject claim that there is something bizarre about the universe, rather than about quantum mechanics as a theory!

Dozens of scientists have concluded that this proves that consciousness, either human or feline, has a direct, *occult* impact on the universe—that the universe, as John Wheeler has written, could not exist unless there was a human being to observe it. Even stranger notions have become quite popular: one, a staple of science fiction, is that *every time* an electron has to "make up its mind" where to jump, a new universe splits off—in one universe the electron jumps one way, in another the other way. This creates new universes as rapidly as Hawking's baby-universe theory, is just as fantastic—and is also based on quantum ideas.

While some scientists' explanations are fantastic, the contradictions in quantum mechanics are quite real. They aren't in outright contradiction with experiment, as with quantum electrodynamics' prediction of an electron's infinite mass. But they do inevitably involve contradictions with other, equally well-verified theories—specifically special relativity. The most striking illustration of this is in a widely noted series of experiments by Alain Aspect of the University of Paris. Aspect polarized pairs of photons—oriented them similarly—and then sent them to two different measuring devices, polarized filters (Fig. 8.5). Using quantum mechanics one can predict the probability that each photon will pass through a filter tilted at a given angle. But, according to the same theory, the probability that one photon will pass through depends on how *both* filters are tilted, because both are measuring the same quantity.

Aspect made sure that the filters were sufficiently far apart, and that their orientation was varied electronically quickly enough, that no signal from one could reach the other in time to affect the second measurement, even if the signal traveled at the

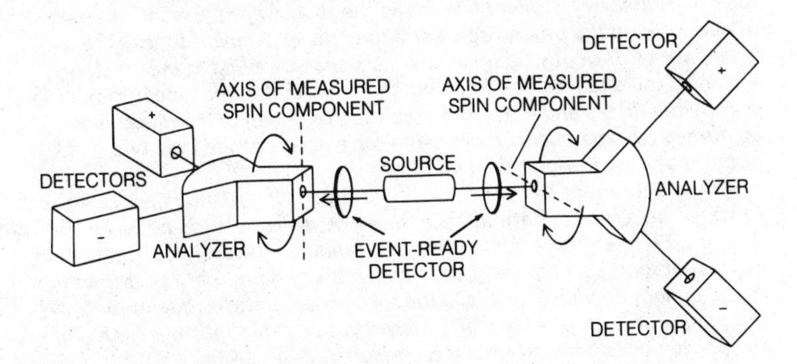

Fig. 8.5a. In Alain Aspect's experiment, two photons, identically polarized, are emitted in opposite directions from a single source. Each passes through a polarized filter whose angle is varied rapidly by electronic means. For a given angle, some of the photons will pass through, others will not. Quantum mechanics predicts that if a photon passes through one filter, there is an effect on the probability that it will pass through the other filter.

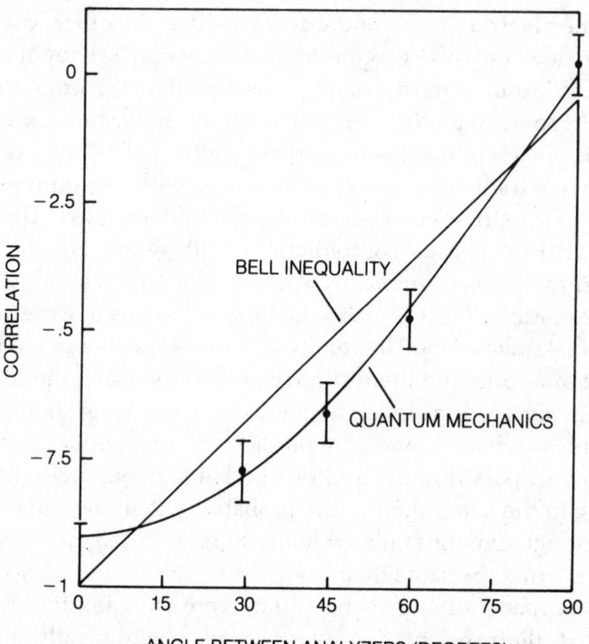

Fig. 8.5b. The curved line in the graph shows the chance that a photon will pass through one filter at a given angle to another, if it has passed through the other filter. The straight line shows the chance if there were no effect of one filter on the other. Aspect arranged the experiment so that the two filters were sufficiently far apart that no signal, traveling at the speed of the light, could reach one filter from the other during the time it took to change the filter's angle. Yet the prediction of quantum mechanics was confirmed by experiment: there was a correlation between the two photons passing through the two filters.

This effect cannot be used for a practical signaling scheme since one has to know the results at both filters to measure the correlation between the photons. Each sequence by itself remains random. However, it appears that a signal of some sort has traveled between the two filters at faster than the speed of light, implying a limit to the theories of relativity, quantum mechanics, or both. The standard interpretation is that the polarization of the photon does not exist until it is measured.

speed of light. Yet the results were just those predicted by quantum mechanics. Apparently a signal *did* travel faster than the speed of light.

This is conventionally interpreted as showing how contrary to common sense the universe is: quantum mechanics is true, relativity is true, and their results are contradictory. Therefore, logic doesn't apply to the quantum world. Such a viewpoint among scientists has served as an open invitation to all sorts of irrationalism and occultism. Quantum mechanics has been used to justify the existence of extrasensory perception, telepathy, and other fantasies.

This can't be considered strange, because the conventional view of quantum mechanics, the view taught in every physics department today, introduces magic into the heart of science. Ultimately it is assumed that quantum phenomena are *acausal*—that there is no cause for the decay of a nucleus at a particular instant, or the emission of a photon. These things "just happen." And when the basic principle of causality—that everything is caused by something else—is abandoned, magic can become quite acceptable. (To be sure, the majority of scientists who use quantum mechanics don't think about these matters very much and don't believe in magic. But the magical nature of quantum transition is central to the standard interpretation of the theory, nonetheless.)

Like the other paradoxes of modern physics, the contradictions of quantum mechanics are, in effect, swept under the rug. Many articles are written about them, but they tend to conclude, "Isn't the universe bizarre?" No articles within the mainstream of physics view these contradictions as somehow implying that, despite the theories' great success, they are limited in some fundamental way.

QUANTUM ORIGINS

What is perhaps most peculiar about this situation is that quantum mechanics itself arose in response to the contradictions of earlier theories, which its founders viewed *not* as indications of the limits of human logical understanding, but simply as limits of a particular theory, and of the need to develop a new one.

The first impetus toward the idea that energy comes in discrete packages, or "quanta," and can behave like particles, came from the study of black-body radiation—the same black-body radiation I discussed early in this book. At the end of the nineteenth century, scientists were studying the spectrum of light emitted by perfect absorbers—essentially black boxes with only a tiny pinhole for the measuring device. They found that the spectrum is always the same in shape, except that it grows in intensity and shifts in frequency as the temperature of the box increases.

Using the well-tested laws of electromagnetism, Lord Rayleigh, a leading authority on light, calculated what the spectrum should be theoretically—and the answer didn't make sense (Fig. 8.6). For long wavelengths the proposed curve agreed well with observation. But for short wavelengths, the theoretical curve kept escalating without limit. The shorter the wavelength, the more intense the light. The total intensity of the light emitted is, therefore, infinite! This impossible result was immediately dubbed the "ultraviolet catastrophe."

This is exactly the sort of contradiction that bedevils present-day physics. The laws of electromagnetism were incontrovertible—they had been tested by millions of experiments—yet they predicted an impossible result, *infinite* light intensity. Similarly, QED today predicts an electron's infinite mass, and quantum

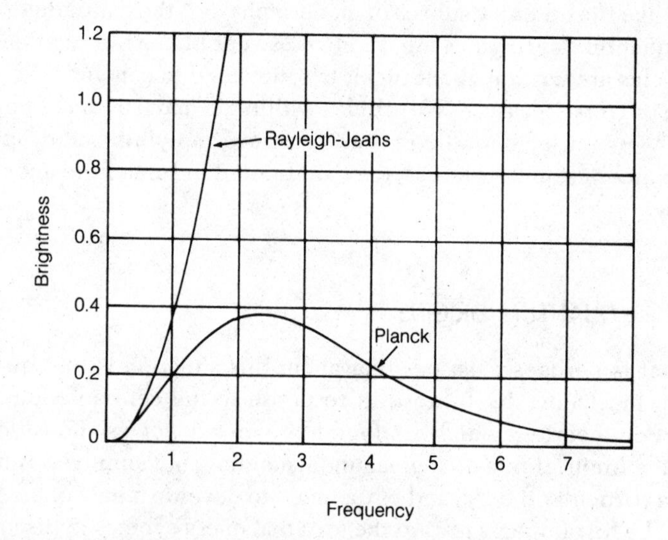

Fig. 8.6. Lord Raleigh's prediction of the spectrum of a black-body increased without limit with higher frequencies, a result called the ultraviolet catastrophe. The true spectrum was quite different (lower curve). The contradiction between the prediction of a well-established theory and experiment led to the development of quantum theory.

mechanics and special relativity have been experimentally shown to contradict each other.

The physicists of the start of this century could have taken the same approach as the physicists of today: "Oh well, the theory is obviously right. Let's just substitute the *real* equation in this case." Had they done so, though, we would have no quantum mechanics and, among other things, no electronics.

Max Planck, however, used Kepler's method: he assumed that the contradiction between theory and experiment showed a limit to the theory, well tested as it was. Something new was required, and he didn't search his mind to find the idea of the quantum! Through trial and error he tried to fit the observed curve with a mathematical formula, much as Kepler tried to fit the positions of Mars. By this method Planck discovered the equation that describes the spectrum. He then asked himself: What physical process could possibly produce a spectrum described by this mathematical equation? Guided by what he already knew of the equation, he came up with the shocking idea that energy is quantized. He showed mathematically that, given this, one can theoretically account for the observed spectrum. Only a single parameter is needed, a quantity relating the energy of a quantum to the frequency of light emitted. This parameter came to be known as Planck's constant, the basic constant of quantum theory.

Quantum mechanics arose as a result of efforts to overcome contradictions of experiment and theory. Yet for the sixty years since the development of QED scientists have evaded such fruitful contradictions. As in cosmology, beginning a decade earlier, quantum theorists moved steadily away from a concern with reality and observation, toward the sterile contemplation of mathematical purity. Over the decades the deductive method became dominant and the effort's underlying philosophy became more pessimistic. No longer was it the aim of science to make sense of the world, but merely to create abstract mathematical theories which had less and less contact with nature.

Although each step rested on earlier ones, each was denounced by the pioneers of the earlier developments. With his cosmological speculations about a homogeneous, symmetrical, closed universe, Einstein had taken the first step in divorcing theory from observation in fundamental science. But when Heisenberg and Bohr went further—interpreting quantum mechanics as proving that nature is incomprehensible to human reason, that only mathematical formulations existed—the pioneers of quantum theory rebelled. Einstein, Schrödinger, Planck, and Louis de Broglie denounced Heisenberg and Bohr's Copenhagen interpretation as a step toward mysticism.

In turn, when renormalization swept the contradictions of point particles out of sight, older theoreticians condemned it as a mathematical trick. Heisenberg later dismissed quarks as nonsense. A further step was taken with the GUTs, whose only verifiable prediction—proton decay—was ignored when it wasn't confirmed by observation. And with string theory, the last tenuous link with reality is broken and the theorists arrive at a hypothesis which makes no predictions about the real world. Again, it is denounced by those who have paved the way for it: Sheldon Glashow, one of the architects of ethereal GUTs, writes, "Contemplation of superstrings may evolve into an activity . . . to be conducted at schools of divinity by future equivalents of medieval theologians. . . . For the first time since the dark ages we can see how our noble search may end with faith replacing science once again."

The result of this divorce of theory and reality has been, as in ancient times, a growing sterility and stagnation of fundamental science. There have been tremendous advances in most areas of physics, such as materials science and hydrodynamics, which remain tied to experiment; but since the development of QED, the discovery of the neutrons and antimatter in 1928 to 1930, there have been no major gains in our understanding of the underlying structure of matter. We still do not know how the nuclear force works. Our progress in nuclear technology is based on a combination of basic quantum mechanics and a vast body of experimental knowledge gained over fifty years—not on any application of theoretical advances.

This stagnation has had a major, if delayed, impact on technology. The theoretical breakthrough of quantum mechanics led thirty years later to the technological breakthroughs of the transistor and the laser. And the subsequent lack of discoveries has contributed to the *cessation* of major technological revolutions in the past thirty years—and in turn, to the stagnation of living standards globally.

In the 250 years since the industrial revolution—and probably in the half millennium since the beginning of the scientific revolution—there hasn't been such a long period of theoretical and technological stagnation.

Despite the well-justified critiques by the pioneers of quantum mechanics and their successors, the philosophy that drove a wedge between observation and theory is as evident in the early part of this period as in the latter—and it is this philosophy that propelled the flight from experimental science. Einstein, in his words, "sought to know God's thoughts." Dirac writes, "It is more important to have beauty in one's equations than to have them fit experiment . . . because the discrepancy may be due to minor features that are not properly taken into account and that will get cleared up with further developments of the theory." This is exactly the reasoning used today by scientists who ignore inconvenient observational results.

Einstein and Dirac, like Kepler before them, were able *in practice* to subordinate their love of mathematical beauty to their even stronger urge to understand reality. Einstein threw out dozens of beautiful equations in his work on general relativity because they did not fit observation, just as Kepler junked his early beautiful Pythagorean solids for the humble ellipse.

Yet later theorists, lacking the commitment to understanding that made the pioneers great scientists, took only Platonic philosophy as their guide, which led them ever farther from experimental reality. Einstein condemned Heisenberg's concept of particles jumping around of their own free will, yet for Heisenberg and those who followed, the logical absurdities of quantum mechanics were irrelevant: only the mathematical equations were real, everything else was mere appearance. From a fascination with mathematical beauty followed the devaluation of the understanding of the reality mathematics is supposed to describe.

A Syracuse University theoretical physicist, Fritz Rohrlich, wrote in 1983 that it is impossible in quantum mechanics even to *ask* such simple questions as how a photon is generated when it is emitted from an atom. "This is a meaningless question in quantum mechanics and it has no answer. The mathematical language does not lend itself to asking such a question."

In the conventional, "Copenhagen" view of quantum mechanics, not only is the real world subordinated to the world of mathematical description, of pure reason, but severe limits are placed on the application of reason itself. In the quantum world the fundamental idea of rationality—that of cause and effect—no longer holds. Events can occur without cause, a particle can simply pop into and out of existence magically.

And if it is possible for electrons to pop into existence without any cause, why wouldn't a whole universe pop into existence without cause? The difference between a virtual particle and the Big Bang is only one of quantity. Indeed, the most recent cosmological theories are based on this quantum acausality—the universe, in the theories of Hawking and others, is one gigantic quantum fluctuation.

As in the ancient world, the revival in the early twentieth century of the duality of idea and appearance, of pure thought and observation, leads to an unstable alliance of rationality and dualism. And as in the ancient world, this alliance tends to fall apart into two trends—one emphasizing rationality and leading back toward the observation of nature, the other emphasizing the supremacy of the spirit and leading away from rationality toward occultism and a magical view of the world.

Ultimately, rationality must be tied to observation of the real world. Without the test of empirical reality, one man's reason can be another's madness. Despite his fond hopes that "pure thought can grasp reality as the ancients dreamed," Einstein never accepted the Platonic duality of idea and reality, nor did he abandon the test of observation. "Pure logical thinking cannot yield us any knowledge of the empirical world," he concluded in 1933. "All knowledge of reality starts from experience and ends in it." Einstein never accepted quantum mechanics' dismissal of causality. He did not merely object to an abandonment of a world of strict determinism. As we saw in Chapter Seven, even strictly causal laws, like that of Newtonian gravity, can lead to indeterministic systems. He objected to the idea that an event can occur with no cause whatsoever.

But the founders of the Copenhagen interpretation, Heisenberg and Bohr, clung to dualism at the expense of rationalism. To them, rational understanding can penetrate only so far, can only predict the average behavior of large numbers of particles. Beyond that lay the Platonic world of ideas, in which quantum particles would come into, and out of, existence in unfathomable ways.

As with Pythagoras and Plato, the line between the worship of reason divorced from observation and irrationality is thin and frequently crossed. Just as Plato combined a belief in reason with a rehabilitation of the discredited Greek gods of Olympus, so theoretical physicists today believe in the magical powers of number and accept electrons that jump around without cause.

QUANTUM AND SOCIETY

This retreat from reality to a mathematical irrationality did not pop out of nowhere. It originated in the chaotic social conditions of post–World War I Germany, where the senseless slaughter of the war had shattered for much of the middle class any belief in rationality and progress. As historian Paul Forman contends, it was above all the efforts of German scientists to accommodate themselves to the rise of irrationalism and occultism in the Weimar Republic that led to their integration of irrationality and acausality into the very foundations of physics.

In the wake of the world war, the pessimistic philosophy of Oswald Spengler's *The Decline of the West* became the fashion among many German intellectuals. To Spengler, the villain of modern society is rationality, which destroys both body and soul —unlike life-bringing mysticism. Its embodiment is the scientist: "The abstract savant, the natural scientist, the thinker in systems whose entire mental existence is founded upon the principle of causality is a late manifestation of the hatred of the powers of destiny, of the incomprehensible."

The revolutions and devastating hyperinflation in the Germany of 1923 only cast further doubt on the power of rationalism —the world seemed to be veering without cause from catastrophe to catastrophe. The flourishing of occultism, as in other periods of social crisis, was a widespread response—a response that paved the way for the rise a decade later of Nazism, with its glorification of the mystical *Volk*. And as Einstein's friend, the Irish writer James Garner Murphy, wrote to him in 1932, "Scientists live in the world just like other people. Some of them go to political meetings and all are readers of current literature. They cannot escape the influence of the milieu in which they live. And that milieu is characterized at the present time largely by a struggle to get rid of the causal chain in which the world has entangled itself."

Paul Forman emphasizes that many German physicists, in order to free themselves from the opprobrium attached to the notion of causality, resolved that they must rid themselves of causality itself. Long before the development of quantum mechanics in 1925—in the years immediately after the German defeat in 1918—the idea of an acausal, fundamentally irrational physical universe had gained great credibility among German scientists.

What began as a rejection of the rigid determinism of classical mechanics (a rejection that had significant scientific justification, as we saw in Chapter Seven) rapidly mutated into a very different idea—a general rejection of causality. From this rejection a new mysticism arose at the heart of the scientific establishment.

The connection between social evolution and the evolution of the ideas of fundamental physics is nowhere better illustrated than in Werner Heisenberg himself. By his own testimony he came to physics primarily through his fascination with Plato's view of a world of perfect ideas. He first read Plato's *Timaeus* while serving as a teenager in the notorious right-wing *Freikorps* militia, which fought against the workers' battalions in Munich in the civil war of 1919. After fighting, Heisenberg used his leisure to read the *Timaeus*, where he learned of Plato's notion of a world reduced to mathematical forms perceptible only to the mind.

In the next few years Heisenberg, just out of university, entered upon the study of physics. As he confessed to his first professor, the distinguished physicist Arnold Sommerfeld, "I am much more interested in the underlying philosophical ideas than in the rest."

Heisenberg was influenced more directly than many of his colleagues by the irrationalist winds blowing across postwar Germany. As a member of Germany's *Wandervogel*, or youth movement, he was taught that there is a mystical unity of man and nature which far transcends rationalism. The *Wandervogel* glorified the German *Volk* and, like the *Freikorps*, later became a prime recruiting ground for the Nazi Party. While many of his older colleagues were probably only paying lip service to growing ideological pressures, Heisenberg himself was drawn to this irrationalist current and was determined in his career as a scientist not to reduce nature to the cold rationality identified with a rotting old society.

For Heisenberg it was imperative "to get away from the idea of objective processes in time and space." As he delved more and more deeply into the then-fluid subject of quantum phenomena this rejection of a rational, objective reality became his guiding principle.

At this time, the experimental facts of quantum mechanics or atoms were well known, but had not been integrated into a single theory. Both Heisenberg and Erwin Schrödinger simultaneously succeeded in doing so. Their solutions are mathematically equivalent but conceptually very different. Heisenberg developed quantum mechanics, in which the real objects of scientific inquiry—atoms and electrons—are replaced by mathematical objects, sets of matrices, which are manipulated according to certain exact laws. In the course of this work, Heisenberg formulated his famous uncertainty principle, which states that it was impossible to measure both the position and velocity (or other specified pairs of properties) of a particle with total accuracy. As one measurement increases in accuracy, the other decreases in accuracy.

In itself, this was not so revolutionary. But Bohr and Heisenberg interpreted the uncertainty principle in an odd way: it is not a limit on our ability to *measure* phenomena; rather, it means that an electron or photon *materializes* in a given spot only when it is measured. How it can do this, and indeed how photons or electrons can come and go in this way, is a mystery, they argued, because humans cannot comprehend the quantum world. The apparent irrationality of a world in which electrons materialize arbitrarily at a given point (limited, of course, by the probabilities derived from quantum laws) is simply to be accepted.

In contrast, Schrödinger had come up with equations that define waves, which he believed to be as objective as the radio waves used routinely in communication or the light waves that allow us to see. For Schrödinger, the position of an electron cannot be precisely defined because there is no such thing as a point-particle electron; it is merely a particular part of the probability wave. Schrödinger admitted that his theory could not fully account for the existence of particles, but he insisted that it described an objective reality, not an occult world of arbitrary actions.

SOCIAL CONSEQUENCES

Although it was Schrödinger's equation that has permitted physicists to use quantum mechanics even to this day, Heisenberg and Bohr's philosophical interpretation of quantum mechanics had a profound impact on the intellectual life of the time. It was welcomed in broad intellectual circles as the scientific confirmation of the role of the irrational in the universe, a justification of occultism. The uncertainty principle was generalized into the belief that rationality is necessarily limited.

Again, Heisenberg is a remarkable example of the relationship between physics and social ideology. In the six years after he had formulated quantum mechanics Heisenberg witnessed Germany's march toward fascism. He had become less political and tried studiously, as he put it, to "ignore the ugly scenes in the street" where workers and fascists fought pitched battles. After Hitler's victory in 1933, though, he could no longer avoid making choices. A Jewish mathematician was dismissed from Heisenberg's department at his university, and some of the faculty urged a mass resignation in protest. Many colleagues in Germany and Italy had already left for America. Should Heisenberg follow their example, denying his aid to what he himself felt to be an odious regime?

After much consideration he decided to stay. He gave his reason to Enrico Fermi when the latter confronted him six years later, in 1939, at a time when war was inevitable. Heisenberg argued that it was impossible to predict the consequences of any act. "There are no general guidelines to which we can cling. We have to decide for ourselves and cannot tell in advance if we are doing right or wrong," he told Fermi. Perhaps if he emigrated a social catastrophe would overcome the U.S. as well—there was just no telling, so any action was morally justifiable.

With these social uncertainty principles in view Heisenberg proceeded down the path leading from smaller compromises, such as ignoring the dismissal of his Jewish colleagues, signing his letters with "Heil Hitler," and collecting money for the Nazi Party, to far larger ones. Otto Hahn had discovered nuclear fission right before the war, and physicists around the world, including Heisenberg, knew that the combatants would race to develop the atomic bomb. By returning to Germany from a visit to America in 1939 he chose to head the German atomic bomb program.

He justified his actions by arguing—*after* the war—that he had believed Germany unable to build an A-bomb before the end of the war, and therefore his efforts could have no real effect. (There is, in fact, no evidence that he believed this before Germany was defeated.)

Heisenberg's morally repugnant actions, and his justification for them, illustrate the huge difference between his idea of an acausal universe and the actual unpredictability that exists in the real world. The outcome of a vast social crisis such as World War II cannot be predicted in advance, yet it is far from uncaused. The defeat of the Axis was the result of the decisions and actions of hundreds of millions of individuals; it was the effect of those myriad causes. To refuse, as Heisenberg did, to look at the effects of one's own actions, to judge the results, to pretend that there is no link between action and effect, is moral bankruptcy.

Bohr also ended up working, unwittingly in his case, on the German A-bomb. Despite the urging of British intelligence agents, Bohr refused to leave Denmark or to stop his nuclear research, which contributed directly to Heisenberg's project. Bohr, blinded by his belief in the abyss between thought and reality, refused to acknowledge that the "pure" research in his laboratory might have devastating technological consequences. He did not recognize the truth until after he had finally been forced to leave a few days before the Nazis rounded up Danish intellectuals.

Heisenberg was right about one thing—the Nazis *were* incapable of developing the A-bomb, and not because of the inherent difficulty of the task. The Manhattan Project started far later than Heisenberg's uranium project, yet managed to succeed only a few months after the end of the European war. Of course, Germany lacked the economic resources that America was able to throw into the project. But the main factor that ensured Germany's failure to develop the A-bomb was its lack of the scientific personnel needed: except for a few like Heisenberg, those with the necessary talent were in America. They had decided not to support a state that was spreading a regime of slavery and murder throughout the globe, and they left to support those who would fight such a monstrous tyranny. They had recognized the unbreakable link between action and effect, between the theories of physics and the reality of technology, the battles of modern society, and the struggle against a modern slavery. And they thereby ensured that Hitler would never be informed, as Roosevelt was by Einstein, of the importance of the A-bomb, and that he would never succeed in building one.

Of course, there is no one-to-one correspondence between a scientist's physical theories and his or her political and moral beliefs. Heisenberg, for one, took very seriously the philosophy that motivated his physics, and acted on it. Many other scientists pay no attention to the philosophy of the Copenhagen interpretation and concern themselves only with the *equations* of quantum mechanics, which in no way depend on philosophy. Many who do think about the paradoxes of quantum behavior do so as a kind of hobby isolated from their general worldview.

But Heisenberg's example is a relevant one, because the social significance of the philosophies of fundamental physics remains great. The great uncertainties of the twentieth century continue to promote a swelling occultism. During the past thirty years, in which technological and material progress has halted, rationalism has fallen into increasing disfavor. We see this today in the rise of mystical religions and fundamentalism throughout the world and in the spread of religious fanaticism.

These tendencies cannot but encourage physicists to portray their work as part of a search for mystical truth, as a pursuit of beauty far from the disdained materialistic pursuit of any knowledge useful in the real technological world. In turn, as in the late twenties, the latest physics theories are used to support the newest occult concepts spreading among the public.

The same pressures, forcing theory further from experimental reality, are present within the institutions of science as well. In the teaching of physics, mathematical manipulation is emphasized at the expense of physical understanding, thus focusing students toward the purely formal aspects of science and away from the outstanding contradictions of the standard approaches.

As in cosmology, slowing technological advance has broken the link between science and technology. Until the end of World War II, the need to understand the nucleus and its forces kept theorists speculating about practical work. But once it became clear, with the development of the A-bomb and the H-bomb, that the existing empirical knowledge of nuclear reactions is adequate for military purposes, the pull of technology ebbed and theory was free to float skyward into a mystical search for unity with what Heisenberg termed "the Central Order."

THE SEARCH FOR AN ALTERNATIVE

It would be satisfying if at this point I could present an alternative theory to fundamental physics which resolves the contradictions of the conventional ideas. Unfortunately, no such theory exists as yet. What do exist, though, are various clues as to how to approach such an alternative, and substantial reason to believe such an alternative is possible.

We can define a set of criteria that a new theory must meet. It must resolve the long-standing contradictions of particle theories. It must explain what particles are and how they exist without blowing themselves apart. How can they be real entities, with size and dimensions, not perfect, infinitesimal particles? How are they created and destroyed? How are they controlled by the waves described in Schrödinger's equations? How can they have a finite, specific mass, and why do they have the mass that they do? Why do some particles decay while others don't, and why do the unstable ones have the lifetimes observed?

Equally important, such a theory must resolve the contradictions among the various theories in use today, especially between quantum mechanics and relativity. Unless one believes in magic, Aspect's experiment clearly demonstrates that some form of communication faster than the speed of light occurs. The fact that it at present seems impossible to put this signaling to use technologically is irrelevant to the physical process taking place —nature, after all, isn't organized for the convenience of human needs. But the clear result of Aspect's experiment is that either special relativity or quantum mechanics or both are limited in some way. A new theory would have to explain this experiment in a self-consistent manner.

As Prigogine points out, a theory of particle dynamics must include the irreversibility of time. In a broader sense we must be able to explain elementary particles, such as protons and electrons, *and* the laws that govern them, as the products of historical processes—just as we today understand atomic nuclei, stars, planets, penguins, and people as historical products.

The method of such a new theory must extrapolate from the known phenomena of the laboratory and of the macroscopic world, not from the a priori invention of perfect laws for the perfect microscopic world. As a result, such a theory, of course, must make clear-cut and comprehensive predictions that can be confirmed or refuted in the laboratory. Only such a theory can form the basis for fundamental technological advances.

Such a theory is possible. A theory *can* avoid the acausal approach of conventional quantum mechanics. As we saw in Chapter Seven, the revulsion of intellectuals and lay people against classical determinism, a world in which everything is fixed from the beginning to the end of time, has continually encouraged a mystical reaction. In physics this reaction fueled the rise of acausality in quantum mechanics. But determinism, which implies, at least in theory, that events can be exactly predetermined, has been thoroughly confused with a quite different notion, causality—the idea that all events occur as a result of some other events, that they are connected by natural processes. When determinism was rejected, causality was rejected along with it.

But Prigogine's work has shown that this is entirely wrong. Completely causal processes, such as the motion of a comet in an orbit around the sun, can also be entirely indeterminate over time. In the past few years, Prigogine and his colleagues have shown concretely how the same can be true for quantum mechanical systems. It is entirely possible for such systems to be governed entirely by causality, in which no event occurs "spontaneously" or instantaneously. Yet because of these systems' inherent instability, it is impossible, as quantum mechanics and experiments agree, to predict exactly a specific event. But this is not due to our inadequate knowledge of the underlying dynamics. Even the most precise (but finite) knowledge of those dynamics will not improve our ability to make exact predictions.

In Prigogine's approach, the basic quantum laws are reformulated so that they are no longer intrinsically time-reversible, but reflect the distinction between past and present. Quantum transitions do occur rapidly, but not instantaneously—there are no discontinuities in space or time and no uncaused events.

Prigogine's ideas have received some support from computer simulations performed by Adrian Patrascioiu of the University of Arizona. Patrascioiu simulated the behavior of a vibrating medium, a mechanical analog of the electromagnetic vibrations in the black-body experiments of the turn of the century. He found that, although the simulations involve no quantum assumptions, but merely classical mechanics of an unstable system, the distribution of energy among vibration frequencies closely resembles the Planck spectrum. Here a supposedly quantum effect emerges from a system whose energy is *not* quantized.

Although Prigogine's preliminary work has shown that it is entirely possible for a fully causal system to obey the nondeterministic laws of quantum mechanics, no full-blown alternative theory has been articulated. He has yet to put forward any interpretation of the Aspect experiment. In a general sense, Prigogine's approach leads toward the idea that quantum processes are inherently nonlocal, influenced by the entirety of the system. But the problem remains that relativity indicates that *no* influences within *any* system can travel faster than the speed of light, yet they appear to do so in Aspect's experiment.

Nonetheless, Prigogine's efforts indicate that a search for alternatives is by no means doomed to failure. There are clues that indicate where such a search might lead. The first and most generally noticed are the form of the equations of quantum mechanics and electromagnetism. Since the nineteenth century it's been recognized that the equations of electromagnetism are almost identical with the equations of hydrodynamics—the equations governing fluid flow. Even more curious, Schrödinger's equation, the basic equation of quantum mechanics, is *also* closely related to equations of fluid flow. Since 1954 many scientists have shown that a particle moving under the influence of random impact from irregularities in a fluid will obey Schrödinger's equation.

More recently, in the late seventies, researchers found another curious correspondence while developing mathematical laws that govern the motion of line vortices—the hydrodynamic analogs of the plasma filaments I have discussed. The governing equation turns out to be a modified form of Schrödinger's equation, called the nonlinear Schrödinger equation.

Generally in science when two different phenomena obey the same or very similar mathematical laws, it means that in all probability they are somehow related. Thus it seems likely that both electromagnetism and quantum phenomena generally may be connected to some sort of hydrodynamics on a microscopic level. But this clue, vague as it is, leaves entirely open the key question of what the nuclear particles are. And what keeps them together? How can fluids generate particles?

Here we can appeal to the laboratory. We know that a magnetized fluid—plasma—does form particlelike structures, which appear naturally on scales ranging from microns to light-years—the plasmoids. Because we've seen that such structures develop with sizes differing by 10^{22} or so, it seems likely that we should be able to extrapolate downward by another factor of a billion to reach the quantum level. Obviously, it's not so simple, because laboratory plasmas are composed of protons and electrons. If plasmalike processes play a role in the structure of the electrons and protons themselves, these processes cannot be the same as ordinary plasma processes.

But the idea of particles formed from vortices in some fluid is certainly worth investigating. (This is a *real* return to Ionian ideas: the idea of reality being formed out of vortices was first raised by Anaxagoras 2,500 years ago!) Probably the first investigator to seriously raise this idea in the modern period was the Soviet physicist Lev Landau in the forties. More recently Winston Bostick and his former student Daniel Wells have developed a fluid vortex model of the electron.

Unfortunately, no such theory has produced much in the way of verifiable, quantitative predictions. However, I think there are additional clues, some developed from my own work, which indicate that plasma processes and quantum mechanical processes are in some way related.

First and foremost are Krisch's experimental results on spin-aligned protons. Qualitatively, the results clearly imply that protons are actually some form of vortex, like a plasmoid. Such vortices interact far more strongly when they are spinning in the same direction—which is certainly the behavior Krisch observed in proton collisions. Because vortex behavior would become evident only in near-collisions, the effects *should* be more pronounced at higher energies and in more head-on interactions—again, in accordance with Krisch's results.

A second clue lies in particle asymmetry (see p. 336). Particles act as if they have a "handedness," and the simplest dynamic process or object that exhibits an inherent orientation is a vortex. Moreover, right- and left-handed vortices annihilate each other, just as particles and antiparticles do.

A third clue concerns the characteristic velocities of plasma filaments. As we saw in Chapter Six, velocity is a scale-invariant quantity in plasma. My own work led to the calculation of the characteristic velocity of ions in a plasma vortex filament. This velocity forms a sort of cosmic speed limit, an upper limit to the orbital velocities of objects even as large as superclusters.

This velocity is important in the macroscopic realm, so might it not have a similar importance in the microscopic? Astronomer Paul Wesson had, in fact, previously noted that the velocities characteristic of astronomical objects are similar to those at the atomic level.

To test this idea I compared the characteristic velocity of a plasma vortex with a quantum equivalent. The closest analog is in the process called the Quantum Hall effect: electrons placed in crossed magnetic and electric fields attempt to circle around the magnetic field lines, but the crossed fields push them in a direction perpendicular to both sets of field lines. The electrons drift with a velocity directly related to Planck's constant.

I had already calculated the typical velocity of an ion in a plasma filament, so it was relatively simple to calculate how fast the average ion drifts down the filament. With both plasma and quantum, this velocity is directly proportional to the effective resistance of the filaments. I compared the two velocities and found them to be identical.

These velocities are based on what appear to be quite different processes—one quantum, the other plasma. They are also based on two different fundamental constants: the plasma velocity is derived from the ratio of proton mass to electron mass, the quantum velocity on Planck's constant. According to conventional theory, there is no inherent connection between these two quantities. Each is known to an accuracy of better than one part in ten million, yet the velocities based on these quantities are identical within this accuracy.

Of course, such identities can be mere coincidence—but agreement to one part in ten million is *quite* a coincidence. My own confidence in this relation increased when new values for both proton mass and Planck's constant were released a year after I had published the calculations. Both values had changed, yet the relationship between the two velocities remained the same. To me this relation suggests that vortices are in some way important at the quantum level.

Obviously, we're a long way from a new theory consistent with the known laws of quantum mechanics and able to explain the existence of particles as well. Such a theory would also have to predict significant features of the weak and strong nuclear forces. From this standpoint, such forces should arise from the close interactions of vortex particles.

To be sure, such a new theory would in no way be a Theory of Everything. If we could, for example, fully explain nuclear, electromagnetic, and gravitational interactions as aspects of the properties of vortices formed in some primitive fluid, we would still have to explain how the properties of the fluid, however simple, arose. But if fundamental physics is viewed as part of the general effort to describe an infinite history of the universe, there is no reason to fear an eventual end to the enterprise.

HOW GOES THE SCIENTIFIC REVOLUTION?

Let's step back from particle physics and look at the larger picture. In three closely related disciplines—cosmology, thermodynamics, and particle physics—our notions of the history of the cosmos, of time and matter, are undergoing a profound change, a change that can be characterized as a general scientific revolution.

However, the transformation is at very different stages in these three fields. It is most advanced in thermodynamics, in which the alternative to the old ideas has been elaborated and is coming under wide discussion. Articles by Prigogine and his colleagues appear in leading journals and are being popularized in a number of books. It remains a minority view in the field but is gaining adherents swiftly.

In cosmology the situation has advanced more slowly. The ideas of the Big Bang have become scientifically untenable. Large-scale clustering and the nonexistence of dark matter contradict all the crucial predictions of the Big Bang. The universe is not homogeneous, and the dark matter is not there. Alternative plasma hypotheses can better explain the helium abundance and the microwave background, *and* they predict new phenomena, such as the supercluster complexes and the absorption of radio waves, which have been confirmed by observation.

Nevertheless, the revolution in cosmology is still in its infancy. The questions of the Hubble expansion remain entirely open, and a general debate over the issue of the Big Bang has barely begun.

With particle physics the revolution hasn't begun at all, although, in my view, it won't be long in coming. As yet, this field lacks its Copernicus, let alone its Kepler. The inadequacies of the existing approach, the general feeling that the theorists are at a dead end, is certainly becoming widespread. But an alternative approach has not yet crystallized.

Cosmologists and particle physicists attend the same conferences. As the debate over the Big Bang increases, the untenability of the current approach in particle theory—the pursuit of perfect symmetry—will become more obvious, and more scientists, especially younger ones entering the field, will begin to concentrate on alternatives. It seems likely that successful new concepts will soon arise.

Certainly the most important stimulus to such new concepts will be new experiments and observations. As I described in the Appendix, probably the most important cosmological observations over the long term will be in the mapping of the universe, which will demonstrate the extent of inhomogeneity and probably even the actual form of the Hubble relation. Equally important to both cosmology and particle physics will be the creation of, and experiments with, ambiplasmas—matter–antimatter mixtures. Such work could sensitively test both the assumptions of

Alfvén's antimatter cosmology and the symmetries and asymmetries of matter and antimatter, which are certainly fundamental to the structure of both.

The extension of Krisch's work is also key. Beginning in 1994, a highly sensitive version of his spin-aligned target will be inserted in the beam of the Soviet Union's new UNK accelerator. This will extend the range of spin-effect studies from 28 GeV to 3 TeV, over one hundred times higher in energy, providing vital high-energy data on spin effects, and giving a clear test to all alternative theories.

But experimentation alone will not suffice; there must also be a freer debate of the results. As I mentioned in Chapter One, with the present peer-review system such debate is difficult at best. For each specialized journal, papers are reviewed by the very people whose pet theories are being challenged. The fact that a paper may appear in another journal is of little account, because the specialization of science fragments debate—astronomers don't read plasma physics journals.

The current system of specialized peer review originated in the late nineteenth and early twentieth centuries, as science became more closely tied to, and supported by, large-scale capitalist enterprise. While inventor-entrepreneurs like Thomas Edison chose for themselves what to research, the later financier-industrialists wanted the "quality of work" guaranteed in advance. So they, together with leading academics, encouraged the idea of peer review—the inspection of scientific work by the "best authorities" in a given field.

At the same time, the growing industrialization of scientific research led to an increasing level of specialization. The older generation of scientists had picked their research topics according to their own interests and often hopped across an entire field (as the best twentieth-century scientists continue to do). But as scientific research became organized in large-scale industrial labs, and as university work fell under the sway of industrial concerns, research came to focus on specific topics of commercial need, and scientists were encouraged to devote their entire career to single specialties.

The combination of growing specialization and the peer-review system have fractured science into isolated domains, each with a built-in tendency toward theoretical orthodoxy and a hostility to other disciplines.

Today, the baneful effects of peer review and hyperspecialization can be overcome only by what geophysicist Juan Roederer has referred to as "interdisciplinarification"—the systematic synthesis of "separate" fields. In Roederer's view this would involve several simultaneous initiatives such as the encouragement of interdisciplinary research programs, and the organization of far more extensive interdisciplinary education at the undergraduate level. Such changes would be simplest to carry out for peer review itself: both papers and funding proposals would be reviewed by referees from more than a single specialty, especially when the papers themselves were interdisciplinary. Clearly a plasma cosmology paper should not be rejected by astrophysicists at one journal and then published in another journal after review by plasma physicists, as occurs today. It would be far better for the first journal to have the paper reviewed by both plasma physicists and astrophysicists. In such a system, split reviews would lead to publication as part of an ongoing debate, not rejection in defense of an isolationist orthodoxy. And because such papers would need to be written more clearly if, say, a geophysicist were to understand an astrophysics paper, the system would also improve the currently abysmal standard of technical writing (and the sloppy thinking it both disguises and perpetuates).

Evidence that "interdisciplinarification" does, in fact, fight orthodoxy and encourage the development of new ideas is in the willingness of the Nobel Prize committees to recognize mavericks like Alfvén and Prigogine. The committees consist of representatives from the whole of a broad field, such as physics or chemistry, and so they do not respect the specific orthodoxies of a given specialty and are far better able to judge a scientist's work on its merit, no matter how controversial it may be.

A REVOLUTION IN TECHNOLOGY

The outcome of the emerging revolution in science depends not only on developments in science, but also on the evolution of society as a whole, as I shall discuss in the last chapter. But it appears likely that if such a revolution does occur it will have nearly as profound an effect on technology as did the first scientific revolution of the seventeenth century. Just as that revolution gave Europeans the technology they needed to cross the oceans and explore the world, so the new revolution may give humanity the means to rebuild our world, abolish want, and explore the universe.

The most promising field for the new work is controlled fusion. Basic research in plasma cosmology and plasma physics can lead to new approaches to harnessing fusion, which could have a profound impact on technology as a whole.

Up to now, the main fusion efforts, especially those focusing on the tokamak, have been hobbled by an approach that ignores or tries to suppress plasma instabilities. As I noted in Chapter Five, this approach tends to be based on MHD (magneto-hydrodynamic) mathematical methods, which make assumptions about plasma that rarely apply. This approach requires very large and extremely expensive magnets to stabilize the plasma, so the resulting reactors will never be cheaper than existing fission reactors.

In addition, no external magnetic device can contain plasma temperatures above 200 to 300 million degrees—incredibly hot by everyday standards, but low by fusion standards. Only one fuel, deuterium-tritium (both isotopes of hydrogen), can fuse at these temperatures, but it produces nearly all of its energy in the form of neutrons. These can generate radioactivity in the structure of a reactor—far less radioactivity than in a fission reactor, to be sure, but enough to require considerable safety precautions.

The result is that tokomak-based fusion reactors appear to be not much more economical than existing fission reactors. Their main advantage, a significant one, is that their fuel is almost inexhaustible: deuterium is present in sea water, and tritium can be generated in fusion reactors from lithium, a relatively common element. In contrast, fission reactors rely on uranium, a rare element.

The alternative approach to fusion *exploits* plasma instabilities, inducing the plasma itself to generate the magnetic fields that compress and heat it, as in quasars and galaxies. The plasma focus and several other devices attempt to use these instabilities by employing the pinching forces of high currents, which are relatively easy and cheap to produce. There is no necessary limit to the strength of the magnetic fields, while there is with those produced by conventional electromagnetism. Thus, there is no limit on the temperatures that can be obtained.

At these higher temperatures the "clean" or neutron-free fusion fuels such as hydrogen-boron may be used. Hydrogen and boron fuse to make helium at a temperature of about 2.5 billion degrees. The reaction releases no neutrons, only charged particles—the helium nuclei. Essentially no radioactivity is produced, so the reaction is safe.

Equally important, if such reactors are practicable, they could produce electricity directly, without steam generators or turbines. Conventional fusion approaches, fission reactors, and fossil-fuel generators all heat water to steam, which then flows through a turbine whose spinning magnets produce electricity. The process is both expensive and inefficient. But neutron-free fusion directly produces electricity, a flow of charged particles—the helium nuclei. Because all of these fusion devices are pulsed, a pulse of electricity can be directly induced in conductors by the pulse of ions just as the electricity in one winding of a transformer induces electricity in another winding.

With no enormous magnets to produce the energy and no steam generators and turbines to transform it into electricity, neutron-free fusion devices could be exceedingly inexpensive and compact.

But it is hard to predict the behavior of large reactors from small laboratory experiments. Here plasma cosmology and astrophysics can be of critical help. As Alfvén emphasized forty years ago, plasma scaling laws allow one to take laboratory data and scale it up to the astrophysical scale, or to take data obtained from astronomical observation, for example, of galaxies and quasars, and use it to predict plasma behavior in the lab. The theories and models now being developed to explain astrophysical phenomena, particularly such events as quasars and the nuclei of galaxies, can thus be used to guide the design of instability-based experiments and eventually of reactors on earth. By imitating natural phenomena in the laboratory, we can, quite possibly, learn how to control plasma to produce thermonuclear energy.

It would be difficult to overestimate the impact of the development of such an ideal energy source. Such reactors would have essentially inexhaustible and cheap fuels (boron is a relatively abundant element). Fusion would lead to a sharp fall in the cost of energy and thus in the cost of nearly all material goods that need significant amounts of energy for production, especially materials like steel and aluminum. (It would also eliminate the recurrent wars fought over Middle Eastern oil.) Raw-material extraction would be transformed, because lower-grade ores could be economically mined. There would, as well, be an enormous impact on the pollution problem. Canadian scientists have already developed, and the Westinghouse Corporation is marketing, a plasma torch which heats pollutants to such high temperatures that they break down into harmless constituents. Organic pollutants are eliminated and metallic contaminants, such as mercury, can be recaptured for further use. At present electricity costs, such torches can be used economically only for concentrated cleanups of toxic waste. With cheap electricity, they could become a ubiquitous means of preventing pollution in the first place, breaking up or recovering potential pollutants before they leave the plant site. At the same time, the replacement of fossil fuel–based energy with fusion-based energy would eliminate the buildup of carbon dioxide in the atmosphere and thus the threat of a greenhouse warming of the earth.

Such controlled-fusion reactors are also probably the only means by which human space travel could become practicable. The vast expense of manned space travel at present is due in part to the limitations of the chemical fuels all rockets use. When burned, these fuels are exhausted from rockets at speeds of at most a few kilometers per second. In order to achieve velocities of more than twenty-five kilometers per second, needed to escape the earth's gravity, it is necessary to use vast amounts of fuel to lift small payloads—typically a hundred tons of fuel per payload ton. This leads to huge and expensive rockets.

With present technology, there is such a premium on reducing the weight carried that it is impossible for manned expeditions to compete economically with robotic spacecraft. At present, unmanned exploration is ten to one hundred times cheaper than manned exploration, and such projects as the colonization of Mars are completely impractical.

However, neutron-free fusion rockets, if they could be developed, would achieve exhaust velocities of up to ten thousand kilometers per second. Even carrying only one pound of fuel for one pound of payload, such rockets could have speeds of nearly one thousand kilometers per second; a flight to Mars would require less than a week. The cost of space travel could be cut by a hundredfold, making human beings, with their inherent flexibility, better and more economical explorers than robots. Over time, such energy sources may make colonies on other planets a practical prospect.

The development of fusion power is the most direct technological application of the new direction in cosmology. If, however, a broad, new approach toward the fundamental structure of matter and energy were to develop, as appears to be happening, far more startling technical innovations could result. The study of matter–antimatter interactions, for example, could eventually lead to the development of efficient means of producing antimatter, and of storing it as an incredibly concentrated source of energy. Antimatter is the only fuel known today that could make interstellar travel practical. An antimatter drive, if it could be developed, would enable spaceships to achieve velocities near the speed of light and thus make possible at least unmanned, and possibly manned, exploration of nearby stars and planetary systems.

Setting aside such speculation, the technologies that naturally arise from the emerging scientific revolution are exactly those that humankind most urgently requires. The elimination of global misery and poverty is simply not possible with available energy sources. To extend the current American living standard (certainly not yet ideal by any means) to the world's population would involve a fivefold increase in global energy consumption. Such an expansion would be both economically and environmentally impossible with present energy sources. Only an advanced form of fusion would provide the wherewithal to decently feed, clothe, and house all of humanity.

Over the longer run, humanity must colonize space, for we will eventually run out of room for a growing population on earth. As explained in Chapter Seven, it is impossible for an evolving system to stop growing—it must grow or die. To achieve the higher level of complexity needed to advance or even to survive, society needs more individuals, just as the biosphere must continually generate more species. A more technological society needs a greater division of labor, and thus, inevitably, more people. And, if society is appropriately organized, more people means more new inventions, new discoveries, and a greater rate of advance. Throughout history, in fact, the greatest rate of social advance has always been linked to periods of rapid population growth.

But the population of the earth cannot grow without limit. Fusion power would essentially eliminate the threat of any energy shortage, and agricultural productivity can certainly be improved far above current levels—but land is limited. Today, perhaps 2 percent of the total landmass is covered by cities and their suburbs. A century from now it may be 20 percent, and in a century and a half, 50 percent. Obviously, before that it will become impossible to maintain an adequate quality of life, however efficiently we manage our resources and waste. A healthy biosphere could not exist without large regions remaining in a natural state, nor would people be satisfied with a total, global city.

Thus, sometime in the next century, humanity will have to develop the means to venture en masse into the universe. And the technology needed for this can come only from enormous advances in basic physical science, from a scientific revolution.

This is not as fantastic as it sounds. From the 1830s to 1960 the technology of human travel increased over a thousandfold—from the horse to the space rocket. If rapid technological progress resumes in the future, it is entirely possible that another century or so will bring another thousand- or ten thousandfold increase in technology, from that of the present-day rocket to means of traveling close to the speed of light.

It is likely that such an imperative to rapid expansion would apply to any civilization, once it has achieved a given level of population and technology. Since evolution, as we've seen, tends toward increasing levels of complexity and faster rates of evolution, one would expect such civilizations to be a common occurrence through the universe. But if this were so, a galaxy-wide society would long since have come into existence. Such a civilization would surely have left obvious evidence of its existence to be glimpsed by means of our telescopes already, even if its members did not care to visit us openly.

The only reasonable explanation would appear to be that we are among the very first to have reached this critical stage of evolution, that at this moment the few pioneers of intelligent life are, at scattered points throughout the universe, including here on earth, either preparing or launching the first stage of cosmic expansion. The cosmos is on the point of discovering itself, of entering a new stage of evolution, and what we do here on this small planet may well echo through unlimited space and time.

■

9 INFINITE IN TIME AND SPACE

All the labor of the ages, all the devotion, all the inspiration, all the noonday brightness of human genius are destined to extinction in the death of the solar system. . . .

—BERTRAND RUSSELL

Unique in this respect among all the energies of the universe, consciousness is a dimension to which it is inconceivable and even contradictory to ascribe a ceiling or to suppose that it can double back on itself. . . . If progress is a myth . . . our efforts will flag. With that the whole of evolution will come to a halt—because we are evolution.

—PIERRE TEILHARD DE CHARDIN, 1938

Just as the Ptolemaic cosmology was yoked to the theology of medieval Catholicism, so the Big Bang is today entangled with religious and theological ideas. It is used to support those concepts, and religion in turn is marshaled in defense of modern cosmology. Once again, as four hundred years ago, some theologians attempt to define which scientific concepts are permissible and which are not.

The new scientific revolution, like the Copernican revolution, is *not* an attack on religion as a whole, but on the entanglement of science and religion—the idea that religious authority can dictate or reject scientific doctrines, or that the evidence of science can be used to bolster religious authority. As the new cosmology becomes known, it is triggering within theology a renewal of the ancient debates over a finite and infinite universe, over the relation between science and faith.

A few decades ago the idea of a potential conflict between science and religion might have seemed an anachronism—a throwback to the dark days of the Scopes "Monkey Trial" of the twenties or the even darker time of Galileo. This is scarcely the case today. As in previous periods, when progress falters, when living standards begin to fall, rationality becomes increasingly suspect as a guide to action, and large sections of the population turn to the irrational certainties of religious fanaticism. In such periods, religious views intrude into all realms of social activity, and religious conflicts begin to pervade politics, art, and science.

Such has been the trend over the past ten or fifteen years. Today blood is spilled in religious strife throughout the world: Catholic against Protestant in Northern Ireland, Christian against Muslim in Lebanon, Jew against Muslim in Israel, Sikh against Hindu in India.

Such events are not limited to the Old World. In the U.S. the two leading bookstore chains stopped selling Salman Rushdie's novel *The Satanic Verses* after it was condemned by the Ayatollah as blasphemous, and its author condemned to death in absentia. The chains restored the book only after mass protests from writers' organizations. During the furor over the book, the *New York Times* printed a letter from the president of the Pakistan League of America, who argued that he had a *constitutional* right to murder Rushdie: "The United States Constitution grants freedom to all religions," he writes. "If my religion calls for the death penalty for blasphemy, wouldn't I be renouncing my religion to deny it?"

Nor is the U.S. free from homegrown religious strife, in some cases explicitly impinging on scientific issues. In the past decade a half-dozen states have passed laws restricting the teaching of evolution in the public schools, and compelling the introduction of "creation science"—the doctrine that the earth was created in seven days a few thousand years ago—into the curriculum.

In a 1987 decision the Supreme Court ruled one of these laws unconstitutional for violating the separation of Church and State, but creationist lobbies have had a major impact on textbooks throughout the nation, causing the entire subject of evolution to be either watered down or eliminated from them. Moreover, the Supreme Court decision was not unanimous, and the dissent, written by Associate Justice Anthony Scalia, was joined by the highest judicial official in the country—Chief Justice William Rehnquist.

In the dissent, the two justices argue that the Louisiana act, which required equal time for creationism and evolution, is constitutional. Creation science is not a religious doctrine, the justices write, but merely a "collection of scientific data supporting the theory that life abruptly appeared on earth." Such a miraculous occurrence does not require a creator, but in any case "to posit a past creator is not to posit the eternal and personal God who is the object of religious veneration." Finally, the justices, generalizing their dissent, conclude by expressing their belief that the constitution, correctly interpreted, does not prohibit laws that have a purely religious purpose, that act only to enhance or encourage a particular religious belief.

▪ COSMOLOGY AND THEOLOGY

So we should not be surprised that today cosmology remains entangled with religion. From theologians to physicists to novelists, it is widely believed that the Big Bang theory supports Christian concepts of a creator. In February of 1989, for example, the front-page article of the *New York Times Book Review* argued that scientists and novelists were returning to God, in large part through the influence of the Big Bang. A character in John Updike's 1987 novel *Roger's Version* is cited as typical of the trend. The character, a computer hacker, says, "The physicists are getting things down to the ultimate details and the last thing they ever expected to be happening is happening. God is showing through, facts are facts . . . God the Creator, maker of heaven and earth. He made it, we now can see, with such incredible precision that a Swiss watch is just a bunch of little rocks by comparison."

Astrophysicist Robert Jastrow echoes the same theme in his

widely noted *God and the Astronomers*: the Big Bang of the astronomers is simply the scientific version of Genesis, a universe created in an instant, therefore the work of a creator. These ideas are repeated in a dozen or more popular books on cosmology and fundamental physics.

Such thinking is not limited to physicists and novelists, who could perhaps be dismissed as amateur theologians. Ever since 1951, when Pope Pius XII asserted that the still-new Big Bang supports the doctrine of creation *ex nihilo*, Catholic theologians have used it in this way. The pope wrote in an address to the Pontifical Academy of Sciences, "In fact, it seems that present-day science, with one sweeping step back across millions of centuries, has succeeded in bearing witness to that primordial '*Fiat lux*' [Let there be light] uttered at the moment when, along with matter, there burst forth from nothing a sea of light and radiation, while the particles of the chemical elements split and formed into millions of galaxies. . . . Hence, creation took place in time, therefore, there is a Creator, therefore, God exists!"

To be sure, these views are by no means unanimous within the Catholic Church. The present pope, John Paul II, is far more cautious in mixing science and religion. In his own address on the subject, he repeatedly apologized on behalf of the Church for the persecution of Galileo and reaffirmed the autonomy of religion and science. Addressing the Pontifical Academy of Sciences in 1981 he paraphrased Galileo, saying that the Bible "does not wish to teach how heaven was made but how one goes to heaven." It is therefore not up to religion, he argues, to judge one or another cosmological theory. Yet in the same address, John Paul II favorably quotes from Pius XII's earlier speech and contends that the question of the beginning of the universe is not one that can be solved by science alone—to do so requires "above all the knowledge that comes from God's revelation." (His ideas on this subject are clearly evolving: in a more recent address to the same group in 1988 he warned against "making uncritical and overhasty use for apologetic purposes of such recent theories as that of the Big Bang in cosmology.")

For many the link between cosmology and theology still exists: scientific theories can support theology. The converse is equally true—many argue that the doctrines of religion and philosophy preclude the idea of a universe infinite in time and space.

INFINITY AND DEITY

As in previous epochs, the question is not one of a battle *between* science and religion, but of parallel conflicts *within* science and religion. There are those today who violently oppose the idea of an infinite universe as blasphemous and contrary to all religious thought. There are others, equally devout, who hold that religion and science are autonomous, and that the question of whether the universe has a beginning is of no religious importance. And there are even those who base their entire theology on the notion of a universe that evolves in an infinite expanse of time.

The idea of an origin of the universe is an alien one to many religions. Hinduism, for example, assumes a cosmos characterized by infinite cycles of development and decay. To such religions the idea of an infinitely existing universe is not a problem. It is in Judaism, with its doctrine of a creator, and more so in Christianity, which extends this with its doctrine of creation *ex nihilo,* that the interaction of religion and cosmology become the sharpest, and it is here we'll concentrate our attention.

To many in the Judeo-Christian tradition, the idea of a universe infinite in time and space is not allowed for the same reasons Augustine argued two millennia ago: infinity is exclusive to the deity, and thus prohibited for the material universe. To say that the universe is unlimited is to obscure a crucial difference between God and nature, and thus to advocate pantheism—the idea that nature itself is inherently divine and, perhaps, needs no God. Thus a belief in an infinite cosmology implies heresy. Such reasoning is intimately linked to the arguments used against Nicholas of Cusa, Copernicus, and Giordano Bruno hundreds of years ago. For many theologians they have lost none of their force today.

One of the leading proponents of such ideas today is Stanley Jaki, a Benedictine priest and widely known philosopher of science. To Jaki, an infinite universe is impossible, ruled out for philosophical reasons by Aristotle. "As soon as you have infinity in mathematics everything breaks down," he contends. But more important, an infinite universe is "a scientific cover-up for atheism." "Physical infinity," he wrote in his 1988 book *The Savior of Science,* "could readily be taken for infinite perfection and from there it was but a step to taking the infinite universe for the ultimate perfect being." Jaki insists that *only* God can be unlimited or infinite, thus any scientific theory asserting the universe to be infinite in *either* space or time is a priori ruled out on religious grounds.

For Jaki, the passions inflamed by Giordano Bruno's advocacy of an infinite universe are still alive. He has written extensively on Bruno, denouncing him as "a muddled dreamer," "an amorous rogue," "a dabbler in magic," and describing his writing as "a heap of dung." For many Bruno was a martyr to free inquiry, but Jaki believes that the burning of Bruno at the stake was "a macabre vindication of basic Christian beliefs."

While few share Jaki's vehemence, his views are scarcely unique. The noted Hebrew scholar Nahum Sarna agrees that a universe infinite in all dimensions "would be like a second divinity"—"a universe must be either finite in space or in time if it is not to rival the deity." Strangely, this restriction is the same as a rule of relativistic cosmology, which allows the universe to be infinite in space or time, but never in both.

Clearly, both Jaki and Sarna believe that there is a conflict between a scientific conception that requires neither beginning nor end to the universe and the dictates of theology, as they see them. The implications of this denial of true infinity of the material universe means far more than just the rejection of plasma cosmology. As we saw in Chapter Seven, Georg Cantor refuted Aristotle's arguments by showing that infinite or "transfinite" numbers can be treated by mathematics, and involve no contradictions. Moreover, he proved that these numbers are just as "real" as other numbers—they describe real aspects of the physical universe and are basic to understanding it.

All of modern analysis, including most of the mathematics underlying modern technology, relies on the concept of continuity—that between any two points in space, there is an infinite number of other points. In the same way, between any two points in time there is an *infinite* number of other moments. Without these assumptions it's virtually impossible to use modern mathematics in a logically consistent way.

Indeed, the idea that space and time are infinitely divisible is vital to explaining the very existence of the irreversible time of the real world. Without such true infinities the world would be a vast digital computer, each instant predetermined by its initial state, without a past, present, or future. So to argue that there is no true infinity, that Aristotle was right, is tantamount to a rejection of modern mathematics, the technology based on the use of that mathematics, and the new discoveries in thermodynamics.

On the other hand, once we accept the continuity of space and time it's hard to object *on philosophical grounds* to an infinite extent of space or time. Cantor proved mathematically that it is possible to compare the size of various transfinite numbers and see which are greater or lesser than, or equal to, others. This involves making a one-to-one correlation between the members of two transfinite sets. On this basis, Cantor demonstrates, for example, that the number of points on a line is larger than the number of numbers that can be counted. But the number of points on *any* line is of the same transfinite magnitude, no matter how long the line is—even if the line is, itself, infinite in extent. So the infinite number of points in a cube one inch on a side is no less than the infinite number of points in an infinite universe. The number of instants between two seconds is as infinite as the number of instants in a span of time with neither beginning nor end.

Thus while the idea of an infinite chain of cause and effect may appear mind-boggling, such an infinite chain exists even in the present, with each passing second—each an infinity of moments. To accept an infinite past is no more or less difficult philosophically than to accept the continuity of time—the infinity of moments in a single second.

While these conclusions are not commonsensical, they are logically consistent. All this, of course, says nothing about whether the universe *is* finite or infinite, which is a scientific question that must be answered by observation.* It does imply that, contrary to Jaki and Aristotle, there is no necessary *contradiction* in a universe of infinite extent. Perhaps more significant, it also implies that, if an infinite universe is rejected philosophically, the infinity of space in a single inch or the infinity of instants in a single second must be rejected as well, along with all the science based on the hypothesis of continuity.

One can still argue on theological grounds that the universe must have some bounds or limits so as not to be equal to an

* Strictly speaking, it is not possible to *prove* scientifically that the universe is infinite. But it is quite possible to claim we have no observational evidence that it is finite.

infinite God. But on this point there is no unanimity among theologians. We've seen that five hundred years ago Catholic theologians like Nicholas of Cusa strongly supported the necessity of an infinite universe. To these thinkers the finite, contained universe of orthodoxy, bounded in space and time, itself implies a limitation of the creator. Surely an infinite God would not create such a petty and limited universe—only an infinite universe befits the supreme divinity and the infinite potential of the human mind.

This same centuries-old debate continues today. "Infinite time wouldn't necessarily bother me," comments James Skehan, a Jesuit priest and director of the Weston Geological Observatory, Boston College. To Skehan, a geologist who has written extensively on the relation of science to Christian creation doctrine, an infinite universe doesn't necessarily give rise to the problems Jaki and others see: "As long as whatever infinite qualities the Deity has are superior to the infinite qualities of the universe, I see no contradiction with Christianity nor support for pantheism."

Thus the revival of the universe infinite in time and space revives ancient theological battles—with much the same arguments still in use. On the one hand are those who, like the medieval orthodox, believe that an infinite universe will challenge the authority of the deity. On the other, their opponents believe only a universe without limits befits an infinite God and human beings in search of infinite knowledge.

▪ THE MOMENT OF CREATION

Some theologians rule out the new cosmology on another ground—it eliminates the moment of creation. Sarna asserts, "A universe without a beginning would pose a problem for the notion of divine creation. From a biblical viewpoint, there has to be some moment in time when the divine will descends upon the process of creation. That is basic to Jewish and Christian theology. If something always existed, this would therefore not be acceptable on the basis of the Bible."

Jaki is less equivocal—he insists not only on a moment of creation, but creation *ex nihilo*: "Creator, God Incarnate, creation out of nothing, immortal soul and human dignity are notions that form a closely knit complex," he writes in *Cosmos and Creator*. Creation out of nothing is the only creation worth considering, and thus without it there can be no God.

Again, this view is scarcely a unanimous one within Judeo-Christian thought. The doctrine of creation *ex nihilo*, as we saw in Chapter Two, is not to be found in the text of Genesis. Sarna himself agrees on this: "It's not possible to deduce from the Hebrew Genesis the idea that the universe was created from nothingness." Other scholars, such as the Protestant theologian Conrad Hyers, agree. "The emphasis in the entire account," he writes, "is on creation as the creation of *order*, which is what cosmological literature is about. . . ." Genesis describes how the orderly cosmos is created, not out of nothingness but out of the "formless void."

"The basic message of Genesis is religious," Skehan emphasizes. "The point is to convey certain *values*: in particular that there is one God, not many, and that he is all-powerful. The creation story is a vehicle for teaching those basic lessons."

The doctrine of creation *ex nihilo* did not become Christian doctrine until the Middle Ages. The first several Church councils, which defined Christian doctrine, such as the Nicene Council, make no mention of it. Today, it is not universally accepted. Hyers, for example, sees the *ex nihilo* doctrine as overly dualistic, emphasizing the unbridgeable gap between creator and creature.

For Hyers and many other theologians the eternal existence of matter and energy in no way contradicts the idea of God the creator. In this case, Hyers writes, "matter and energy would still be seen as proceeding from God, governed by God and fashioned by God. . . . God would be seen as eternally creating, for Divine creativity is not restricted to a finite stretch of time, or to the past, but is a continuing activity." In this view God's creation describes the entire, infinite process by which order emerges from chaos—a process that God guides through the workings of natural laws.

Even some theologians who adhere to the creation *ex nihilo* doctrine, such as Patrick Byrne of Boston College, do not believe it necessarily implies that the universe is finite in duration. "The doctrine is really about God, not about the universe," Byrne contends. "It says he is not dependent on anything else or limited by anything else. No view of the universe would contradict this idea of the unconstricted nature of God."

But to Sarna, an eternally creative God modeling an eternally existing universe still poses several problems: "God would simply not be necessary to create the universe if it always existed. If we can explain each state of the universe as a natural evolution from earlier states, what need is there for God?" An eternally evolving universe is thus autonomous, explicable in all respects as a natural process. While God could be seen as in some sense guiding this natural process, he would not be essential for it to take place.

This argument is tied up with the millennia-old effort to develop logical proofs for the existence of God. Such proofs assert that it is impossible to understand the universe without believing in the existence of God. Science alone cannot account for the universe, in this view, and logic demands a God—for example, to create a world with a beginning in time.

Obviously, if a proof of God's existence depends on the assumption that without revelation science is unable to explain the natural processes of the universe, including its evolution, there is a conflict with an eternally evolving universe. Such a universe does not preclude God's existence, but does negate any logical proof of that existence. The existence of God is, then, purely a matter of faith.

But to many believers, the existence of, and necessity for, God derive not from philosophical logic or from the inadequacy of science to explain the development of the universe, but from the need to provide meaning and purpose to life and to give humankind clear moral guidelines. To Byrne, for example, faith in God is necessary for a belief in the moral teachings of Christianity—"Without a faith in God it makes no sense, for example, to 'love thy neighbor as thyself,'" he says. As Pope John Paul II says, religion teaches one "how to go to heaven," what behavior is demanded of man—something clearly beyond the purview of scientific knowledge (although, to be sure, science can clarify the consequences of various behaviors).

Here again, as with the issue of the infinite universe, we find the idea of an eternal universe, one without a beginning, is viewed by some theologians as unacceptable and incompatible with their religious views, and by others an acceptable scientific concept, a manifestation of an eternally creative deity.

A WORLD WITHOUT END

The universe of plasma cosmology and of the new thermodynamics lacks not only a beginning but an end as well. Here is another point of conflict with some Christian theologians. To Jaki, for example, the transience or temporality of the material universe, the idea that it has both a beginning and an end, is clear evidence of its subordination to God. Moreover, since all material accomplishments must, in the end, crumble to dust as the universe runs down to cataclysm or dissipation, the meaning of life must be obtained from contemplation of the supernatural world.

Pope Pius XII made the same point, explicitly relating the Christian doctrine of an end to all things, a last judgment, to the idea of increasing entropy. "Through the law of entropy," he writes, "it was recognized that the spontaneous processes of nature are always accompanied by a diminution of free and utilizable energy. In a closed material system, this conclusion must lead eventually to the cessation of processes on a macroscopic scale. This unavoidable fate, which . . . stands out clearly from positive scientific experience, postulates eloquently the existence of a Necessary Being." He concludes that a flagging universe necessarily must come to an end, but more significantly, requires something outside itself to imbue it with order at the beginning —a direct link between the idea of ever-increasing disorder and Christian theology.

Here again Big Bang cosmology serves to support a specific theology. Remember, it was the philosophical implications of the second law of thermodynamics that led Lemaître—later the director of the Pontifical Academy of Sciences—to formulate the first version of the Big Bang (see Chapter Four). This is something of a circular argument.

What is particularly curious about the philosophical idea of a mortal universe is that it provides a common meeting ground for one type of Christian faith and a pessimistic existentialism that rejects all religion. The idea of an inevitable universal end is profoundly pessismistic. Conventional cosmology postulates that there will come a moment in the remote future when the last living being in the universe dies—either scorched by the unrelenting heat of a Big Crunch, or frozen as the last star flickers out of existence in an ever-expanding cosmos. *All* human effort—not just the history we know, but the expansive future our descendants will witness—all the magnificent achievements of our species (and probably others too!) will be reduced to nothing.

For many this all proves that the meaning of the universe resides in a progress toward God to be achieved in the last judgment. But to many existentialists (and physicists) this vision is one of complete meaninglessness. Bertrand Russell, for example, writes: "All the labor of the ages, all the devotion, all the inspiration, all the noonday brightness of human genius are destined to extinction in the death of the solar system—all these things, if not quite beyond dispute, are yet so nearly certain that no philosophy which rejects them can hope to stand." Cosmologists such as Edward Harrison describe a similar end: "The stars begin to fade like guttering candles and are snuffed out one by one. Out in the depths of space the great celestial cities, the galaxies cluttered with the memorabilia of ages, are gradually dying. Tens of billions of years pass in the growing darkness . . . of a universe condemned to become a galactic graveyard." Paul Davies, another cosmologist, writes: "No natural agency, intelligent or otherwise, can delay forever the end of the universe. Only a supernatural God could try to wind it up again."

By positing an end to all things, conventional cosmology necessarily implies one of two philosophical stances: either a blind existential pessimism, humanity condemned to a meaningless existence, or a dualistic faith like that of the Middle Ages, which finds meaning only in the world beyond.

BLOOM COUNTY BY BERKE BREATHED

From a scientific standpoint we have seen that these pessimistic conclusions are false. Cosmologically, a universe with as little matter as ours will never collapse. Nor does thermodynamics even demand that the universe run down: Prigogine has demonstrated that there is no inherent limit to the order the universe will attain, or to its increasing energy flows. Our universe is speeding away from the "heat death" of total equilibrium.

Of course, it is true that in *five billion* years or so, our sun will exhaust its supply of hydrogen and expand into a red giant. Yet can it really be believed that by five billion years from now humanity won't have developed the technology to move on to another star? Nor can the universe as a whole run out of energy. Energy, after all, is indestructible—when we use energy, it only undergoes various transformations. The level of order of any system depends on the rate at which energy flows through it, which has little to do with the rate at which entropy increases—the two quantities can move in different directions. It is clear that on earth over the last six hundred million years, the *dissipation* of energy, the rate of increase in entropy, has *decreased*—we know this because the earth is unquestionably colder today than it was in the past. Yet the biosphere's rate of energy flow, the use and reuse of energy by living organisms and its flow through the atmospheric system, has greatly increased. *The earth is making more use out of less energy.*

In human society, the recycling of energy so characteristic of living things is not yet important. However, the first movements to improve this recycling are taking place. For example, most energy in our society is used for transportation, and most is dissipated in a single use. Obviously, gasoline accelerates a car; but a car's motion generates heat through both air friction and the friction of the car's brakes. Yet more advanced means of transportation, such as magnetic levitation trains, which would reuse energy many times over, are now under development. One design, to be operated in a vacuum (an evacuated tunnel, for example), would use electrical energy to accelerate to speeds of thousands of miles per hour and then use the train's momentum to generate nearly the same amount of electrical energy. Its energy consumption would be tiny compared with a car's, yet its speed would be far higher.

The ability of human society to make increasingly better use of energy flows by increasing the level of technology would preclude both an end to life and even an end to the growth of life. Cosmic pessimism is unsupported by science.

There is a conflict of science with *particular* theological ideas, not with religion per se. We again find that the conflict is within religion, that the idea of a necessary end to the universe is not a unanimous view even within Christianity, let alone other religions. In fact, the notion of an eternally evolving universe is the bedrock of the philosophy of Pierre Teilhard de Chardin, one of the most influential of modern Catholic thinkers. Born in 1881 in Paris and ordained a Jesuit priest in 1912, during the twenties and thirties he became a leading paleontologist, specializing in the study of human evolution. From his work, he developed a Christian philosophy based on the idea of evolution as the fundamental process in the universe.

In his most comprehensive work, *The Phenomenon of Man*, written in 1938, Teilhard de Chardin rejects on scientific grounds the idea of a universe doomed to decay. He shows that universal history tends toward higher levels of complexity and accelerating evolution. He sees this process as having occurred through three stages or modes of evolution: prebiological, biological, and now social evolution. In this view, man is not an isolated unit lost in the cosmic solitude "but the axis and leading shoot of evolution." Evolution, for Teilhard de Chardin, is antientropic, moving away from equilibrium—a direct anticipation of Prigogine's approach thirty years later.

Teilhard de Chardin rejects the dualism that assumes two distinct existences, that of the spirit and that of matter. Human consciousness, he argues, cannot be excluded from the realm of scientific study. "An interpretation of the universe," he writes, "remains unsatisfying unless it covers the interior as well as the exterior of things; mind as well as matter." Humanity, including consciousness, must be studied as a natural phenomenon. But if consciousness is considered part of the natural world, then, like the world's other aspects, it must have evolved, gradually and continuously—it didn't spring magically into being in a single instant. This, in turn, implies that incipient consciousness is present to some degree in all matter.

Teilhard de Chardin goes beyond this to identify consciousness with the tendency toward evolution—the development of new and more complex relations among processes in the universe. Specifically, he hypothesizes that there are two forms of energy. One, the "tangential," links an entity—whether animal or human—with existing processes in the universe; the other, "radial," is identified with the creation of new relations and higher orders of complexity. The first is ordinary energy; the second is connected with the rate of growth of new energy flows. This second form of energy Teilhard de Chardin identifies with consciousness—the more rapid the rate of change, of evolving toward greater complexity, the greater the radial energy, the greater the degree of consciousness. Thus in simple entities, such as atoms or simple organisms, the rate of change is so slow that any consciousness can be ignored as infinitesimal. But in higher animals, especially humans, in which new behaviors and relationships evolve moment by moment as new things are learned, consciousness becomes the dominant phenomenon.

Thus, for Teilhard de Chardin, physical evolution and spiritual evolution, or the evolution of the mind, are one and the same process. The role of man, the basis of moral imperatives, is to further that progress. Like Nicholas of Cusa five centuries earlier, he sees no limit to humanity's quest for knowledge and progress. "Is not the end and aim of thought that still unimaginable farthest limit of a convergent sequence, propagating itself without end and ever higher? . . . Unique in this respect among all the energies of the universe, consciousness is a dimension to which it is inconceivable and even contradictory to ascribe a ceiling or to suppose that it can double back on itself." In viewing the progress of evolution and the human mind as unending, Teilhard de Chardin specifically rejects not only the "law of increasing disorder" but the first law of thermodynamics—the law of conservation of energy—as well: he cogently argues that new relations and processes bring into existence new forms of energy as the universe evolves.

Scientifically, Teilhard de Chardin anticipated many of Prigogine's more concrete ideas. But equally important, he argues that only this prospect of an unlimited future can be the basis for human morality, even for human activity—the only prospect that can prevent humanity from despairing. If mankind came to believe that progress would halt, then "mankind would soon stop inventing and constructing for a work it knew to be doomed in advance. And stricken at the very source of impetus that sustains it, it would disintegrate from nausea or revolt and crumble into dust. . . . If progress is a myth . . . our efforts will flag. With that the whole of evolution will come to a halt—because we are evolution."

During his life some of his superiors considered his work heretical or at least unsound, and prohibited him from publishing it. However, it was published posthumously and his ideas have become part of a broad and influential stream of Catholic thought. To Skehan, for example, "It is Teilhard de Chardin's concept of a unified spiritual and physical evolution that makes the idea of a universe without a beginning appealing and consistent." He did not write theology as such, but his work shows that the idea of an eternally evolving universe, of which humankind is part, remains a powerful concept for both religion and science.

THE DESIGN OF THE UNIVERSE

If the universe is evolving over an infinite span of time, and if this evolution toward greater complexity and higher energy flows is explicable in terms of natural processes, then the creation of order out of chaos is comprehensible. Moreover, as both Prigogine and Teilhard de Chardin emphasize, humankind is not alienated from nature: our existence is not a meaningless accident in an indifferent universe doomed to extinction, but the cutting edge of a process of universal evolution.

Here again the new ideas in cosmology and thermodynamics conflict with theological aspects of the old cosmology. For in conventional cosmology, evolution is a freak affair in a decaying universe. The development of order—spiral galaxies, myriad stars, life, humanity, the lawfulness of the universe itself—can be explained in only one of two ways: in existentialist terms, as a series of colossal accidents, or in Platonic-Christian terms, as an intentional, elegant *design*. Both approaches emphasize seemingly inexplicable coincidences or chance events that permit our existence, either as mere accidents, the vagaries of an absurd universe, or as so improbable that they *cannot* be accidents, rather must be the signature marks of God. Both approaches limit reason and both are contradicted by a science that can explain these apparent "coincidences" or "accidents" as misinterpretations of a general evolutionary tendency in the cosmos, of which order and progress are the concrete results.

On the most general level of this argument, Jaki contends that because God designed the universe, there *must* be a mathematical blueprint—a Theory of Everything—which can be defined by a single set of equations describing all physical phenomena. The alternative—that such laws describe only processes that are themselves evolving—is theologically unacceptable. But Jaki rules out the idea that the Theory of Everything can be derived by pure reason, because it will inevitably contain arbitrary constants or mathematical forms that cannot possibly be further explained. These fundamental anomalies will stand as clear-cut evidence of the God that designed them into the structure of the universe.

Since such a theory has not been discovered, such arguments about its formal qualities are purely speculative. As we saw in Chapter Eight, there is no good reason to believe that such a final law, an end to the search for knowledge, will ever exist. At every point in the past when that final goal was thought to be in sight, new and wholly unexplained phenomena were soon discovered. But Jaki and many others already interpret what they believe to be inherently arbitrary facts—either observed phenomena or requisites of various theories—as evidence of God's design. It's worth examining some of these.

Several concern alleged coincidences in the fundamental constants of the universe, such as the universal gravitational constant. According to the argument, only a slight change would make life, even the existence of the universe as we know it, impossible. Therefore, these constants must have been chosen by God in order to generate the universe.

Jaki points to the quantity of matter in the universe at the time of the Big Bang, as expressed by the density parameter, omega. As we've noted earlier in this book, according to the Big Bang theory, if this ratio had differed from unity by one part in 10^{50}, the universe would have collapsed or exploded into nothingness in an instant. With good reason Jaki rejects the idea of missing mass, so, he asserts, observationally omega is not unity. In the beginning it must have been .999999999. . . . This peculiar value is evidence of divine foresight.

Another example is cited by Paul Davies and Brandon Carter. Davies points out that the properties of stars such as our sun can be roughly calculated knowing their structure and certain basic constants of nature, such as the gravitational constant (G), electron charge, and Planck's constant. According to Davies, Carter calculated that if gravity's force were to differ from what it is by only one part in 10^{40}, stars like our sun could not exist—only giant stars, which presumably would not support planets with life. Again such a fine tuning of fundamental constants is attributable to divine foresight.

A final such case of fine-tuned constants comes from Fred Hoyle, who points out that if a certain energy level in the oxygen nucleus—in essence, how fast the nucleus vibrates—were .5 percent higher than it is, all the carbon produced in a star's core would burn immediately to oxygen, and none would be left over to make up living organisms.

However, in each of these cases the "astonishing coincidence" is, in fact, evidence only of the faulty theories involved, a basic misunderstanding of statistics, or a simple mix-up of the facts. The first example, that of the incredibly accurate omega, we've seen before. It is one of many indications that Big Bang theory is inadequate—eliminate the Big Bang, and the incredible fine-tuning of omega is no longer needed. In the case of the "delicate balance" needed for the sun, the idea that G is fine-tuned to one part in 10^{40} has no basis in Dr. Carter's actual calculations. In fact, the electrical force would have to be twice as strong or the force of gravity six thousand times stronger than it is before stars like the sun would be impossible.

For the third example, the case of the missing carbon, it's necessary to briefly discuss some basic ideas of statistics. To calculate the probability of an event's occurring, one must ensure that one has defined the event correctly. For example, many of us have unexpectedly run into someone we know while far from home, perhaps in another country. The odds of meeting that particular person at that particular time and place are so tiny as to be virtually impossible. *But* if one tries to estimate the chance of bumping into an acquaintance far from home at *any* point in our lives, the chances become so great that such an event is no more surprising than it is common.

Similarly, if one looks through all the various energy states that are relevant to some nuclear reaction in a star's core, it is not at all unlikely to find one or two that are close to critical values—values that would make some reaction very improbable. Hoyle's case in itself is not improbable at all, since the presence of carbon on earth does not depend on a specific energy level, but only on its being below a critical value—something that a priori has a fifty-fifty chance of occurring.

There are also quite a few examples in which the extraordinary accident is based on a wrong theory. Astronomer Owen Gingerich, for example, cites the extraordinary luck of having an asteroid or comet hit our planet sixty-five million years ago, knocking off the dinosaurs and clearing the way for mammals and eventually humans. Yet paleontologists are strongly divided on whether the asteroid theory is right. For one thing, it has failed to explain why, at the same time the dinosaurs died off, there was a gradual retreat of shallow seas from all the continents—an event that can scarcely be caused by an asteroid collision.

A third type of these arguments is the most interesting, for they are directly based on the denial of the evolutionary tendencies of the universe. Gingerich cites the relative constancy of the earth's temperature even as the sun's radiation has gradually increased with time. The corresponding changes in the atmosphere, the decline in heat-trapping carbon dioxide, Gingerich argues, is both inexplicable by natural mechanisms and too coincidental to be chance. More generally, authors such as Fred Hoyle have cited the evolution of life as extremely improbable. Hoyle contends that *random* chemical reactions of the primitive oceans could never have arrived by mere chance at the complex molecules we see in the most primitive forms of life. Finally, even such experts on evolution as Stephen Jay Gould view the evolution of human beings as exceedingly improbable—"the accidental result . . . of an enormous concatenation of improbabilities." Other writers use this "extreme improbability" to argue that only divine guidance, not science, can explain humanity's development.

What all ignore, and what is emphasized in the new view of cosmology and thermodynamics, is the natural tendency of all matter, both animate and inanimate, to evolve continuously toward higher rates of energy flow, toward the capture of greater currents of energy. At the simplest level, laboratory experiments have shown that simple molecules important to life, such as amino acids, necessarily form when chemical mixtures similar to those of the primitive oceans are exposed to bursts of electromagnetic energy. Why do they form? Because they most efficiently trap the energy briefly available to the system. Experiments have not yet demonstrated how the next steps toward a living system actually took place, but as Prigogine emphasizes, the new structures arising from instabilities set the stage for more complex instabilities, further capture of energy, and further elaboration of structure. Life did not arise as an accidental, wildly improbable leap from molecules to cells or even to viruses, but through a step-by-step evolution, just as humans did not evolve in a single leap from one-celled creatures.

James Lovelock and others have shown entirely plausible natural mechanisms whereby the biosphere as a whole, through feedback responses, adjusts the components of the earth's atmosphere to favor a greater biomass and a faster overall rate of evolution. No particularly delicate balance is needed, either—the temperature of the earth has fallen over the past six hundred million years.

Finally, the idea that the evolution of humankind is purely an accident, divinely engineered or otherwise, ignores the vast mass of evidence that there are long-term trends in biological evolution. Over these millions of years there has been an irregular but unmistakable tendency toward adaptability to a greater range of environments, culminating in human adaptation to virtually any environment. Over this period the intelligence of the most developed animals on earth has risen with increasing speed, from trilobites, to fish, to amphibians, to the dinosaurs, to mammals, to primates, to the hominid apes and the direct ancestors of humankind.

Of course, through this long period there have been many chance events, many zigs and zags, advances and setbacks, which determined the exact timing and mode of the development of a creature capable of social evolution. Yet this unpredictability in no way erases the long-term tendency that makes the development of higher levels of intelligence, and eventually something resembling human beings, all but inevitable—as inevitable as the development of amino acids in a primal chemical soup.

Historical processes don't evolve linearly from advance to advance, as I have emphasized. But it is entirely wrong to thus conclude that there are no trends in evolution at all. As Prigogine's work and indeed the entire history of the biosphere and the cosmos as a whole show, the development of intelligent life is but the latest phase in a long acceleration of evolution itself.*

Thus we find that the apparently improbable accidents of the universe are neither the products of a random and incomprehensible cosmos nor evidence for a designing creator. Rather, they are misinterpretations of the general evolution of the universe.

The old cosmology and the old physics leave humanity with a choice between despair at contemplating a purposeless cosmos and abandonment of the scientific project and the ascription to the deity of all that science cannot explain. In either case a gap is created between a rational humanity and a fundamentally irrational, incomprehensible nature—whether or not it is guided by God. In the old view humanity lives, as Jacques Monod writes, "on the boundary of an alien world. . . . Man knows that he is alone in the universe's unfeeling immensity, out of which he emerged only by chance." It is only a small change to ascribe the chance events to a divine loading of nature's dice. Nature remains alien and irrational.

By contrast, the new cosmology, the new physics, posits a pro-

* Stephen Jay Gould's major argument against the existence of any progressive trend in evolution, while expertly documented in his book *Wonderful Life*, doesn't prove his case in the least. Gould shows that the number of animal phyla (basic body plans such as the vertebrates) was greater soon after multicellular life developed six hundred million years ago than it is today. Therefore, he contends, the diversity of life has not increased—even though there are many more species alive today, there are fewer fundamentally different body types. He asserts that diversity has not increased, that there are no long-term trends in evolution of any sort. As a corollary of this, Gould writes, the development of any intelligent life is purely an accident of evolutionary history—as unpredictable as the development of any given species.

But this is a giant non sequitur. Not only is there a huge difference between the contingencies that lead to the evolution of a particular species and a long-term trend in evolution, such as toward greater adaptability or intelligence, but Gould rests his case on facts that are an example of just such a trend! Over time, evolution has tended to concentrate more and more on specific modes of development. Nearly all chemical elements were in existence ten billion years ago or more. The types of compounds vital to life—DNA, RNA, proteins, and so on—were all present on earth some four billion years ago. The main kingdoms of life—animals, plants, fungi, and bacteria—have existed for two billion years; there have been no new ones in that time. As Gould shows, the main phyla have existed for six hundred million years, and the major orders (a lower grouping) for about four hundred million years.

As evolution has sped up, it has become more and more specific, and the earth has been transformed by the social evolution of a single species, our own. This is exactly the sort of long-term trend that Gould, despite his great contributions to evolutionary theory, is ideologically determined to ignore. Yet it exists, as does the trend toward intelligence.

gressively evolving, comprehensible universe. Such a view, to be sure, does not demand a creative God and is perfectly compatible without one. But nor does it preclude a creative deity. As Teilhard de Chardin demonstrates, a passionate faith in the intelligibility of the universe through science can exist side by side with a passionate faith in God. He wrote of this dual faith in "How I Believe": "If as a result of some interior revolution, I were to lose in succession my faith in Christ, my faith in a personal God, and my faith in spirit, I feel that I should continue *to believe* invincibly *in the world*. The world (its value, its infallibility and its goodness)—that, when all is said and done is the first, the last and the only thing in which I believe. It is by this faith that I live."

What is precluded by the new cosmology is the allocation of the mysteries left by bad science to the charge of a deity so clumsy as to leave his calling cards as incomprehensibilities written across the galaxies or in the equations of physics. It assumes that the universe is intelligible and that the scientific method can push back the frontiers of ignorance, so that no mystery will remain forever unexplained.

SCIENCE AND RELIGION

The debates of four centuries ago are thus reviving. What is involved is not only two views of the universe, but two views of the relation of science and religion. As in Galileo's day there are today those like Stanley Jaki who believe that religious authority can dictate the limits of scientific inquiry. Indeed, to Jaki, science itself is impossible outside a Christian context. For him and others science is incapable of comprehending the universe—certain mysterious cosmic coincidences must remain closed to it.

Not surprisingly, those who would limit scientific inquiry limit the universe as well: they perceive an infinite and eternal universe as a heretical threat to the authority of religion and to the transcendence of God.

Others, though, affirm the intrinsic autonomy of both science and religion. To James Skehan and many others, any attempt to determine from religious authority what concepts are permitted to science would threaten the entire basis of scientific research.

"To allow a priori ideas, religious or otherwise, to constrain science," Skehan warns, "is to enter the threshold of a new and even gloomier Dark Age."

Such a view in no way excludes a dialogue between science and theology. It is inevitable that religious beliefs and theories will influence scientific thought, as does every other aspect of society. Equally, theology will continue to ponder the implications of new developments in science. But for either to dictate the other's purview or possibilities is unacceptable. As Pope John Paul II put it, "Religion is not founded on science nor is science an extension of religion. Each should possess its own principles, its pattern of procedures, its diversities of interpretation, and its own conclusions . . . neither ought to assume that it forms a necessary premise for the other."

Religion is threatened neither by unlimited scientific inquiry nor by an unlimited universe—nor by any other conclusions of science. But the new cosmology does imperil any religious or philosophical doctrine that seeks to impose itself on science, to limit the bounds of human reason, or to dictate a cosmic pessimism. The failure of the Big Bang has subverted such doctrines, which claim conventional cosmology as their main support.

With its emphasis on observation, the new scientific revolution brings with it a revival of a scientific outlook not dependent on, or entangled with, religious doctrines. Like any worldview, such an approach will have philosophical implications. But its philosophical premises are only a cosmos knowable to the senses and governed by cause and effect. The empirical conclusions of this new science show a cosmos without beginning or end, one whose fundamental characteristic is progress. Any philosophy or theology that assumes a contrary reality will inevitably fight the new science, as the medieval church fought Galileo and Copernicus long ago.

■

10 COSMOS AND SOCIETY

In the late sixteenth and early seventeenth centuries—the time of the first great scientific revolution—European society was in deep crisis. The Spanish conquest of America had brought back to the continent no real wealth, only a river of gold which inflated prices and forced down real wages. Under the harsh rule of Spain and its allies, the European peasantry suffered increasing rents and feudal exactions. Living standards fell by over 50 percent. People looked back longingly to the days of their grandfathers, who at least had enough to eat and good beef on Sundays. In the new colonies of Spanish America, things were far worse. The reduction of the entire population to bondage had destroyed the native system of agriculture. Weakened by toil and starvation, the population succumbed to alien European diseases—in a century 80 percent of the indigenous population of Mexico and Peru was wiped out.

The old society—the society of lord and serf—had reached its limits, and the new society of merchant, manufacturer, worker, and free peasant had not yet been born. In Prigogine's terms, European society had reached a bifurcation point, where either an old mode of existence is superseded by a new one or it collapses back into chaos. Either there was to be a social revolution or Europe would suffer the fate of ancient Rome.

The scientific revolution had a critical role in deciding which way society would go. The new ways of thinking of Copernicus and Digges, Galileo and Kepler showed both to the small educated class and, in popularized form, to the mass of the population that the old society was not divinely ordained. The scientific method challenged authority with the evidence of observation, and in the process undermined the authority of the lords of the earth. Perhaps most important, it provided a potent antidote to oppression and increasing immiserization—it brought hope and the prospect of technical progress. The scientific revolution counterposed to the finite, fixed, and limited medieval cosmos the vision of an infinite universe and the abolition of all limits to the achievements of mankind. The new science gave the benefits of technology to those fighting for a new society.

Armed with this technology and these ideas, the partisans of social revolution were victorious. Sixteenth- and seventeenth-century Europe did not crumble in despair, but gave birth to the most vibrant advance of human culture ever seen—the epoch not only of Galileo and Kepler, but of Shakespeare and El Greco. A new society triumphed with the English and Dutch revolutions and by its triumph ensured the victory of the scientific revolution as well.

Today, another scientific revolution is beginning, one that may change our view of the cosmos as radically as the last. And today it again seems likely that the effects of this revolution, both social and scientific, will be profound.

A SOCIETY IN CRISIS

One wouldn't guess it from reading the self-congratulatory editorials and essays that abound in the American press, but world society has begun to retreat. According to the writers of these pieces, western capitalism is triumphant—it is only the socialist or formerly socialist countries that face economic and social crisis. In a widely praised article "The End of History?" in the magazine *The National Interest*, State Department functionary

Francis Fukuyama argued that in present western society humankind's evolution has ended, achieving "the final form of human government," a perfectly egalitarian "classless society" in which economic prosperity is permanently assured.

As in cosmology, in political and economic thought there is often an abyss between theory and observation. The actual state of society, so evident in socioeconomic statistics or a glance at any urban area, is quite different from the utopia of the editorialists and essayists. Society is retreating, living standards are falling *throughout* the world—including the developed and developing market economies that constitute the bulk of the world's population and wealth.

In the United States, Western Europe, and Japan, real wages, adjusted for inflation, are stagnant or falling. Writers like Fukuyama may see the American economy as utopia, but the American population has seen average real wages fall by 17 percent since their peak in 1973, a retreat to the levels of the late fifties (Fig. 10.1). Despite the vast influx of women in the work force, the median American family with two wage earners earned less

Fig. 10.1.

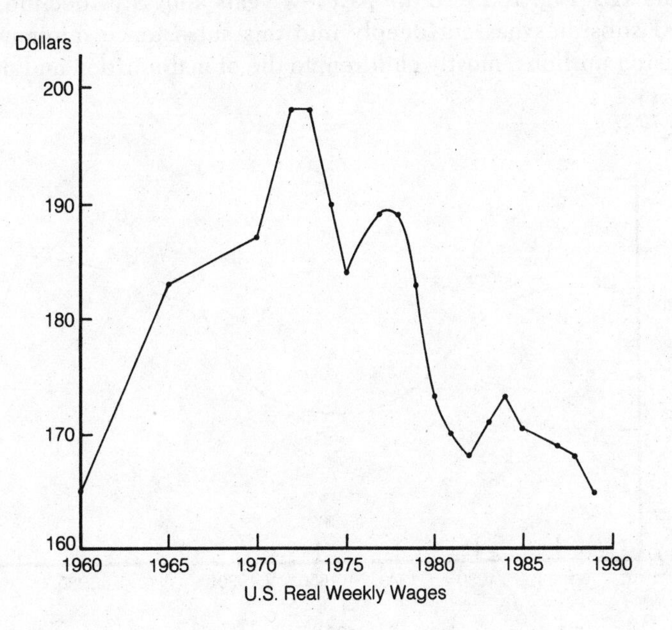

U.S. Real Weekly Wages

real income in 1989 than a single-income family did twenty years earlier. While wages have fallen, the number of hours of work have risen. According to a Harris poll in 1973, the average American adult had 26 hours of leisure time a week, but now has only 16 hours.

In Europe unemployment rates have remained near or above 10 percent for nearly a decade. Factories around the globe are shutting down, even in a "boom" period, and production of basic goods has fallen. In the market economies as a whole (excluding the socialist nations) steel production per capita has dropped 44 percent in the past fifteen years, and energy consumption per capita has fallen as well.

The developing countries of Latin America, Africa, and Asia are suffering far more. There, living standards have declined by 25 percent or more over the past decade, according to the United Nations. For these countries, the best measure of living standards is the world food supply: since 1971 the annual per capita supply of cereals in the market economies has fluctuated at about 340 kilograms, a level that leaves two billion malnourished and over a billion near starvation. The annual per capita supply level is only 10 percent above that achieved before World War I, seventy years ago (Fig. 10.2). In the past few years a massive decline in food subsidies has cut deeply into this subsistence minimum, causing millions, mostly children, to die of malnutrition and dis-

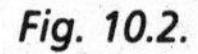
Fig. 10.2.

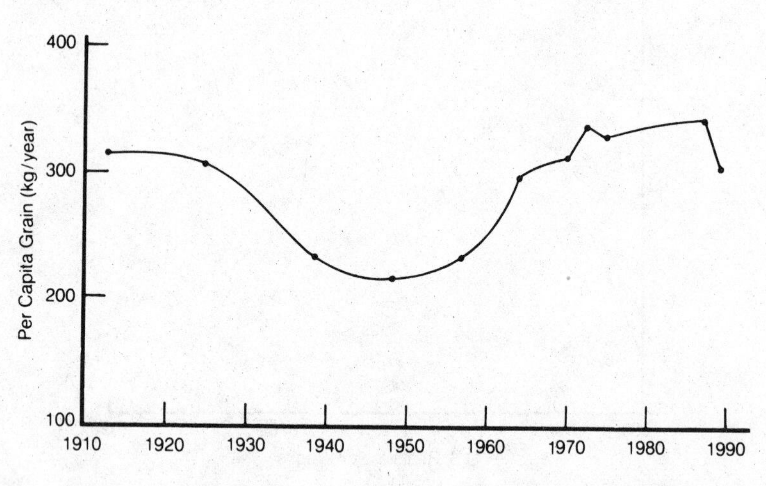

ease. In the poorest countries of Africa the weakened population is falling victim to AIDS, which has taken a million lives; ten times that many are infected with the AIDS virus.

Compared objectively with the crisis in the market economies, the "triumphant west," the far more public crises in the socialist countries seemed until recently almost mild. In the Soviet Union, Eastern Europe, and China, food supplies in 1989 were almost double what they were before World War I and housing and health care had increased similarly, although the levels remained well below that of the United States. But here too growth had slowed to a crawl or even ceased over the past decade, leading to the present crisis. It was, however, only the move toward market economies that has led to drastic falls in production and living standards. With this move the reversal of social advance has become truly global and all-encompassing.

Simultaneous with declining economic activity and living standards has come environmental degradation. The past decade has seen a massive acceleration of deforestation, the depletion of atmospheric ozone, an increase in air and water pollutants, and the growth of acid rain.

As in past epochs, a sharp drop in living standards has led toward a halt in population growth. In the advanced countries birth rates have plummeted: in the sixties the average American family had three or four children and the average European family two or three, but today even two children per family is atypical. In Germany the population has begun to fall, and in Europe as a whole its increase has effectively ended for the first time in three hundred years (Fig. 10.3). If present fertility rates continue, Europe's population will decline by 15 percent in the next generation. The great explosion of population that began with the capitalist epoch has, it appears, come to an end.

The advance of science and technology has radically slowed too. While biology remains vibrant, physical technology has been limited to mere quantitative advance for nearly thirty years, again a situation unprecedented in over two hundred years (Table 10.1).

This halt in the material advance, which progressed so rapidly during the nineteenth century, cannot be considered an inevitable phenomenon, an ultimate limit to human achievement. As is the case with any system, as was the case with the European

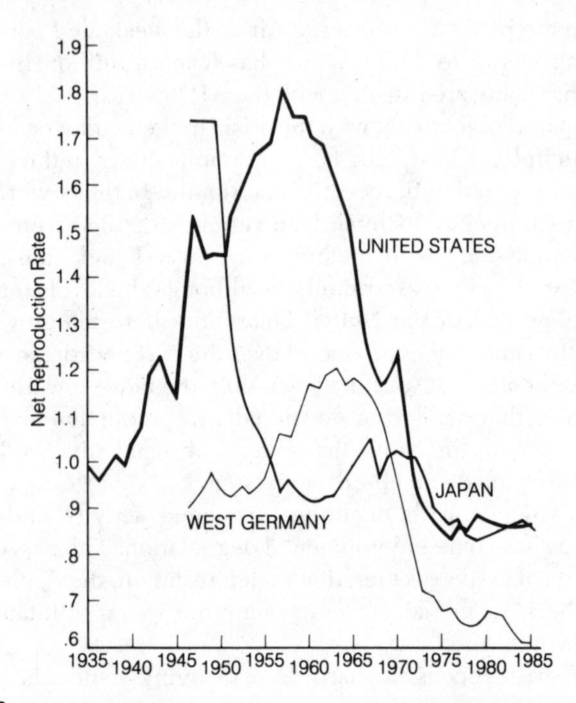

Fig. 10.3.

society of the sixteenth century, modern capitalist countries have reached the limits of expansion. Ancient Rome needed new slaves and feudal kingdoms new lands—capitalism has always required new markets. At the turn of the century the limits of the market expansion were first reached as the entire globe was incorporated into the various colonial empires. The global economy stopped growing and stumbled headlong into the catastrophes of World War I, the Depression, and World War II. After a generation of recovery, the same limits were encountered again in the early seventies. Once more, in the past twenty years, we have been faced with the paradox of a gigantic unfilled need for goods, for food, clothing, and housing, side by side with a "lack of markets," a saturation of the market for goods that can be sold *at a profit*. While children in Latin America lack clothing, clothing manufacturers are closing in the United States. While cities fall into decay and millions go homeless, steel mills are dismantled as unprofitable.

TABLE 10.1
MAJOR TECHNOLOGICAL INNOVATIONS IN THE PHYSICAL SCIENCES

1710	Newcomen steam engine
1720	
1730	
1740	
1750	
1760	
1770	Watt steam engine
1780	
1790	gas lighting
1800	battery
1810	steamboat
1820	electromagnet, photography
1830	electric motor, telegraph, locomotive
1840	
1850	Bessemer steel process
1860	typewriter
1870	electric dynamo, electric light, microphone, telephone
1880	phonograph
1890	automobile, motion picture, radio, tape recorder, X-rays
1900	airplane
1910	
1920	television
1930	
1940	computer, radar, transistor, nuclear energy
1950	laser, satellite, space probe
1960	
1970	
1980	
1990	

As the market has ceased to grow, investment in new production has become futile. Instead, new investment has been diverted almost entirely to nonproductive purposes, activities that add nothing to the world's real wealth. Rather than building new factories, owners have transferred plants from high-wage to low-wage areas. While there is no money for new housing, a trillion dollars has been poured into an armaments buildup that dwarfs any before it. While new plants cannot be profitably built, billions are spent buying existing ones: in the past few years nearly $200 billion per year, an amount well in excess of U.S. corporate profits, has poured into this Wall Street takeover spree. The profits made in such takeovers are recycled only in further speculation. The stock market has become a sort of financial black hole into which the wealth of nations is disappearing, leaving only a vast growth of debt. In the course of the past decade total debt has increased by nearly 50 percent in real dollar terms —interest payments in the U.S. alone consume over $300 billion per year. This debt has postponed the inevitable collapse of financial speculation and served as a convenient avenue of investment for billions that cannot be invested productively at a profit.

The debt has pushed back the final reckoning only by terrifically aggravating the crisis. In order to pay off debt generated in takeovers, corporations have closed plants, laid off workers, slashed wages, and cut back on research and development. The government's entire deficit now equals the interest on the national debt. Social services are eliminated purely in order to pay this interest.

In the Third World, the trillion dollars owed to the banks of the advanced countries has become a millstone dragging the impoverished population toward disaster. In order to pay off the interest on these titanic debts, Third World countries have exported food their own population desperately needs, axed investment in new internal development, and eliminated health and education services, all with catastrophic effects. The heaviest burden has fallen on the world's children. As the 1989 report of UNICEF put it:

> Three years ago, former Tanzanian President Julius Nyerere asked the question, "must we starve our children to pay our debts?" That question has been answered in practice. And the answer has been "Yes." In those three years hundreds of thousands of the world's children have given their lives to pay their countries' debts, and many millions more are still paying the interest with their malnourished minds and bodies.

Largely due to the titanic debt, UNICEF reports, incomes in the Third World have fallen by a quarter, spending on health by

50 percent and on education by 25 percent. Some of the world's poorest countries have in grim paradox become net food exporters to gain foreign exchange for debt payments. In Uganda, where one person in eight is infected with AIDS, a national budget imposed by the World Bank allocates nearly all income to debt payments, leaving less than one dollar a year per capita for all medical services.

The debt has also enormously aggravated environmental destruction. As Brazilian economists have documented, the devastation of the Amazon rain forest has been mainly motivated by the need to generate revenues through agricultural exports, such as beef from cattle grown on cleared lands. The financing for such export projects has come directly from international financial organizations such as the World Bank.

Eastern Europe, particularly Poland, has been pulled into this web of debt as well. The most enthusiastic supporters of the current program in Poland, which slashed wages by 40 percent, are not communist bureaucrats but their capitalist creditors and the pro-capitalist Polish government.

If we look at the evidence, there can be no doubt that the development and advance of global society has halted, that the current dominant society, capitalist society, has reached its ultimate limits. Clearly, the Stalinist model that collapsed in the Soviet Union, China, and Eastern Europe offers no alternative. The choice facing humanity is either to develop a new, as yet untried form of society or suffer a collapse of culture and civilization, which sooner or later must surely end in nuclear war. The question is whether this current stagnation is merely a pause between two stages of development or, as was the case two thousand years ago, the end of an entire period of human advance. Are we on the verge of a new Renaissance or a new Dark Age?

THE CONSEQUENCES OF COSMOLOGY

It is in the context of this social crisis that the effects of the scientific revolution and of the concomitant debate in cosmology must be judged. What we find is that the old ideas in cosmology tend to reinforce, and are reinforced by, society's dominant ideas, but the new ideas undermine that ideology with an alternative set of conceptions, as subversive in their own way as was the Copernican cosmos in its time. These ideas point to a new society and a renewal of progress.

Conventional cosmology encourages the prevailing ideas in a number of ways. First, it gives a scientific veneer and cosmic endorsement to the black pessimism that characterizes so much of today's intellectual climate. It is hard to find a generation in history when leaders in the fields of art, music, literature—the "leading thinkers" generally—have evinced such a profound pessimism and nihilistic despair as at the end of the eighties. Both the visual arts and, to a lesser extent, the literary ones are pervaded by a general sense of purposelessness, the existential absurdity of life. The sudden popularity in intellectual circles of Fukuyama's essay on the end of history is symptomatic of this growing intellectual despair. For only slightly beneath the superficial smugness of this article, which hails the final victory of the west, is a hopeless denial of progress, which has been central to western civilization for three hundred years.

Fukuyama writes that history has ended because western democracy is the final form of human society; all alternatives—communism and fascism—have been defeated or discredited. There is nothing beyond what we have now: all the misery and injustice is the best that can possibly be achieved, the endpoint of human evolution. There is no sense in decrying what occurs in the west, Fukuyama implies, nothing can be done about it. Fukuyama's bleak view becomes clear at his closing: "The end of history will be a very sad time. . . . In the post-historical period there will be neither art nor philosophy, just the perpetual caretaking of the museum of human history."

This fashionable intellectual pessimism pervades society as a general aimlessness and despair, which becomes obvious in an explosion of greed and selfishness. When things fall apart, when there seems no clear way forward, when no alternative is apparent, the response is to save oneself from the general debacle: "I'll get mine" and "The devil take the hindmost" have, with examples from Washington and Wall Street, become the slogans of the decade. When society is advancing, or when it seems possible that it can in the future, human beings have proven themselves capable of the highest heroism and self-sacrifice for a common cause. But when that hope is gone, social bonds degenerate into an orgy of greed, the child of hopelessness.

In a more acute crisis than exists at present, such despair and selfishness can be transformed into the unbridled evil of fascism, as Spengler's fashionable pessimism of the twenties paved the way for Hitler's vision of a *Götterdämmerung* where only the strongest and most ruthless would prevail.*

Such currents of despair arise in times of social crisis and retreat. In the sixteenth century, though, science provided a potent antidote to such hopelessness by putting forward the alternative idea of unlimited human progress. Today, in contrast, the concepts of conventional cosmology and the allied ideas in fundamental physics add credibility to the counsels of despair, as did Augustine's cosmology in fourth-century Rome.

Conventional cosmology today envisions a universe that is on a one-way street from an explosive start to an inevitable, ignominious end—a universe wound up twenty billion years ago and now running down. This, we are told, is not only the cosmic fate in the distant future, but the tendency everywhere and always, including here and now. In such a universe, progress or evolution is at best an accident or a miracle, contrary to the overall tendency of the cosmos. It should be no surprise, then, that human progress on this tiny planet has now come to an end—the accident is over and there is nothing to be done about it. Decay has simply caught up with us. Cosmologists will discover within a decade the Theory of Everything, which will mark the end of our quest for knowledge. Just as pundits like Fukuyama predict the end of history, so conventional cosmology predicts the end of science.

But the new ideas of the emerging scientific revolution bring an entirely different outlook. If the universe is evolving from an infinite past to an infinite future, if human development is only the latest stage of continual progress stretching through the unlimited reaches of time, then the very idea of an "end to history" is ludicrous, an unfunny joke. History can no more have an end

* A few decades earlier, biologist Ernst Haeckel emphasized the pessimistic, Malthusian interpretation of evolution as a vicious struggle for available resources, giving pseudoscientific justification for a virulent racism that portrayed human society as a war of "higher"—Aryan—races against "lower" ones. Haeckel's Malthusian Darwinism encouraged the widespread intellectual acceptance of anti-Semitic racism even before the rise of fascism.

than time itself. If human development pauses or retreats, then it is only because some specific form of society has reached its limits. If there is something wrong, then it must be fixed.

Such an outlook offers a hope of renewed progress. It asserts that scientific advance and technological development are not at an end, but could be starting a new period of explosive growth. It provides the motivation human beings need to join together in collective efforts rather than to fragment into self-centered anarchy. It shows that the technology exists to eradicate want on earth and open the path outward into an infinite universe.

Probably even more important than the effects of the content of the cosmological debate are the effects of the differing notions of scientific method on the current growth of irrationalism and occultism.

When society retreats, when progress is halted, rationality is discredited and many turn to the supernatural. Today occultism and religious fanaticism are reviving through both the developed and developing worlds. Wall Street investors consult astrologers for financial advice, while millions buy books describing the latest revelation of reincarnated prophets from the ice age, or earnestly seek healing from rock crystals and plastic pyramids.

All such occultism and fanaticism rejects the testimony of experience and places faith in authority. In the early stages of a crisis, like today's, such authority may seem innocuous—the quaint guru whose followers are at worst a bit gullible. But in the past such belief in the occult has in times of more desperate crises been channeled into a mass irrational worship of race, land, and *Volk*. In Germany the same people who quaintly worshiped nature in the *Wandervogel* a decade later expressed their mystical unity with Germany as storm troopers. The seemingly innocuous authority of the local guru was replaced with the authority of the national leader—the Führer.

No facile analogy is intended here—history does not repeat itself and there are many differences between Weimar Germany and nineties America. But whenever large groups adopt irrational beliefs they become easy targets of political manipulation, for they have no sure guide to differentiate truth from falsehood or wisdom from madness. The appeal to an irrational authority has always served as the basis for forced labor, slavery, and serfdom. It should be remembered that the last such systems, American slavery and Russian serfdom, were wiped out only two lifetimes ago; and the last serious effort to reestablish slavery, in Nazi Germany, was destroyed only a half-century ago.

The antidote to such irrationality has always been the appeal to observation, the evidence of the senses, and reason. Traditional science and the new cosmology, both rooted in such well-founded methods, can provide a scientifically educated population with the means to judge the claims of authority, and to reject those claims if they prove unfounded. Not only do such methods undercut a belief in the occult, they also discredit all blind reliance on authority, the willingness to "let the experts decide." A population that understands in practice how a theory's final test is its results in practice would, for example, be unlikely to accept a prescription of wage cutting and austerity to promote future prosperity, when every such policy in the past has led only to further falls in living standards. For this reason, the scientific method, the inductive method, remains as in Galileo's day a subversive force.

By contrast, the authority of deductive science is used to defend the rule of the "experts" in society. For example, Aleksei N. Boiko, an economic advisor to then–Soviet President Gorbachev, contemptuously dismissed a proposal for a referendum on price increases and a market economy as akin to "trying to solve a problem in higher mathematics by voting on it." What need is there for voting when the correct policies can be derived through the pure reason of those in authority? (Similar arguments have been used in the United States to defend the decisions of the Federal Reserve Bank from the interference of the electorate.) Moreover, the methods and assumptions of conventional cosmology and fundamental physics reject scientific method, and thereby disarm science in the face of irrationality and occultism. Here, after all, is a world of electrons, photons, even entire universes popping into existence without cause, a world governed by neither causality nor logic. The criterion of truth is not fidelity to observation or experiment but the subjective beauty of equations, the symmetries plumbed from the mind of the scientists or "the mind of God"—ultimately the authority of established physicists and of established ideas.

It is no coincidence that the ideas of such science are now used to lend credence to the "channeling" of departed spirits and other such occult matters. Jane Roberts, communicating the opinions of the ethereal being "Seth," uses the multiple world hypothesis of quantum philosophy to justify "paranormal" phenomena. Other occultists assert that the notion that observation can affect events that happened billions of years in the past (such as the emission of a photon from a distant quasar) is evidence for psychokinesis. Such inferences are by no means absurd. If one accepts the illogic of current ideas about quantum mechanics, one is ill prepared to reject such claims.

Cosmologists and fundamental physicists may scorn the use of their theories by popular writers on the occult and the irrational. But their own methods, by rejecting the tests of observation, by rejecting the inductive method of science, leave them defenseless against the assault of the irrational. In a similar way German scientists like Heisenberg, by accepting acausality, negated science as an intellectual counterweight to the rise of irrationalism.

To the extent that the "deductive" methods of cosmology and particle theory remain the most popularized view of science, science will remain impotent in the face of a new irrationality and its political consequences.

THE WAY AHEAD

The debate in cosmology will thus have a significant impact on today's intellectual climate, on the pessimism and irrationality that will sustain the old society to its end, or the hope and empirical judgment that may transform it. The new ideas may have another effect in outlining the direction that society must move in.

The core of the new cosmology is the belief that theory and experiment, mind and hand, must be unified. This unity is vital to the scientific enterprise and to the development of society as well. If we look back to the chart on p. 317 of Chapter Seven, we can see that those societies in which the direction of economic activity is, to the greatest degree, in the hands of those who actually carry on society's work are the ones that advance most swiftly. By contrast, those societies in which work is furthest separated from thought and social planning are the most stagnant.

Without doubt, the period of the most rapid human development relative to population size was the period of neolithic revolution and the immediately following urban revolution; this period saw the invention of the technologies basic to civilization —agriculture, animal husbandry, writing, mathematics, spinning, potting, weaving, and metallurgy. They emerged from a society that had not yet ossified into separate classes, where communal agriculturists were free to develop new techniques of immediate benefit to their community and themselves. Next was the brief Ionian period, when small craftsmen and merchants developed new ways of writing and thinking, before the rise of chattel slavery. Two millennia later came the late Renaissance, from 1550 to 1650, when the small population of Europe gave rise to genius after genius; Digges's "mechanics," artisans, small manufacturers, reading of the latest in scientific developments, gave birth to an explosion of new technology before the resurgence of aristocratic power. Finally, there was the nineteenth century, when inventors, entrepreneurs, skilled workers, and scientists—the Edisons, Marconis, Maxwells, and Faradays—combined to transform the world while Beethoven and Brahms composed, Monet and van Gogh painted, Tolstoy, Dickens, and Conrad wrote.

Each of these periods gave rise to a greater or lesser degree of political democracy, but all were characterized by a close link between hand and mind, by a democratization of at least major sectors of the economy.

By contrast, the periods of slowest development were in the later Bronze Age, the slave societies of 300 B.C. to A.D. 700, and to a lesser extent the feudal society of the Middle Ages. These societies squandered the minds of the population, reducing them to mere tools of a small ruling class.

From the standpoint of the underlying theory of evolution it's reasonable that this relationship should hold. The freest societies are those that most directly allow the individual to make changes in the mode of production and thus in the society as a whole, that most readily encourage innovation. In Prigogine's terms these are the most "unstable" and therefore the fastest-evolving.

Today the economic democracy essential to progress has almost disappeared, and progress cannot resume unless those who do the work decide what work is to be done and how it is to be done. Neither a few thousand immensely wealthy capitalists nor a few thousand party bureaucrats have the wisdom to run the vast and complex world economy. That task can be accomplished only by those who work, by the people themselves.

The issue of who shall rule is today posed most sharply in Eastern Europe and the former Soviet Union, where the old Stalinist regime has self-destructed. But what will take its place? One answer, the obvious one, is capitalism. This is the answer advocated by the governments of Eastern Europe and the Soviet Union. But this would be a change of masters, if even that, rather than a triumph of self-rule. The people of these nations are not gaining control of the factories. The factories are to be sold to either multinational corporations or, amazingly, to the very bureaucrats who managed them under the old regimes. Yeltsin, like Gorbachev, is proposing to convert a section of the old bureaucracy into a new capitalist class, at the expense of the working population. The steps toward a market economy, as in Poland, have brought prosperity to the few but a 40 percent decline in real wages and mass unemployment to the many—soup lines where there were bread lines. What is replacing the Stalinist regimes is not democracy but a return to the rabid nationalism of the interwar period, the threatened Balkanization and disintegration of all East Europe and the Soviet Union. In Poland political leaders again blame the Jews; in Romania statues are being erected to Antonescu, the wartime fascist dictator, and throughout the area old ethnic enmities are being rekindled. In the meantime, the national wealth is being sold off, at bargain prices, to foreign investors.

A return to capitalism in what was once the Soviet Union threatens more chaos, with collective farmers thrown off privatized land, industrial workers out of jobs, and the country torn into hostile, nuclear-armed fragments.

The alternative is for the peoples of the nominally socialist states to gain democratic control over their own economies and their own fates, to decide for themselves if they need more austerity or more production, faster payment of foreign debt or more meat on the family table. The alternative to bureaucratic planning is not the autocratic rule of private capital but democratic planning—the working people deciding themselves, through truly democratic institutions, what should be produced and how, from the level of the factory to that of the nation, and thus gaining control of the socialized industry that is rightfully theirs.

The question of economic democracy is a real one in the west as well. In the "egalitarian" United States, which, according to thinkers like Fukuyama, has achieved a classless society, economic power is concentrated to an unprecedented degree. Despite all the rhetoric of free enterprise, nine-tenths of the economy is in the hands of the top five hundred companies, and 80 percent of that nine-tenths is in the hands of the top one hundred. As of 1980, before the recent spate of takeovers further concentrated power, a U.S. Senate report showed that controlling ownership of the stocks of all these companies lay in the hands of two dozen top financial institutions—banks, insurance companies, and pension funds. The ownership or control of these institutions in turn lies in the hands of each other. Over a third of the shares of Citibank, for example, were held by twenty-four of its leading "competitors."

The five hundred or so individuals who sit on the boards of directors of these powerful institutions directly control through corporate stock ownership dominant interests in all these institutions, so the shareholders they are "responsible to" are primarily each other. Since many of these individuals sit on the boards of several of the top industrial corporations in America, they constitute the majority of these boards as well. Those who own finance and industry are the same small group—a financial elite.

These same men indirectly elect the majority of the Federal Reserve Board, the world's most powerful economic institution, which decides the interest rates and credit policies that can send the nation or the world into recession or inflation.

Of course, the federal government, with its huge budget, wields immense economic power and is controlled by elected officials. Yet here, too, practice falls somewhat short of the perfect "final form of human government." In the ideal democracy of today's America, no less than 98 percent of all incumbents in the House of Representatives are reelected, a performance that in any other country would be scorned as indisputable evidence of rigged elections. Such representatives need not fear the wrath of the impotent electorate. Instead, they answer to those who finance their campaign—the infamous Political Action Committees of various moneyed special interest groups and corporations.

Similar mechanisms of control, different in detail, exist in Western Europe and Japan. In the estimate of *Fortune* magazine, not more than one thousand individuals own or control 90 percent of the world's economic capacity.

This is the small economic elite whose banks can demand the repayment of debt even if it causes the collapse of entire national economies, the immiserization of whole peoples, who can shut plants and lay off hundreds of thousands of workers, who can dictate the expenditure of billions on armaments and eliminate billions for housing, health, and education.

The whole history of social evolution shows that there is no way forward unless the world economy can be liberated from this tyranny of greed, unless the people themselves can gain control of that economy. There is no blueprint that explains how that control can be won or exercised—it has never been done before.

Yet events are already posing the problem. The American financial system—lynchpin of the capitalist world—is disintegrating, large chunks falling bankrupt into the hands of the federal government. The finance sector is being nationalized piecemeal. Who shall run such a nationalized finance sector and in whose interest? Will it be, as at present, by and for those who have stolen the most from their depositors, who have looted one bank and fled to another while the government pays off their victims with the taxpayers' money? Or will the entire system instead be taken over in bulk to benefit the entire population, so that interest rates can be slashed and money redirected from speculation to production?

Who should decide—the experts or the people? Should $300 billion a year in interest be drained from the economy, or should it be diverted back to consumption and production, eliminating the federal deficit and increasing family income? Should the Third World debt be repaid at untold human cost, or should it be written off, so that the impoverished South can import the products made by workers in the North? These questions aren't academic—they will be decided politically in the coming few years and they come down to the same issue—who shall rule the economy, and in whose interest? By the people and for the people, or by the experts and for the greedy few?

The end of the Cold War poses another question for the overgrown defense industry. Should hundreds of billions still be squandered each year on gold-plated weapons with no conceivable use? If not, should this capacity be wasted on equally useless civilian projects, like a supersonic business jet (as some corporations have proposed)? Or, as many have advocated for years, should this, the most advanced sector of the American economy, be regeared to meet the needs of people?

The $100 billion now spent on armaments could be used to rebuild American industry, to create two to three million new jobs a year, to eliminate in less than a decade the housing crisis, and to rebuild the crumbling cities of America. Will this be done? Who will decide what the defense industry should produce? The people or the corporations and their elected friends?

Again and again war threatens in the Middle East. Should American and Arab blood be spilled to fatten the profits of American oil and defense companies, of Arab sheiks and dictators? Should the world continue to squander hundreds of billions of dollars a year for oil? Or should some of those billions be used to develop fusion energy, which could replace oil with cheap and clean power?

It is clear that the people don't make these decisions now. To change this would take a great struggle, a radical transformation of politics—the development of new parties, new institutions for control of the economy from the factory to the nation.

Yet the results will also be on a grand scale. The technology and productive capacity now lying idle or wasted on armaments can, if directed by the people for their own needs, provide within a few decades adequate food, decent housing and clothing, clean water, good health care, and quality education for the entire world population. The elimination of want and misery on earth can, in turn, set the stage for the expansion in the next century of human society into the universe.

The question remains, Will it happen? Whether such a transformation takes place depends in large measure on what people believe is possible. According to polls, three out of four Americans think that a decade from now there will be more unemployment, more inflation, more homelessness, more divorce, more crime, more drugs. Those who think that this is inevitable, part of the "way things are" will do nothing. Only those who still believe it is possible for the future to be better than the past will demand change. Those who think decisions must be left to experts will do nothing, but those who trust the evidence of their own eyes will demand the right to real self-rule.

A CROSSROADS OF HISTORY

In every system, be it physical or social, there are times when small events determine the course of ensuing events—either forward to new and higher plains or backward, cut off by the barriers of the old. Society stands at such a crossroads now, at the end of the second millennium. If it falls backward then the present embryonic scientific revolution will be swept away in the general retreat, becoming, like Aristarchus' heliocentric theory, a mere footnote to a debacle.

What happens now depends in part on what people think and this, in turn, rests on how they view the universe and how they believe truth can be known. To know the truth about the universe and our own society, must we rely on the authority of experts who can fathom the mysteries of pure reason? Or does our own freedom rest on the ability of all to define truth through observation and the testing of ideas in practice?

Must thought and action be forever separate, the rulers and the ruled? Or can those who do the work decide what work is to be done?

Do we live in a finite universe doomed to decay, where humans are insignificant transitory specks on a tiny planet? Or are we instead the furthest advance of an infinite progress in a universe that has neither beginning nor end? Will our actions today have no meaning in the end of all things, and are we now being swept into that inevitable decay? Or does what we do here and now permanently change the cosmos, a change that will echo through a limitless future?

How we answer these questions affects which road we take: whether we fall back despairing within a finite world or move forward into the infinite universe.

▪

APPENDIX

WHAT CAUSES THE HUBBLE SHIFT?

There are several theories that attempt to explain the Hubble shift without recourse to the Big Bang's assumptions—which, as we saw in Chapters One and Six, are refuted by observation.

First is Alfvén's antimatter theory, described in Chapter Six. The theory faces some definite problems, at least one of which appears, in my own view, to be fairly serious but not insurmountable.

First, it's clear that antimatter explanations are not needed to explain DeVaucouleur's limit on the velocities of astronomical objects. As we've seen, this can be explained more simply by the plasma processes that form the objects. But three problems remain. How does ambiplasma acquire sufficient energy to overcome the Leidenfrost layers that separate matter and antimatter? The layers are characterized by velocities of around 50,000 km/sec, yet, as we've seen for the most part, astronomical objects don't acquire velocities higher than about 1,000 to 2,000 km/sec. This is at least six hundred times less energy than is contained in the Leidenfrost layers and thus seems unlikely to lead to instabilities—the sort of turbulence that would break up the layers, permitting the explosive mixing of matter and antimatter.

There may be a simple solution. In the plasmoids that form at the center of a galaxy, leading to the creation of active galactic nuclei and quasars, magnetic pinching generates high velocities, far higher than can be contained by gravitational fields—about 10,000 km/sec. Within the high-velocity regions it's possible that the Leidenfrost layers will be torn apart, leading to a massive release of energy. This would, in turn, heat surrounding plasma enough to rupture its Leidenfrost layer, setting off a chain reaction.

Obviously, this process does not happen in all quasars and galactic nuclei. So the model would require that matter and antimatter be generally separate on the galactic scale, with only some galaxies having ambiplasma mixtures.

A second problem is that matter–antimatter explosions would tend to blow structures apart, rather than preserving the hierarchical ordering of the universe. If one-quarter of a galactic cluster's mass, for example, is annihilated, after several billion years the ex-clusters remaining would be smeared across most of the observable universe. Clearly, if such clusters and other objects are to survive their own explosion, some mechanism must automatically determine that the violence of a given explosion decreases with the size of the objects involved. Thus a supercluster complex a few hundred million light-years across should explode with a maximum velocity of some 10,000 km/sec, so it will not disintegrate in ten billion years. A cluster would have an even gentler explosion, but the metagalaxy as a whole would bear the full brunt of an explosion with top velocities near that of light.

Fortunately, there's a rather natural reason to expect that this hierarchy of explosions would occur. As we've seen, the universe is organized in such a way that the space between objects increases as the objects' size decreases. If matter-antimatter explosions occur only when objects of different composition collide, larger objects should collide more frequently than smaller ones —thus stars within a galaxy almost never collide, galaxies collide relatively rarely, but if clusters move at all randomly within a supercluster complex, collisions would be frequent. For the complexes themselves, if they have substantial random velocities, collisions would be inevitable. Thus for smaller objects matter-antimatter collision will only slowly add energy to the intergalactic medium and force a gradual expansion. But for the larger entities and for metagalaxies as a whole, a collision of its constituents—the supercluster complexes—would be common and the energy release far greater.

But the third and most serious problem arises when this scheme is considered at the level of the metagalaxy as a whole. Simply put, the metagalaxy would tend to blow itself apart before enough energy was released to fuel the Hubble expansion observed. The rate of energy released by a matter-antimatter collision is proportional to the density of the plasma, and the time needed for an object to blow itself apart is proportional to its radius. So the larger the product of density times radius (nR), the greater the energy release per particle and the more violent the explosion. But the nR of objects tends to drop with increasing size. This is a consequence of the DeVaucouleur limit, which limits the product of density and area. (If $n \times r^2$ is a constant, as it is in the DeVaucouleur relationship, then n is proportional to $1/R^2$ and thus $n \times r$ is proportional to 1/R.)

Let's assume that the universe we now observe, the metagalaxy, was one-third its current size and had an orbital velocity somewhat above the DeVaucouleur limit—2,000 km/sec. The annihilation rate would be sufficiently low that the explosion would generate only maximum velocities of a few thousand kilometers per second before the universe expanded significantly. Only if the metagalaxy was at some point far more compact, only ten or twenty million light-years across, would it develop sufficient energy sufficiently fast to create a Hubble-like expansion.

But that would pose two difficulties. Such a compact explosion would blow apart objects as large as the supercluster complexes, and the initial body would have an orbital velocity of 50,000 km/sec. It's difficult to see how such a compact object could have formed. We have good reason to think that such objects don't exist today, even temporarily, and that the angular momentum generated during contraction prevents them from far exceeding DeVaucouleur's speed limit.

One way around this problem is to suppose that galaxy-size plasmas collide and annihilate one another. The difficulty here is that such collisions would not occur very frequently in the metagalaxy, as I just pointed out.

It is certainly not impossible to construct a scenario that leads to a consistent result. For example, perhaps dense plasma filaments paired together with one matter and one antimatter filament could approach and annihilate each other quickly, but with the resulting energy spread along the tangled length of the filaments. Here, the dynamics that led to the creation of such pairs would have to be carefully examined. But for the antimatter theory to be plausible, it would have to simultaneously account for the large velocities observed in the Hubble expansion and the large-scale structures that were obviously not blown apart in that expansion.

It's important to emphasize here that the metagalaxy may be half antimatter *even if* matter-antimatter annihilation is not the source of the Hubble expansion. These are, in essence, two separate, if related, questions. It's entirely possible that matter and antimatter have remained effectively separated by Leidenfrost layers even at the supercluster complex level, so that their interaction is always small. There are two pieces of evidence in favor of this idea. First, there seems to be no way to produce matter without antimatter, as is detailed in Chapter Eight. This by itself would make a mixed universe the probable one. Second, the Laurent-Carlqvist calculations of the X-ray and gamma ray spectrums expected from matter-antimatter annihilation match observations well. This agreement with observation is again good evidence that antimatter exists, unless another equally plausible explanation of the X-ray/gamma ray background can be found.

A second set of theories about the Hubble relation assumes that the universe is not expanding, but that the redshift is created as light travels through space. According to Paul Marmet and Grote Reber (a co-initiator of radio astronomy), quantum mechanics indicates that a photon gives up a tiny amount of energy as it collides with an electron, but its trajectory does not change. As the photon travels, its energy declines, shifting its frequency to the red. Marmet has calculated this effect for our own sun, showing that it explains a long mysterious redshift between the limb and center of the sun. (The effect is not significant at the high densities of the earth's atmosphere.)

Potentially this effect could explain the high redshifts of some quasars, since light traveling through the outer atmosphere of the quasar could be redshifted before leaving it.

But for explaining galactic redshifts there is a fatal flaw. Here, Marmet assumes that the light is redshifted in traveling through intergalactic space. Yet his calculations imply that a density of ten thousand atoms per cubic meter is required to achieve the observed redshifts. This is *far* more than the one-tenth of an atom per cubic meter that has been observed locally or even the ten atoms required by dark matter theories. Such a high matter density would have enormous gravitational effects that simply aren't observed. Unless the matter were far more smoothly distributed than that observed thus far, supercluster complexes would, with this density, produce galaxy velocities close to the speed of light. In fact, according to general relativity, such a high density would create a closed universe with a radius of only a few hundred million light-years.

Two other hypotheses appear to be less easily refuted. The first is another version of the idea of "tired light"—the loss of energy as the light travels. In this version, however, J. P. Vigier has hypothesized a new term in the equations of quantum mechanics which cause the vacuum itself to absorb the energy. As is the case with the cosmological constant explanation, this theory is relatively difficult to refute, but it involves somewhat ad hoc hypotheses.

Similar problems afflict the final main alternative. This is the oldest one, first suggested by quantum pioneer Paul Dirac in 1938. Dirac proposed that instead of the space between the galaxies expanding, as general relativity predicts, all space is expanding because the basic scale of all objects from electrons to galaxy clusters grows with time, due to an unknown physical law. While the form of this hypothesis proposed by Dirac predicts that fundamental constants such as the force of gravity will change—something that doesn't seem to happen—other versions allow all the constants to remain the same. That is, the size of everything —objects and the space between—evenly expands, so distant objects only appear to be redshifted; but no real expansion is taking place since the density of the universe remains unchanged, everything expanding in place.

Advocates of all these hypotheses need to make far more precise predictions before they can be thoroughly tested and confirmed or refuted. The Vigier and Dirac proposals, as well as the cosmological constant idea, involve new laws of physics. In the first two cases, the new physical laws should, at least in principle, be possible to test in the lab, and neither would be considered particularly convincing unless it provided concrete predictions other than the Hubble relationship.

Alfvén's hypothesis alone has the advantage of using exclusively known laws of physics. But it, too, is extrapolated far beyond experimental results; simulations to test how antiplasma might behave, and ideally lab experiments, are essential.

Fortunately, there is one astronomical test that could help to sort out the various Hubble proposals. Each predicts a slightly different relation between distance and redshift, especially at high redshifts. If galaxies at large redshifts could be plotted as Tully and Fisher plotted nearby ones, local areas of concentration should appear, as they do in our part of the cosmos. If the redshift were converted to distance using the "wrong" relation, the concentrations would be characteristically distorted. They would all tend to be elongated toward the earth in some maps, or flattened in our direction in others. Only in a map using the "correct" law would the shapes be oriented without respect to earth.

Over time, such large-scale mapping will define observationally what the true Hubble relation of redshift and distance is. Then, I believe, it will be possible to concentrate on those theories that actually yield such a relationship.

Alternatively, work on solving the problems in fundamental physics outlined at the end of Chapter Eight may well lead naturally to a sound explanation of the Hubble shifts. In any case, the nature of this shift and the nature of the elementary particles remain two outstanding unsolved problems of physics. It would be tidy if they had a common solution—but the universe isn't always tidy!

BIBLIOGRAPHY

1. The Big Bang Never Happened

Alfvén, H., "Cosmology in the Plasma Universe," *Laser and Particle Beams*, vol. 16 (Aug. 1988), pp. 389–98.

———, "On Hierarchical Cosmology," *Astrophysics and Space Science*, vol. 89 (1983), pp. 313–24.

Lerner, Eric J., "Magnetic Vortex Filaments, Universal Scale Invariants and the Fundamental Constants," *IEEE Trans. Plasma Sci.* vol. PS-14 (Dec. 1986), pp. 609–702.

———, "Plasma Model of the Microwave Background and Primordial Elements," in *Laser and Particle Beams*, vol. 6 (Aug. 1988), pp. 457–69.

Lilly, S. J. "Discovery of a Radio Galaxy at a Redshift of 3.395," *Astrophysics Journal*, vol. 333 (Oct. 1988), pp. 161–67.

Peratt, A. L., "Are Black Holes Necessary?" *Sky and Telescope*, vol. 66 (July 1983), pp. 19–22.

———, "Simulating Spiral Galaxies," *Sky and Telescope*, vol. 68 (Aug. 1984), pp. 118–22.

Shaver, P. A., "Possible Large-Scale Structure at Redshifts of 0.5," *Nature*, vol. 326 (1987), pp. 773–75.

Tully, R. Brent, "More About Clustering on a Scale of .1 c," *The Astrophysical Journal*, vol. 303 (1986), p. 25.

Tully, R. Brent, and J. R. Fischer, *Atlas of Nearby Galaxies* (Cambridge: Cambridge University Press, 1987).

Valtonen, Mauri, and Gene Byrd, "Redshift Asymmetries in System of Galaxies and the Missing Mass," *The Astrophysical Journal*, vol. 303 (1986), pp. 523–34.

Yusef-Zadeh, F., and M. Morris, "The Linear Filaments of the Radio Arc Near the Galactic Center," *The Astrophysical Journal*, vol. 322 (1987), p. 721.

2. A History of Creation

Augustine of Hippo, *The Confessions, The City of God, On Christian Doctrine* (Chicago: William Benton, 1952).

Breasted, James H., *A History of Egypt* (New York: Bantam Books, 1967).

* particularly recommended

Butterworth, G. W., trans., *Clement of Alexandria* (London: William Heinemann; New York: G. P. Putnam's Sons, 1919).

* Childe, V. Gordon, *What Happened in History* (Baltimore: Penguin Books, 1942).

Heath, Sir Thomas, *Aristarchus of Samos: The Ancient Copernicus, A History of Greek Astronomy to Aristarchus Together with Aristarchus's Treatise on the Sizes and Distances of the Sun and Moon: A New Greek Text with Translation and Notes* (Oxford: Clarendon Press, 1913).

Farrington, Benjamin, *Head and Hand in Ancient Greece: Four Studies in the Social Relations of Thought* (London: Watts & Co, 1947).

Ferguson, John, *Pelagius: A Historical and Theological Study* (Cambridge: W. Heffer & Sons Ltd., 1956).

Hanson, Robert W., ed., *Science and Creation: Geological, Theological, and Educational Perspectives* (New York: Macmillan, 1986).

Hyers, Conrad, *The Meaning of Creation: Genesis and Modern Science* (Atlanta: John Knox Press, 1984).

Lilla, Salvatore, *Clement of Alexandria: A Study in Christian Platonism and Gnosticism* (Oxford: Oxford University Press, 1971).

Norris, R. A., Jr., *God and World in Early Christian Theology* (New York: The Seabury Press, 1965).

*Pagels, Elaine, *Adam, Eve, and the Serpent* (New York: Random House, 1988).

Plato, *Republic* (Oxford: Oxford University Press, 1945).

———, *Timaeus* (New York: Liberal Arts Press, 1949).

Roberts, Alexander, and James Donaldson, *Ante-Nicene Christian Library* (Tertullian) vol. XV (Edinburgh: J. & J. Clark, 1870).

Roberts, Robert E., *The Theology of Tertullian*, Th.D. thesis (London: Epworth Press, 1924).

Russell, Bertrand, *A History of Philosophy* (New York: Simon and Schuster, 1945).

Sarton, George, *A History of Science: Ancient Science Through the Golden Age of Greece* (Cambridge, Mass.: Harvard University Press, 1960).

3. The Rise of Science

Bett, Henry, *Nicholas of Cusa* (London: Methuen, 1932).

Brophy, James, and Henry Paolucci, eds., *The Achievement of Galileo* (New York: Twayne Publishers, 1962).

Cassirer, Ernst, *The Individual and the Cosmos in Renaissance Philosophy*, Mario Domandi, trans. (New York: Harper & Row, 1963).

Chestnut, Roberta C., *Three Monophysite Christologies: Severus of Antioch, Philoxenus of Mabbug, and Jacob of Sarug* (Oxford: Oxford University Press, 1976).

Crombie, A. C., *Robert Grosseteste and the Origins of Experimental Science 1100–1700* (Oxford: The Clarendon Press, 1953).

Easton, Stewart C., *Roger Bacon and His Search for a Universal Science* (New York: Russell & Russell, 1952).

Geymonat, Ludovico, *Galileo Galilei: A Biography and Inquiry into His Philosophy of Science*, Stillman Drake, trans. (New York: McGraw-Hill, 1965).

Hespel, Robert, *Sévère d'Antioche: la polémique antijulianiste* (Louvain: Secretariat du Corpus SCO, 1968).

Islam, Philosophy and Science. Four public lectures organized by UNESCO, June 1980 (Paris: The UNESCO Press, 1981).

Johnson, Francis R., *Astronomical Thought in Renaissance England: A Study of the English Scientific Writings from 1500 to 1645* (Baltimore: The Johns Hopkins Press, 1937).

Kant, Immanuel, *Universal Natural History and Theory of the Heavens* (Edinburgh: Scottish Academic Press, 1981).

Mason, Stephen F., *A History of the Sciences* (New York: Collier Books, 1962).

McInerny, Ralph, *St. Thomas Aquinas* (Boston: Twayne Publishers, 1977).

Nasr, Seyyed Hossein, *An Introduction to Islamic Cosmological Doctrines* (Cambridge, Mass.: The Belknap Press, 1964).

Omar, Saleh Beshara, *Ibn al-Haytham's Optics: A Study of the Origins of Experimental Science* (Chicago: Bibliotheca Islamica, 1977).

Poupard, Paul Cardinal, ed., *Galileo Galilei: Toward a Resolution of 350 Years of Debate—1633–1983*, Ian Campbell, trans. (Pittsburgh: Duquesne University Press, 1987).

Said, Hakim Mohammed, *Ibn al-Haitham*, Proceedings of the Celebrations of 1000th Anniversary Held under the Auspices of Hamdard National Foundation, Pakistan.

Sambursky, G., *The Physical World of Late Antiquity* (Princeton, N.J.: Princeton University Press, 1962).

Shapere, Dudley, *Galileo: A Philosophical Study* (Chicago: University of Chicago Press, 1974).

Sigmund, Paul E., *Nicholas of Cusa and Medieval Political Thought* (Cambridge, Mass.: Harvard University Press, 1963).

Singer, Dorothea Waley, *Giordano Bruno, His Life and Thought with Annotated Translation of His Work, on the Infinite Universe and Worlds* (New York: Abelard-Schuman, 1950).

Steenberghen, Fernand Van, *Thomas Aquinas and Radical Aristotelianism* (Washington, D.C.: The Catholic University of America Press, 1980).

Westman, Robert S., and J. E. McGuire. *Hermeticism and the Scientific Revolution,* papers read at a Clark Library Seminar, March 9, 1974 (Los Angeles: University of California Press, 1977).

Wildberg, Christian, trans. *Philloponus: Against Aristotle, on the Eternity of the World* (Ithaca, N.Y.: Cornell University Press, 1987).

4. The Strange Career of Modern Cosmology

Alfvén, Hannes, "How Should We Approach Cosmology?" in *Problems of Physics and Evolution of the Universe* (Yerevan: Academy of Sciences of Armenian SSR, 1978).

Alpher, R. A., H. Bethe, and G. Gamow, "The Origin of Chemical Elements," *Physical Review,* vol. 73, no. 7 (April 1948), pp. 803–4.

Bondi, H., and J. Gold, "The Steady-State Theory of the Expanding Universe," *Monthly Notices of the Royal Astronomical Society,* vol. 108 (July 1948), pp. 252–70.

Burbidge, E. Margaret, G. R. Burbidge, William A. Fowler, and F. Hoyle. "Synthesis of the Elements in Stars," *Reviews of Modern Physics,* vol. 29, no. 4 (October 1957), pp. 547–650.

Commoner, Barry, *The Poverty of Power: Energy and the Economic Crisis* (New York: Alfred A. Knopf, 1976).

Dicke, R. H., P. J. E. Peebles, P. G. Noll, and D. T. Wilkinson, "Cosmic Black Body Radiation," *The Astrophysical Journal,* vol. 142 (1965), pp. 414–19.

Eddington, A. S., *The Nature of the Physical World* (New York: Macmillan, 1929).

———, "The End of the World: From the Standpoint of Mathematical Physics," Supplement to *Nature,* no. 3202 (March 1931), pp. 447–53.

Einstein, Albert, *Essays in Science* (New York: Philosophical Library, 1934).

"The Evolution of the Universe," Supplement to *Nature,* No. 3234 (Oct. 1931).

Ferris, Timothy, *Coming of Age in the Milky Way* (New York: William Morrow, 1988).

Gamow, George, *The Creation of the Universe* (New York: Mentor, 1952).

———, "Expanding Universe and the Origin of Elements," *Physical Review,* vol. 70, pp. 572–57.

———, *One, Two, Three Infinity* (New York: Mentor, 1947).

Gribbin, John, *In Search of the Big Bang* (New York: Bantam, 1986).

Hawking, Stephen, *A Brief History of Time* (New York: Bantam, 1988).

Hoyle, F., "On the Formation of Heavy Elements in Stars," *Proceedings of the Physical Society of London,* vol. 49 (1947) pp. 942–48.

———, "The Synthesis of the Elements from Hydrogen," *Monthly Notices of the Royal Astronomical Society,* vol. 106 (1947) pp. 343–48.

Jeans, James, "The Physics of the Universe," Supplement to *Nature,* vol. 122, no. 3079 (Nov. 1928), pp. 689–700.

Lemaître, Georges, *The Primeval Atom: An Essay on Cosmology,* Betty H. and Serge A. Korff, trans. (New York: D. Van Nostrand Company, Inc., 1950).

Pais, Abraham, *Subtle Is the Lord* (New York: Oxford University Press, 1982).

Piaggio, H. T. H., "Science and Prediction," Supplement to *Nature,* no. 3202 (March 1931), p. 454.

Penzias, A. A., and R. W. Wilson, "A Measurement of Excess Antenna Temperature at 4080 Mc/s," *Nature,* vol. 142, no. 1 (1965), pp. 419–21.

Poe, Edgar A., *Eureka: A Prose Poem* (New York: P. Putnam, 1848).

Ryle, M., and R. W. Clarke, "An Examination of the Steady-State Model in the Light of Some Recent Observations of Radio Sources," *Monthly Notices of the Royal Astronomical Society,* vol. 122, no. 4 (1964), pp. 349–62.

Silk, Joseph, "Cosmic Black-Body Radiation and Galaxy Formation," *The Astrophysical Journal,* vol. 151 (Feb. 1968), pp. 459–71.

Wagoner, Robert V., William A. Fowler, and F. Hoyle, "On the Synthesis of Elements at Very High Temperatures," *The Astrophysical Journal,* vol. 148 (1966), pp. 3–50.

Weinberg, Steven, *The First Three Minutes* (New York: Basic Books, 1977).

5. The Spears of Odin

Alfvén, Hannes, "A Cosmic Cyclotron as a Cosmic Ray Generator?" *Nature,* vol. 138 (Oct. 31, 1936), p. 76.

———, "Double Layers and Circuits in Astrophysics," *IEEE Transactions in Plasma Physics,* vol. 14 (Dec. 1986), p. 779.

———, "Existence of Electromagnetic Hydrodynamic Waves," *Nature,* vol. 150 (Oct. 3, 1942), pp. 405–6.

———, "On the Motion of Cosmic Rays in Interstellar Space," *The Physical Review,* vol. 55, no. 5 (March 1, 1939), 2d ser., pp. 425–29.

———, "Origin of Cosmic Radiation," *Nature,* vol. 131 (April 29, 1933), pp. 619–20.

———, *Plasma Physics, Space Research, and the Origin of the Solar System,* Nobel lecture (Stockholm: Kungl. Boktryckeriet P. A. Norstedt & Soner, 1971).

———, "Tentative Theory of Solar Prominences," *Arkiv for Matematik, Astronomi Och Fysik* (Sweden), vol. 27A (1940), pp. 1–10.

———, and Gustaf Arrhenius, "Structure and Evolutionary History of the Solar System, III," *Astrophysics and Space Science,* vol. 21 (1973), pp. 117–76.

———, and Per Carlqvist, "Interstellar Clouds and the Formation of Stars," *Astrophysics and Space,* vol. 55 (1978), pp. 487–509.

———, and Carl-Gunne Fälthammar, *Cosmic Electrodynamics: Fundamental Principles* (Oxford: Clarendon Press, 1963).

———, and N. Herlofson, "Cosmic Radiation and Radio Stars," *Physical Review,* vol. 78 (April 17, 1950), p. 616.

Dessler, A. J., "The Evolution of Arguments Regarding the Existence of Field-Aligned Currents," *Magnetospheric Currents,* ed. T. Potemera (Washington, D.C.: American Geophysical Union, 1984), pp. 22–28.

Devik, O., *Blant Fiskere* (Oslo, Norway: Aschehoug, 1971), cited in A. Egeland and E. Leer, "Professor Kr. Birkeland," *IEEE Transactions in Plasma Science,* vol. PS-14 (Dec. 1986), p. 666.

Johannesson, Olof, *The Great Computer: A Vision,* Naomi Walford, trans. (Stockholm: Albert Bonniers, 1966).

6. The Plasma Universe

Alfvén, Hannes, *Cosmic Plasma* (Holland: D. Reidel, 1981).

———, "Cosmology and Recent Developments in Plasma Physics," *The Australian Physicist,* vol. 17 (Nov. 1980), pp. 161–65.

———, "Hubble Expansion in a Euclidean Framework," *Astrophysics and Space Science,* vol. 66 (1979), pp. 23–37.

———, "Plasma Physics Applied to Cosmology," *Physics Today,* vol. 24 (Feb. 1971), pp. 28–33.

———, *Worlds-Antiworlds: Antimatter in Cosmology* (San Francisco and London: W. H. Freeman and Company, 1966).

———, and Oskar Klein, "Matter-Antimatter Annihilation and Cosmology," *Arkiv For Fysik,* vol. 23 (1962), pp. 187–94.

DeVaucouleur, G., "The Case for a Hierarchical Cosmology," *Science,* vol. 167 (Feb. 27, 1970), pp. 1203–13.

Lerner, Eric J., "The Big Bang Never Happened," *Discover* (June 1988), pp. 70–80.

———, "Galactic Model of Element Formation," *IEEE Transactions in Plasma Science,* vol. 17 (April 1989), pp. 259–63.

———, "Magnetic Self-Compression in Laboratory Plasma, Quasars and Radio Galaxies," *Laser and Particle Beams,* vol. 4 (1986), pp. 193–222.

———, "Radio Absorption by the Intergalactic Medium," *The Astrophysical Journal*, vol. 361 (Sept. 20, 1990), pp. 63–68.

Peratt, Anthony L., "Evolution of the Plasma Universe," *IEEE Transactions in Plasma Science*, vol. 14 (Dec. 1986), pp. 639–60, 763–78.

———, and James Green, "On the Evolution of Interacting, Magnetized, Galactic Plasmas," *Astrophysics and Space Science*, vol. 91 (1983), pp. 19–33.

———, *Physics of the Plasma Universe*, (New York: Springer-Verlag, 1992).

Rieke, G. H., and M. J. Rieke, "Stellar Velocities and the Mass Distribution in the Galactic Center," *The Astrophysical Journal*, vol. 33 (July 1, 1988), pp. L33–37.

7. The Endless Flow of Time

Einstein, Albert, *Ideas and Opinions* (New York: Crown, 1954).

Lovelock, J. E., *Gaia* (New York: Oxford University Press, 1979).

*Prigogine, Ilya, "Time, Structure, and Fluctuations," *Science*, vol. 201 (Sept. 1, 1978), pp. 777–85.

———, and R. Lefever, "On Symmetry-Breaking Instabilities in Dissipative Systems, II," *The Journal of Chemical Physics*, vol. 48 (Feb. 15, 1968), pp. 1695–1700.

———, and G. Nicolis, "On Symmetry-Breaking Instabilities in Dissipative Systems," *The Journal of Chemical Physics*, vol. 46 (May 1, 1966), pp. 3542–50.

———, G. Nicolis, R. Herman, and T. Lain, "Stability, Fluctuations and Complexity," *Collective Phenomena*, vol. 2 (1979), pp. 103–9.

———, and Isabelle Stengers, *Order Out of Chaos* (New York: Bantam Books, 1984).

Glansdorff, P., and I. Prigogine, *Thermodynamic Theory of Structure, Stability and Fluctuations* (London: Wiley-Interscience, 1971).

Gunzig, Edgard, Jules Gehenian, and Ilya Prigogine, "Entropy and Cosmology," *Nature*, vol. 330 (Dec. 17, 1987), pp. 621–24.

8. Matter

Aspect, Alain, Philippe Grangier, and Gerard Roger, "Experimental Tests of Realistic Local Theories via Bell's Theorem," *Physical Review Letters*, vol. 47, no. 7 (Aug. 17, 1981), pp. 460–63.

Bohm, D., and J. P. Vigier, "Model of the Causal Interpretation of Quantum Theory in Terms of a Fluid with Irregular Fluctuation," *Physical Review*, vol. 96, no. 1 (Oct. 1, 1954), pp. 208–15.

Davies, Paul, *Other Worlds: Space, Superspace and the Quantum Universe* (New York: Simon and Schuster, 1980).

Forman, Paul, "Weimar Culture, Causality, and Quantum Theory, 1918–1927: Adaptation by German Physicists and Mathematicians to a Hostile Intellectual Environment," *Historical Studies in Physical Sciences,* vol. 3 (1971), pp. 1–115.

Heisenberg, Werner, *Physics and Beyond* (New York: Harper and Row, 1971).

Krisch, Alan D., "Collisions between Spinning Protons," *Scientific American,* vol. 257, no. 2 (Aug. 1987), pp. 42–50.

Prigogine, I., and Tomio Y. Petrosky, "An Alternative to Quantum Theory," *Physica,* vol. 147A (1988), pp. 461–86.

Roederer, Juan G., "Tearing Down Disciplinary Barriers," *EOS,* vol. 66 (Oct. 1, 1985), pp. 681, 684–85.

Rohrlich, Fritz, "Facing Quantum Mechanical Reality," *Science,* vol. 221 (Sept. 23, 1983), pp. 1251–55.

*Zee, A., *Fearful Symmetry: The Search for Beauty in Modern Physics* (New York: Macmillan, 1986).

9. Infinite in Time and Space

Bruno, Giordano, *The Ash Wednesday Supper,* Stanley L. Jaki, trans. (The Hague: Morton, 1975).

Davies, P. C. W., *God and the New Physics,* (New York: Simon and Schuster, 1983).

Gould, Stephen Jay, *Wonderful Life: The Burgess Shale and the Nature of History* (New York: W. W. Norton, 1989).

Jaki, Stanley L., *The Savior of Science* (Washington, D.C.: Gateway Editions, 1988).

John Paul II, *Scientific Research and Man's Spiritual Heritage,* Address of Pope John Paul II to the Pontifical Academy of Sciences (Oct. 3, 1981).

Pius XII, "Modern Science and the Existence of God," *The Catholic Mind,* vol. 49 (March 1972), pp. 182–92.

*Teilhard de Chardin, Pierre, *The Phenomenon of Man* (New York: Harper and Row, 1959).

10. Cosmos and Society

Fukuyama, Francis, "The End of History?" *The National Interest* (Summer 1989).

UNICEF, *The State of the World's Children 1989* (Oxford: Oxford University Press).

ILLUSTRATION CREDITS

Fig. 1.2, page 17: Courtesy M. Seldner, B. L. Siebers, E. J. Groth, and P.J.E. Peebles, *Astronomical Journal* 82, 249 (1977)

Fig. 1.3b, page 19: Courtesy Mount Wilson and Palomar Observatory

Fig. 1.4a, page 21: Courtesy R. Brent Tully, University of Hawaii

Fig. 1.4b, page 22: Courtesy R. Brent Tully, University of Hawaii

Fig. 1.5, page 25: Courtesy Alex Szalay, Johns Hopkins University

Fig. 1.6, page 30: NASA

Fig. 1.10, page 43: Courtesy ROE/ANT Board, 1979

Fig. 1.12, page 46: Courtesy Anthony L. Peratt, Los Alamos National Laboratory

Fig. 1.13, page 47: Courtesy Anthony L. Peratt, Los Alamos National Laboratory

Fig. 2.1, page 59: From J. H. Breasted, *A History of Egypt* (New York: Charles Scribners, 1937)

Maps page 61: From *What Happened in History* by Gordon Childe (Penguin Books, 1942). Copyright © 1942 by the Estate of V. Gordon Childe. Reprinted by permission of Penguin Books Ltd., London.

Fig. 2.2, page 71: From *The Crime of Galileo* by de Santillana. Copyright © 1955 University of Chicago. Reprinted by permission of the University of Chicago Press.

Fig. 3.1, page 106: From *The Cambridge Encyclopedia of Astronomy,* edited by Dr. Simon Mitton; Copyright © 1977 by Trewin Copplestone Publishing Ltd.; reprinted by permission of Crown Publishers

Fig. 5.3, page 174: Courtesy Dr. Syun Akasofu, Geophysical Institute, University of Alaska; Copyright © 1977

Fig. 5.4, page 176: Courtesy Anthony L. Peratt, Los Alamos National Laboratory

Fig. 5.5, page 177: Courtesy Anthony L. Peratt, Los Alamos National Laboratory

Fig. 5.6, page 184: Courtesy Anthony L. Peratt, Los Alamos National Laboratory

Fig. 5.7, page 187: Courtesy National Solar Observatory/Sacramento Peak

Fig. 5.8, page 189: Courtesy Hannes Alfvén

Fig. 5.9, page 195: Courtesy Hannes Alfvén

Fig. 5.10, page 202: Courtesy Hannes Alfvén

Fig. 5.11, page 207: NASA

Figs. 5.12a&b, page 210: Courtesy Hannes Alfvén

Fig. 5.13, page 211: Courtesy Rob Wood/SRW/Copyright © 1988 Discover Magazine

Fig. 6.1, page 217: Courtesy Hannes Alfvén

Fig. 6.2, page 218: Courtesy Hannes Alfvén

Fig. 6.3, page 219: Courtesy Hannes Alfvén

Fig. 6.4, page 221: Courtesy Hannes Alfvén

Fig. 6.5, page 224: Courtesy Hannes Alfvén

Fig. 6.6, page 231: Courtesy Anthony L. Perratt, Los Alamos National Laboratory

Fig. 6.7, page 233: Courtesy Anthony L. Perratt, Los Alamos National Laboratory

Fig. 6.10, page 239: Courtesy Anthony L. Perratt, Los Alamos National Laboratory

Fig. 6.11, page 241: Courtesy Anthony L. Perratt, Los Alamos National Laboratory

Fig. 6.12, page 243: Courtesy V. Nardi, Stevens Institute of Technology

Figs. 6.13a,b,c, pages 244, 245: Courtesy V. Nardi, Stevens Institute of Technology

Fig. 6.17, page 270: Courtesy National Radio Astronomy Observatory

Fig. 7.2, page 294: From *Order Out of Chaos* by Ilya Prigogine and Isabelle Stengers. Copyright © 1984 by Ilya Prigogine and Isabelle Stengers. Foreword copyright © 1984 by Alvin Toffler. Used by permission of Bantam Books, a division of Bantam, Doubleday, Dell Publishing Group, Inc.

Fig. 7.4, page 305: D. H. Erwin, J. W. Valentine, and J. J. Sepkoski, Jr., *Evolution*, vol. 41 (1987), pp. 1177–1186; from data of J. J. Sepkoski, Jr.; reprinted by permission of the authors and the Society for the Study of Evolution

Fig. 7.7, page 323: Courtesy Warren Brown, UCLA; Robert Chapman, University of Rochester; Erol Basar, Institute of Physiology, Lubeck, West Germany

Fig. 8.3, page 347: Property of the author

Fig. 8.4, page 351: From *Lectures on Physics*, by Richard P. Feynman; Copyright © Addison-Wesley Publishing Company; reprinted by permission of Addison-Wesley Publishing Company

Figs. 8.5 a&b, pages 353 and 354: Illustration by Jerome Kuhl from "The Quantum Theory and Reality" by Bernard d'Espagnat. Copyright © 1979 by Scientific American, Inc. All rights reserved

Fig. 8.6, page 356: From *Introduction to Quantum Mechanics*, by Robert Dicke and James Wittke; Copyright © 1960 by Addison-Wesley Publishing Company; reprinted by permission of Addison-Wesley Publishing Company

INDEX